Springer Series in
Surface Sciences

Editor: Gerhard Ertl   21

Springer Series in **Surface Sciences**

Editors: G. Ertl and R. Gomer   Managing Editor: H. K. V. Lotsch

Volume 1: **Physisorption Kinetics**   By H. J. Kreuzer, Z. W. Gortel

Volume 2: **The Structure of Surfaces**   Editors: M. A. Van Hove, S. Y. Tong

Volume 3: **Dynamical Phenomena at Surfaces, Interfaces and Superlattices**
Editors: F. Nizzoli, K.-H. Rieder, R. F. Willis

Volume 4: **Desorption Induced by Electronic Transitions, DIET II**
Editors: W. Brenig, D. Menzel

Volume 5: **Chemistry and Physics of Solid Surfaces VI**
Editors: R. Vanselow, R. Howe

Volume 6: **Low-Energy Electron Diffraction**
Experiment, Theory and Surface Structure Determination
By M. A. Van Hove, W. H. Weinberg, C.-M. Chan

Volume 7: **Electronic Phenomena in Adsorption and Catalysis**
By V. F. Kiselev, O. V. Krylov

Volume 8: **Kinetics of Interface Reactions**   Editors: M. Grunze, H. J. Kreuzer

Volume 9: **Adsorption and Catalysis on Transition Metals and Their Oxides**
By V. F. Kiselev, O. V. Krylov

Volume 10: **Chemistry and Physics of Solid Surfaces VII**
Editors: R. Vanselow, R. Howe

Volume 11: **The Structure of Surfaces II**
Editors: J. F. van der Veen, M. A. Van Hove

Volume 12: **Diffusion at Interfaces: Microscopic Concepts**
Editors: M. Grunze, H. J. Kreuzer, J. J. Weimer

Volume 13: **Desorption Induced by Electronic Transitions, DIET III**
Editors: R. H. Stulen, M. L. Knotek

Volume 14: **Solvay Conference on Surface Science**   Editor: F. W. de Wette

Volume 15: **Surfaces and Interfaces of Solids**   By H. Lüth

Volume 16: **Theory of the Atomic and Electronic Structure of Surfaces**
By M. Lannoo, P. Friedel

Volume 17: **Adhesion and Friction**   Editors: M. Grunze and H. J. Kreuzer

Volume 18: **Auger Spectroscopy and Electronic Structure**
Editors: G. Cubiotti, G. Mondio, K. Wandelt

Volume 19: **Desorption Induced by Electronic Transitions, DIET IV**
Editors: G. Betz, P. Varga

Volume 20: **Semiconductor Surfaces**   By W. Mönch

Volume 21: **Surface Phonons**   Editors: W. Kress, F. W. de Wette

Volume 22: **Chemistry and Physics of Solid Surfaces VIII**
Editors: R. Vanselow, R. Howe

W. Kress   F. W. de Wette   (Eds.)

Surface Phonons

With Contributions by
G. Benedek   J. E. Black   V. Celli
W. Kress   A. A. Maradudin   L. Miglio
D. L. Mills   G. I. Stegeman   J. P. Toennies
S. Y. Tong   F. W. de Wette

With 160 Figures

Springer-Verlag
Berlin Heidelberg New York
London Paris Tokyo
Hong Kong Barcelona
Budapest

Dr. Winfried Kress

Max-Planck-Institut für Festkörperforschung, Heisenbergstrasse 1,
W-7000 Stuttgart 80, Fed. Rep. of Germany

Professor Dr. Frederik W. de Wette

Department of Physics, The University of Texas at Austin,
Austin, TX 78712, USA

Series Editors

Professor Dr. Gerhard Ertl

Fritz-Haber-Institut der Max-Planck-Gesellschaft, Faradayweg 4–6,
1000 Berlin 33, Fed. Rep. of Germany

Professor Robert Gomer, Ph. D.

The James Franck Institute, The University of Chicago, 5640 Ellis Avenue,
Chicago, IL 60637, USA

Managing Editor: Dr. Helmut K. V. Lotsch

Springer-Verlag, Tiergartenstrasse 17,
W-6900 Heidelberg, Fed. Rep. of Germany

ISBN-13: 978-3-642-75787-7   e-ISBN-13: 978-3-642-75785-3
DOI: 10.1007/978-3-642-75785-3

Library of Congress Cataloging-in-Publication Data. Surface phonons / W. Kress, F. W. de Wette (eds.).
p. cm.–(Springer series in surface sciences ; v. 21) Includes bibliographical references and index. ISBN
0-387-52721-4 (alk. paper) 1. Phonons. 2. Surfaces (Physics) I. Kress, W. (Winfried), 1941– . II. Wette, F. W.
de (Frederick W.) III. Series: Springer series in surface sciences ; 21. QC176.8.P5S87   1991   530.4'16–
dc20   90-10345

54/3140-543210 – Printed on acid-free paper

Preface

In recent years substantial progress has been made in the detection of surface phonons owing to considerable improvements in inelastic rare gas scattering techniques and electron energy loss spectroscopy. With these methods it has become possible to measure surface vibrations in a wide energy range for all wave vectors in the two-dimensional Brillouin zone and thus to deduce the complete surface phonon dispersion curves. Inelastic atomic beam scattering and electron energy loss spectroscopy have started to play a role in the study of surface phonons similar to the one played by inelastic neutron scattering in the investigation of bulk phonons in the last thirty years. Detailed comparison between experimental results and theoretical studies of inelastic surface scattering and of surface phonons has now become feasible. It is therefore possible to test and to improve the details of interaction models which have been worked out theoretically in the last few decades. At this point we felt that a concise, coherent and self-contained guide to the rapidly growing field of surface phonons was needed.

The present book contains a series of contributions by experts in the field of surface phonons which review the fundamental concepts of surface waves, describe the most successful methods of calculating the dynamics of crystal surfaces (the slab and the Green's function methods), outline the experimental techniques of atomic beam spectroscopy and electron energy loss spectroscopy, together with the recent developments of the appropriate scattering theory, and give a survey of the measured and calculated surface phonon dispersion curves.

We dedicate this volume to the memory of Heinz Bilz, who made essential contributions to the better understanding of solid state physics, in particular in the fields of lattice dynamics, infrared absorption, light scattering and displacive phase transitions. To the authors of this volume he was more than just an open-minded partner for very many fruitful, encouraging, and extremely stimulating discussions; he was a wonderful human being and a very dear friend.

Stuttgart and Austin
November 1990

Winfried Kress, Frederik W. de Wette

Heinz Bilz
(18 May 1926 – 26 June 1986)

Contents

1. **Introduction**
   By *W. Kress* and *F.W. de Wette* 1

2. **Surface Acoustic Waves**
   By *A.A. Maradudin* and *G.I. Stegeman* (With 10 Figures) 5
   2.1   Surface Acoustic Waves on Various Media 6
         2.1.1   Elastic Media 6
         2.1.2   Piezoelectric Media 16
   2.2   Generation and Detection of Surface Acoustic Waves 24
         2.2.1   Electrical Generation and Detection
                 in Piezoelectric Media 25
         2.2.2   Generation on Non-piezoelectric Media 27
         2.2.3   Acousto-optic Detection
                 of Generated Surface Acoustic Waves 28
         2.2.4   Brillouin Scattering from Surface Waves 28
   2.3   Some Applications of Surface Acoustic Waves 30
         2.3.1   Applications to Electronic Signal Processing 30
         2.3.2   Measurements of Intrinsic Material Properties 32
   References 34

3. **The Green's Function Method
   in the Surface Lattice Dynamics of Ionic Crystals**
   By *G. Benedek* and *L. Miglio* (With 10 Figures) 37
   3.1   Outline of the Time-Independent Green's Function Method .. 37
   3.2   The Green's Function Method in Surface Dynamics 42
   3.3   The Intrinsic Perturbation for a Semi-Infinite Lattice 46
   3.4   The Electronic Contribution to Surface Dynamics
         in the Framework of Shell Models 50
   3.5   Surface Phonon Polaritons 53
   3.6   Surface Vibrations in Alkali Halides 55
   3.7   Further Developments:
         The Study of Surface Phonon Anomalies 62
   References 65

4. Study of Surface Phonons by the Slab Method
By *F.W. de Wette* (With 20 Figures) 67
4.1 Formalism 69
    4.1.1 Slab Dynamics 69
    4.1.2 Use of Symmetry 73
    4.1.3 Slab Vibrational Modes
          and Their Dispersion Curves 74
    4.1.4 Macroscopic and Microscopic Surface Modes 78
    4.1.5 Attenuation Curves 78
    4.1.6 Systematic Features of Surface Modes 80
4.2 Computational Considerations 81
    4.2.1 Diagonalization Techniques 81
    4.2.2 Surface Brillouin Zone Sampling 82
4.3 Interaction Models 85
    4.3.1 Simple Pair Potentials (Molecular Crystals) 85
    4.3.2 Shell Models (Ionic Crystals) 86
    4.3.3 Force Constant Models (Semiconductors, Metals) ... 88
4.4 Results 89
    4.4.1 Relaxation and Dynamics
          of the (001) Surfaces of Alkali Halides 89
    4.4.2 Results for fcc and bcc Metals 92
    4.4.3 Layered Structure – Graphite 95
4.5 Derived Physical Quantities 97
    4.5.1 Mean-Square Amplitudes of Vibration
          and Vibrational Correlation Functions 97
    4.5.2 Surface Debye Temperature 103
    4.5.3 Surface Thermodynamic Quantities 104
4.6 Concluding Remarks 107
References 108

**5. Experimental Determination of Surface Phonons
by Helium Atom and Electron Energy Loss Spectroscopy**
By *J.P. Toennies* (With 30 Figures) 111
5.1 Theoretical Background 112
5.2 Kinematics 116
5.3 Helium Scattering 123
    5.3.1 General Considerations 123
    5.3.2 Helium Nozzle Beam Source 126
    5.3.3 Target Chamber 131
    5.3.4 Detector 132
    5.3.5 Typical Measurements and Resolution 133
5.4 Electron Scattering 140
    5.4.1 Apparatus 140
    5.4.2 Typical Measurements 142

5.5	Intensities	142
	5.5.1 Helium Atom Scattering	144
	5.5.2 Electron Scattering	150
	5.5.3 Comparison of Experimental Intensities	153
5.6	Discussion of Experimental Results and Summary	154
	References	162

6. Theory of Helium Scattering from Surface Phonons

	By *V. Celli* (With 5 Figures)	167
6.1	Kinematics	167
	6.1.1 Kinematic Focusing	170
6.2	Dynamical Theory: General Considerations	171
	6.2.1 Box Normalization and Scattering Geometry	171
	6.2.2 Cross Sections and Reflection Coefficients	172
	6.2.3 State-to-State Cross Sections vs Differential Cross Sections	173
6.3	One-Phonon Exchange Processes	173
	6.3.1 The Distorted Wave Born Approximation	174
	6.3.2 The Phonon Matrix Elements	176
	6.3.3 The Atom–Surface Matrix Elements	177
	6.3.4 The Differential Reflection Coefficient	178
	6.3.5 Relation to Phonon Density of States and Correlation Functions	179
6.4	The Inelastic Atom–Surface Interaction	180
	6.4.1 The Static Repulsive Potential	181
	6.4.2 The Static Attractive Potential	183
	6.4.3 The Total Static Potential	185
	6.4.4 The Dynamic Repulsion and the Cutoff Factor	186
	6.4.5 Dynamical Effects of the Attractive Potential	187
	References	189

7. The Study of Surface Phonons by Electron Energy Loss Spectroscopy: Theoretical and Experimental Considerations

	By *D.L. Mills, S.Y. Tong* and *J.E. Black* (With 2 Figures)	193
7.1	A Brief Review	194
7.2	The Surface Phonon Excitation Mechanism in the Impact Regime	197
7.3	The Green's Function Approach to Spectral Density Calculations	200
7.4	Calculations of the Cross Section for Surface Phonon Excitation	204
7.5	Concluding Remarks	206
	References	206

8. **Vibrational Properties of Clean Surfaces:**
   Survey of Recent Theoretical and Experimental Results
   By *W. Kress* (With 82 Figures) 209
   8.1 Ionic Crystals 211
       8.1.1 Alkali Halides (Rock Salt Structure) 211
       8.1.2 Metal Oxides 233
       8.1.3 Refractory Compounds 236
       8.1.4 Perovskite Structure Compounds 248
   8.2 Metals 254
       8.2.1 Body Centered Cubic Metals 255
       8.2.2 Face Centered Cubic Metals 257
   8.3 Miscellaneous 277
       8.3.1 Diamond Structure Crystals 277
       8.3.2 Zinc-Blende Structure Crystals 280
       8.3.3 Layered Structure Crystals 281
   References 287

Subject Index 293

Contributors

Benedek, G.
  Universita degli Studi di Milano, Dipartimento di Fisica
  Via Celoria 16, I-20133 Milan, Italy

Black, J. E.
  Department of Physics, Brock University
  St. Catharines, Ontario, Canada

Celli, V.
  Department of Physics, University of Virginia
  Charlottesville, VA 22901, USA

Kress, W.
  Max-Planck-Institut für Festkörperforschung
  Postfach 80 06 65, W-7000 Stuttgart 80, Fed. Rep. of Germany

Maradudin, A. A.
  Department of Physics, University of California
  Irvine, CA 92717, USA

Miglio, L.
  Universita degli Studi di Milano, Dipartimento di Fisica
  Via Celoria 16, I-20133 Milan, Italy

Mills, D. L.
  Department of Physics, University of California
  Irvine, CA 92717, USA

Stegeman, G. I.
  Optical Sciences Center, University of Arizona
  Tucson, AZ 85721, USA

Toennies, J. P.
  Max-Planck-Institut für Strömungsforschung
  Postfach 28 53, W-3400 Göttingen, Fed. Rep. of Germany

Tong, S. Y.
    Department of Physics, The University of Wisconsin-Milwaukee
    Milwaukee, Wisconsin, USA

de Wette, F. W.
    Department of Physics, University of Texas
    Austin, TX 78712, USA

1. Introduction

W. Kress and F.W. de Wette

Surface dynamics may be considered as a subject now over one hundred years old, since it was in 1887 that Lord Rayleigh first discussed the occurrence of localized waves at the surfaces of isotropic elastic media. Lattice dynamics of solids had its beginnings in the pioneering work of Born and von Kármán and Debye in 1913, and for the next half a century investigations concentrated mainly on the dynamics of infinite solids. It was only about a quarter of a century ago, that surface phenomena began to draw the attention of solid state physicists, and that continuum mechanics and lattice dynamics began to be applied to the dynamics of surfaces. The subject of elastic surface waves has continued to play an important role, first because it describes the continuum or long-wavelength limit of the microscopic description, but equally importantly because it has formed the foundation of the development of a variety of technological applications. For this reason the editors felt that this volume on surface phonons should appropriately begin with a chapter on surface acoustic waves. This is presented by Maradudin and Stegeman (Chap. 2), who lead us from a presentation of the underlying formalism, via the principal phenomena in piezoelectric and non-piezoelectric materials to a discussion of modern applications in acoustic and electrical signal processing.

The main emphasis of this volume is on the theory and experimental investigation of surface vibrations at all wavelengths (not just long wavelengths) which require microscopic, that is, lattice-dynamical descriptions. This immediately raises the question of the relation between dynamics of the bulk crystal, and dynamics of the surface.

The development of inelastic neutron scattering in the mid-fifties opened up the field of bulk lattice dynamics to detailed experimental study. The measurements of complete phonon dispersion curves, including the details of the particle motion (polarizations), enabled a full characterization of the vibrational states of the solid. The rapid accumulation of experimental neutron scattering data on a large variety of solids stimulated theorists to develop lattice dynamical models to interpret these data and subsequently, and where feasible, to develop first-principle theories.

While these experimental and theoretical studies of bulk dynamics rapidly developed both in breadth and in depth, theorists soon turned to the study of surface dynamics. Adapting bulk dynamical theories to surface studies raised questions such as: what are the similarities between bulk and surface dynamics; what are the differences, and most importantly, what can one learn from these

Springer Series in Surface Sciences, Vol. 21   **Surface Phonons**

Editors: W. Kress · F.W. de Wette   © Springer-Verlag Berlin, Heidelberg 1991

surface studies. Because of technical difficulties in the relevant scattering techniques, experimental surface-dynamical methods had a somewhat slower start than the theory, but steady progress was made during the seventies, leading to well-developed methods for inelastic surface scattering – electron energy loss spectroscopy and helium-scattering – in the early eighties. It became possible to measure complete dispersion curves for low-lying surface phonon excitations. In other words, experimental surface dynamics was reaching a level of development comparable to that of bulk neutron studies during the fifties. The rapid expansion of these experimental studies gave rise to renewed theoretical activity along earlier developed lines, but also stimulated the development of more basic approaches to surface dynamics.

In principle, there are two main approaches to surface dynamics: straightforward conventional lattice dynamics applied to a system with surfaces, such as a slab-shaped crystal, and the Green's function method. The latter has been particularly suited for the description of the lattice dynamics of defects, and since a surface is a very special kind of defect, it is natural that at an early stage, the method was adapted for surface dynamical calculations. Benedek and co-workers have taken the lead in this development and it is therefore appropriate that the third chapter, by Benedek and Miglio, deals with the Green's function method. The method is particularly suited for the calculation of densities of phonon states, phonon resonances and correlation functions, which are key ingredients for inelastic scattering calculations. In many cases such calculations are indispensable for the interpretation of the experimental data.

The fourth chapter on the slab method deals with the straightforward application of the method of lattice dynamics in the quasi-harmonic approximation to crystals with two parallel surfaces. Because of its direct approach, this method is particularly suited for the broad study of the phenomena and systematics of surface dynamics carried out by de Wette and a series of co-workers. This chapter by de Wette is partly of tutorial nature and is intended to be of interpretive support for the surface phonon review by Kress. Unlike the Green's function method, the slab method gives very direct access to the microscopic details of the surface vibrations: dispersion curves, polarization vectors, attenuation of surface-mode amplitudes, etc. In addition, the availability of complete phonon information, allows the evaluation of surface thermodynamic functions. While the slab method is in principle not as well suited to the evaluation of phonon densities of states (including surface-projected densities) as is the Green's function method, this disadvantage is rapidly being overcome through the use of supercomputers.

The consideration of crystal slabs is also extremely useful for the study of surface relaxation and surface reconstruction. Such calculations require the knowledge of interaction potentials rather than force constants, because the interparticle forces have to be known over finite ranges of the interparticle distances. Depending on the class of material in question, surface relaxation/reconstruction can have important effects on the surface dynamics. While on the one hand, in alkali halides, because of the closed shell interactions, the effect of surface relaxation on the surface dynamics is generally quite small (except in the heavy alkali halides), in perovskites on the other hand, surface relaxation/reconstruction and surface dynamics are strongly coupled; in fact, surface reconstruction caused by a soft surface mode may provide the triggering mechanism for the bulk transition. A class of materials still more difficult to treat from a fundamental point of view are the covalent crystals, where dangling bounds at the surface strongly modify particle interactions. Finally, metal surfaces are equally or even more difficult to treat, because of the possible strong modifications in the conduction electron states near the surface. In this sense approaches using simple force constants are the least satisfactory for metals, although one has nonetheless had to rely on them most heavily. Recent, more basic approaches to the lattice dynamics of metals such as the dielectric response theory, density functional theory and frozen phonons, are only mentioned in passing in this volume.

Surface scattering experiments provide both the stimuli and the ultimate check for the theory. As mentioned above, inelastic helium scattering and inelastic electron scattering have made giant strides during the past decade. This development has been helped by the fact that, compared to neutron scattering, helium and electron scattering are relatively cheap methods. In Chap. 5 Toennies gives an overview of these methods, with an emphasis on helium scattering, understandable because of the pioneering work of Toennies and his associates in this field.

While the early helium scattering experiments on alkali halides confirmed previous calculations of the surface dynamics of these compounds, it is the measurements on the surfaces of the noble metals Cu, Ag and Au which have stimulated further developments in the theory of surface dynamics of metals. Because of the importance of the conduction electron response in metals, this development has of necessity put increased emphasis on first-principles approaches.

In contrast to electron scattering, helium scattering has the convenient property that the helium atoms are scattered exclusively by atoms in the surface layer so that in many cases a straightforward kinematical description allows one to extract most of the surface dynamical information contained in the experimental data. Nevertheless, frequently more extensive surface scattering calculations are required. The theory of atom–surface scattering is reviewed by Celli in Chap. 6. Despite the strong similarities between neutron scattering and helium scattering on the level of simple kinematics, atom scattering theory goes well beyond a kinematical description because of the basically complicated nature of the atom–surface interaction.

The primary objects of studies with electron energy loss spectroscopy have been the bare and adsorbate covered metal surfaces. Electrons can probe the entire surface region, not just the outer surface layer as does helium scattering. This has the obvious advantage that properties below the surface can be studied, but it complicates the scattering description significantly, and with it, the interpretation of the experimental data. The scope of this problem is discussed by Mills, Tong and Black in Chap. 7.

The volume concludes in Chap. 8 with a review of recent theoretical and experimental results on surface phonons compiled by Kress. Since the amount of published material is undergoing a rapid expansion, a rather strict selection of the available material had to be made to stay within the bounds of this volume.

2. Surface Acoustic Waves

A.A. Maradudin and G. I. Stegeman

With 10 Figures

Surface acoustic waves have been a subject of scientific investigation for more than hundred years. *Lord Rayleigh* [2.1] showed in 1887 that a semi-infinite, isotropic, elastic medium, bounded by a single stress-free planar surface, can support propagating modes that are wavelike in directions parallel to the surface. These excitations are identified as surface waves because they are characterized by acoustic fields that decay exponentially with distance into the medium from the surface, and by a frequency that is linearly proportional to a two-dimensional wave vector, which lies in the plane of the surface. The displacements of the medium lie in the sagittal plane, which is defined by the normal to the surface and the direction of propagation of the wave. In honor of their discoverer, they are known as *Rayleigh waves*.

In the following years, surface acoustic waves continued to be investigated theoretically on the basis of elasticity theory. Until the mid-1960's the principal interest was in seismological waves propagating in the earth's crust. Analyses were extended to Raleigh waves propagating in layered systems, to waves guided by an interface between two media, to surface waves polarized normal to the sagittal plane, and to special cases of acoustically anisotropic media. Indeed, the contributions made by geophysicists to the theory of surface acoustic waves have been many and diverse, but often unknown to physicists and engineers who have, in some cases, rediscovered them many years later.

The recent surge of activity in the field of surface waves has been driven by their application to high frequency signal processing. By using interdigital transducers and piezoelectric media, an electrical signal can be converted efficiently to a surface acoustic wave, the propagation properties of surface waves are used for signal processing purposes, and the acoustic signal can then be converted back to an electrical signal. Although the key technological advance was the invention of the interdigital transducer for the efficient interconversion of electrical and acoustic energy, the most important property of surface acoustic waves is their extremely low speed and hence small wavelength, about 10^{-5} that of electromagnetic waves of comparable frequency. For signal processing operations which depend on wavelike properties of a signal at a given frequency, an acoustic wave device is much smaller than the corresponding electromagnetic device. Furthermore, surface waves can be easily modified or trapped via surface structures, an important advantage over bulk acoustic waves for signal processing purposes. A number of surface wave devices have already found applications in both the commercial and military sectors.

Springer Series in Surface Sciences, Vol. 21   **Surface Phonons**   5
Editors: W. Kress · F.W. de Wette   © Springer-Verlag Berlin, Heidelberg 1991

The rapid evolution of surface wave device technology has stimulated interest in both the basic physics of surface waves and their application to the study of surface or thin film phenomena which affect the surface wave propagation properties. Experimentally the keys have been the development of transducers for efficient generation and pick-up of high-power surface waves, and the application of techniques such as light scattering for detecting and quantifying surface waves. For example, superconductivity, phase transitions, and nonlinear interactions of surface acoustic waves have been investigated by these methods.

The theoretical study of surface acoustic waves is also actively pursued today because the use of continuum mechanics makes possible the solution of certain kinds of problem that are either very difficult or impossible to solve at the present time if one proceeds from a lattice theory of surface phonons. These include the propagation of surface waves along the interface between two different solids or between a solid and a liquid, over curved or rough surfaces, over substrates whose material properties vary with distance from the surface, and the interaction of surface waves with material and/or geometrical discontinuities in their propagation path. Although the results are valid only in the limit of long wavelengths and low frequencies, they provide useful insights into what might be expected from more microscopic treatments, when the latter become possible.

In the present chapter we present the basic elements of the theory of a variety of different types of surface acoustic waves, describe methods for their excitation and detection, and survey various properties possessed by these waves. Relevant experimental results will be presented, and possible technological applications of surface acoustic waves will be described as well.

2.1 Surface Acoustic Waves on Various Media

Surface acoustic waves can propagate along the surfaces of, and interfaces between, several different kinds of media, viz. elastic, piezoelectric, or magnetic media. In this section we present the theory for such waves for elastic and piezoelectric media, and discuss the distinctive features imparted to them by the nature of the medium over which they propagate.

2.1.1 Elastic Media

We begin with the theory of surface acoustic waves on a semi-infinite elastic medium bounded by a stress-free planar surface, the Rayleigh waves. We point out other types of surface acoustic waves, besides Rayleigh waves, that can exist on such media, viz. generalized Rayleigh waves and pseudosurface (or leaky) surface waves.

We consider an elastic medium occupying the upper half-space $x_3 > 0$. The surface $x_3 = 0$ is assumed to be stress-free. Within the framework of the linear theory of elasticity the equations of motion of the medium are

$$\varrho \frac{\partial^2}{\partial t^2} u_\alpha = \sum_\beta \frac{\partial T_{\alpha\beta}}{\partial x_\beta} \ , \qquad \alpha = 1, 2, 3 \ , \tag{2.1}$$

where ϱ is the mass density of the medium, $u_\alpha(\boldsymbol{x}, t)$ is the α Cartesian component of the displacement of the medium at the point $\boldsymbol{x}$ and time t, and $T_{\alpha\beta}(\boldsymbol{x}, t)$ is the stress tensor. The latter is given by Hooke's law,

$$T_{\alpha\beta} = \sum_{\mu\nu} C_{\alpha\beta\mu\nu} \frac{\partial u_\mu}{\partial x_\nu} \ , \qquad \alpha, \beta = 1, 2, 3 \ , \tag{2.2}$$

where the $\{C_{\alpha\beta\mu\nu}\}$ are the elements of the elastic modulus tensor. The elements of the tensor $C_{\alpha\beta\mu\nu}$ are symmetric in α and β, in μ and ν, and in the pairs $\alpha\beta$ and $\mu\nu$. This makes it possible to express them equivalently in a two subscript notation, $C_{\alpha\beta\mu\nu} \rightarrow c_{ij}$, according to the scheme

$$\begin{array}{cccccc} 11 & 22 & 33 & 23 = 32 & 31 = 13 & 12 = 21 \\ 1 & 2 & 3 & 4 & 5 & 6 \end{array} \ . \tag{2.3}$$

The resulting 6×6 matrix representation of the elastic modulus tensor is referred to as the contracted, or Voigt, representation of this tensor.

When we substitute (2.2) into (2.1) we obtain the equations of motion of the medium in the form

$$\varrho \frac{\partial^2 u_\alpha}{\partial t^2} = \sum_{\beta\mu\nu} C_{\alpha\beta\mu\nu} \frac{\partial^2 u_\mu}{\partial x_\beta \partial x_\nu} \ , \qquad \alpha = 1, 2, 3 \ . \tag{2.4}$$

The equations of motion (2.4) have to be supplemented by the conditions expressing the fact that the stresses acting on the surface $x_3 = 0$ vanish:

$$T_{\alpha 3}\big|_{x_3=0} = 0 \ , \alpha = 1, 2, 3 \ , \tag{2.5}$$

or

$$\sum_{\mu\nu} C_{\alpha 3\mu\nu} \frac{\partial u_\mu}{\partial x_\nu}\bigg|_{x_3=0} = 0 \qquad \alpha = 1, 2, 3 \ . \tag{2.6}$$

To solve (2.4) subject to the boundary conditions (2.6) for an arbitrary direction of propagation on an arbitrary surface of a crystal of arbitrary symmetry we rotate the crystal in such a way that the direction of propagation of the surface acoustic wave is along the x_1 axis. The necessary transformation of the elastic moduli is carried out in the following way. When the elastic medium is subjected to an arbitrary real, orthogonal transformation S that sends a point $\boldsymbol{x}$ in it into a point $\boldsymbol{x}'$, where both $\boldsymbol{x}$ and $\boldsymbol{x}'$ are defined with respect to a set of Cartesian axes fixed in space (active convention), according to

$$\boldsymbol{x}' = S\boldsymbol{x} \ , \tag{2.7}$$

where S is the 3×3 real, orthogonal matrix that represents the transformation S, the elastic modulus tensor of the transformed medium is given by

7

$$C'_{\alpha\beta\mu\nu} = \sum_{\alpha'\beta'\mu'\nu'} S_{\alpha\alpha'} S_{\beta\beta'} S_{\mu\mu'} S_{\nu\nu'} C_{\alpha'\beta'\mu'\nu'} \, . \tag{2.8}$$

If the transformation S is one of the elements R of the crystal class G to which the elastic medium belongs, the medium is sent into itself, and we can remove the prime from the right-hand side of (2.4). The transformation law for the elements of the elastic modulus tensor under the operations of G is therefore

$$C_{\alpha\beta\mu\nu} = \sum_{\alpha'\beta'\mu'\beta'} R_{\alpha\alpha'} R_{\beta\beta'} R_{\mu\mu'} R_{\nu\nu'} C_{\alpha'\beta'\mu'\nu'} \, . \tag{2.9}$$

A displacement field of the form of

$$u_\alpha(\boldsymbol{x}, t) = U_\alpha e^{ikx_1 - k\beta x_3 - i\omega t} \quad \text{Re } \beta > 0 \, , \tag{2.10}$$

describes a straight crested wave that propagates in the x_1 direction, whose amplitude decays exponentially with increasing distance into the medium from the surface. Setting $\omega = vk$, where v is the phase velocity of the wave, and substituting (2.10) into (2.4) yields the set of homogeneous equations

$$\sum_\mu \left[\Gamma_{\alpha\mu}(\beta) - \delta_{\alpha\mu} v^2 \right] U_\mu = 0 \tag{2.11a}$$

for the amplitudes $\{U_\alpha\}$, where

$$\Gamma_{\alpha\mu}(\beta) = \frac{1}{\varrho} \left[C_{\alpha 1\mu 1} + (C_{\alpha 1\mu 3} + C_{\alpha 3\mu 1}) i\beta - C_{\alpha 3\mu 3}\beta^2 \right] = \Gamma_{\mu\alpha}(\beta) \, . \tag{2.11b}$$

In order that (2.11) have nontrivial solutions, the determinant of the coefficients must vanish. For each value of the phase velocity v, setting this determinant equal to zero yields a sixth-order equation for the decay constant β. There are therefore six values of β for each value of v. These six roots are either real or members of complex conjugate pairs. Since we are concerned here with surface waves, it is only in the values of β whose real parts are positive that we are interested. In the general case there are three such roots, and we will denote them by $\beta_j(v)$ $(j = 1, 2, 3)$. The amplitudes $\{U_\alpha\}$ for any β_j are related by

$$\frac{U_1^{(j)}}{C_1^{(j)}} = \frac{U_2^{(j)}}{C_2^{(j)}} = \frac{U_3^{(j)}}{C_3^{(j)}} = K_j \, , \quad j = 1, 2, 3 \, , \tag{2.12}$$

where the $\{K_j\}$ are constants and $C_\alpha^{(j)}(v)$ $(\alpha = 1, 2, 3)$ are the cofactors of the elements in the first row of the matrix $\Gamma_{\alpha\mu}(\beta_j(v)) - \delta_{\alpha\mu} v^2$. $C_\alpha^{(j)}(v)$ is therefore a function of $\beta_j(v)$ and of v:

$$C_1^{(j)}(v) = \left[\Gamma_{22}(\beta_j(v)) - v^2 \right] \left[\Gamma_{33}(\beta_j(v)) - v^2 \right] - \Gamma_{23}^2(\beta_j(v)) \tag{2.13a}$$

$$C_2^{(j)}(v) = -\Gamma_{21}(\beta_j(v)) \left[\Gamma_{33}(\beta_j(v)) - v^2 \right] + \Gamma_{23}(\beta_j(v))\Gamma_{31}(\beta_j(v)) \tag{2.13b}$$

$$C_3^{(j)}(v) = -\Gamma_{31}(\beta_j(v)) \left[\Gamma_{22}(\beta_j(v)) - v^2 \right] + \Gamma_{32}(\beta_j(v))\Gamma_{21}(\beta_j(v)) \, . \tag{2.13c}$$

The boundary conditions (2.6) in the present case take the forms

$$
\left[\left(C_{1311} \frac{\partial}{\partial x_1} + C_{1313} \frac{\partial}{\partial x_3} \right) u_1 + \left(C_{1321} \frac{\partial}{\partial x_1} + C_{1323} \frac{\partial}{\partial x_3} \right) u_2 \right.
$$
$$
\left. + \left(C_{1331} \frac{\partial}{\partial x_1} + C_{1333} \frac{\partial}{\partial x_3} \right) u_3 \right]_{x_3=0} = 0 , \tag{2.14a}
$$

$$
\left[\left(C_{2311} \frac{\partial}{\partial x_1} + C_{2313} \frac{\partial}{\partial x_3} \right) u_1 + \left(C_{2321} \frac{\partial}{\partial x_1} + C_{2323} \frac{\partial}{\partial x_3} \right) u_2 \right.
$$
$$
\left. + \left(C_{2331} \frac{\partial}{\partial x_1} + C_{2333} \frac{\partial}{\partial x_3} \right) u_3 \right]_{x_3=0} = 0 , \tag{2.14b}
$$

$$
\left[\left(C_{3311} \frac{\partial}{\partial x_1} + C_{3313} \frac{\partial}{\partial x_3} \right) u_1 + \left(C_{3321} \frac{\partial}{\partial x_1} + C_{3323} \frac{\partial}{\partial x_3} \right) u_2 \right.
$$
$$
\left. + \left(C_{3331} \frac{\partial}{\partial x_1} + C_{3333} \frac{\partial}{\partial x_3} \right) u_3 \right]_{x_3=0} = 0 . \tag{2.14c}
$$

To satisfy these conditions we superpose the solutions corresponding to the three $\beta_j(v)$ to obtain

$$
u_\alpha(\boldsymbol{x}, t) = \sum_{j=1}^{3} C_\alpha^{(j)}(v) K_j e^{i k x_1 - k \beta_j(v) x_3 - i v k t} . \tag{2.15}
$$

Substitution of this form for the displacement field into (2.14) leads to a set of homogeneous linear equations for the $\{K_j\}$. The condition for a nontrivial solution is that the determinant of the coefficients of $\{K_j\}$ vanish,

$$
|M_{\alpha j}(v)| = 0 , \tag{2.16a}
$$

where

$$
M_{\alpha j}(v) = \frac{1}{\varrho} \sum_\mu \left[C_{\alpha 3 \mu 1} + i C_{\alpha 3 \mu 3} \beta_j(v) \right] C_\mu^{(j)}(v) , \tag{2.16b}
$$

$$
\alpha = 1, 2, 3 ; \quad j = 1, 2, 3 .
$$

Equation (2.16) is the equation for the phase velocity v of the surface acoustic wave. In general it has to be solved numerically.

The preceding analysis simplifies considerably for propagation of a surface acoustic wave along a direction of high symmetry on a surface of high symmetry [2.2]. Thus, for example, if $\Gamma_{12}(\beta) = \Gamma_{21}(\beta)$ and $\Gamma_{23}(\beta) = \Gamma_{32}(\beta)$ vanish in (2.11a), the latter system of equations breaks up into the pair

$$
\begin{pmatrix} \Gamma_{11}(\beta) - v^2 & \Gamma_{13}(\beta) \\ \Gamma_{31}(\beta) & \Gamma_{33}(\beta) - v^2 \end{pmatrix} \begin{pmatrix} U_1 \\ U_3 \end{pmatrix} = 0 \tag{2.17}
$$

and

$$
\left[\Gamma_{22}(\beta) - v^2 \right] U_2 = 0 . \tag{2.18}
$$

From (2.11b) this requires the vanishing of six elements of the elastic modulus tensor, viz. those that contain the index 2 only once:

$$C_{1121} \equiv c_{16} = 0; \quad C_{1123} \equiv c_{14} = 0; \quad C_{1321} \equiv c_{56} = 0;$$
$$C_{1323} \equiv c_{54} = 0; \quad C_{2133} \equiv c_{63} = 0; \quad C_{2333} \equiv c_{43} = 0. \tag{2.19}$$

This will be the case, for example, when x_2 is parallel to a two-fold rotation axis of the crystal or is perpendicular to a reflection plane. This result can be obtained from the transformation law for the elastic modulus tensor under an operation from the point group of the crystal with the use of

$$\boldsymbol{R} = \begin{pmatrix} -1 & 0 & 0 \\ 0 & 1 & 0 \\ 0 & 0 & -1 \end{pmatrix} \tag{2.20a}$$

in the case that R is a rotation through $180°$ about x_2, and

$$\boldsymbol{R} = \begin{pmatrix} 1 & 0 & 0 \\ 0 & -1 & 0 \\ 0 & 0 & 1 \end{pmatrix} \tag{2.20b}$$

in the case that R is reflection in the $x_1 x_3$-plane.

The boundary conditions (2.14) also decouple in this case:

$$\left[\left(c_{15} \frac{\partial}{\partial x_1} + c_{55} \frac{\partial}{\partial x_3} \right) u_1 + \left(c_{55} \frac{\partial}{\partial x_1} + c_{35} \frac{\partial}{x_3} \right) u_3 \right]_{x_3=0} = 0, \tag{2.21a}$$

$$\left[\left(c_{13} \frac{\partial}{\partial x_1} + c_{35} \frac{\partial}{\partial x_3} \right) u_1 + \left(c_{35} \frac{\partial}{\partial x_1} + c_{33} \frac{\partial}{\partial x_3} \right) u_3 \right]_{x_3=0} = 0 \tag{2.21b}$$

and

$$\left[\left(c_{46} \frac{\partial}{\partial x_1} + c_{44} \frac{\partial}{\partial x_3} \right) u_2 \right]_{x_3=0} = 0 \tag{2.22}$$

Let us focus our attention for the moment on the equations for $u_2(\boldsymbol{x}, t) = 0$. From (2.18), (2.11b) and (2.19), we find that for U_2 not to be identically zero we must have

$$\frac{1}{\varrho} \left[c_{66} + 2c_{46} \mathrm{i}\beta - c_{44}\beta^2 \right] - v^2 = 0. \tag{2.23}$$

The solution of this equation that has a positive real part is

$$\beta = \frac{1}{c_{44}} \left[c_{44}(c_{66} - \varrho v^2) - c_{46}^2 \right]^{1/2} + \mathrm{i}\frac{c_{46}}{c_{44}} \equiv \beta_{\mathrm{R}}(v) + \mathrm{i}\beta_{\mathrm{I}}, \tag{2.24}$$

provided that

$$c_{44}(c_{66} - \varrho v^2) - c_{46}^2 > 0. \tag{2.25}$$

It is only when (2.25) is satisfied that a surface wave can exist. The 2 component of the displacement field therefore has the form

$$u_2(\boldsymbol{x}, t) = U_2 e^{ikx_1 - k(\beta_{\mathrm{R}}(v) - i\beta_{\mathrm{I}})x_3 - ivkt} \ . \tag{2.26}$$

The boundary condition (2.23) yields the equation for determining v:

$$[c_{46} + c_{44}(i\beta_{\mathrm{R}}(v) - \beta_{\mathrm{I}})]U_2 = 0 \tag{2.27a}$$

or, in view of (2.24),

$$\beta_{\mathrm{R}}(v)U_2 = 0 \ . \tag{2.27b}$$

Thus, the satisfaction of this equation requires that (a) $U_2 \neq 0$, $\beta_{\mathrm{R}}(v) = 0$; (b) $U_2 = 0$, $\beta_{\mathrm{R}}(v) \neq 0$; or (c) $U_2 = 0$; $\beta_{\mathrm{R}}(v) = 0$. None of these alternatives corresponds to a surface acoustic wave.

The only nontrivial case, viz. $\beta_{\mathrm{R}}(v) = 0$, $U_2 \neq 0$, corresponds to a bulk transverse acoustic wave propagating parallel to the surface of the elastic medium. Although it is not a surface acoustic wave it has nevertheless been employed successfully in the construction of devices that have been configured into bandpass filters, delay lines, oscillators, and resonators [2.3].

We have therefore reached an important conclusion: a surface acoustic wave propagating in the x_1 direction in the plane $x_3 = 0$, when the x_2 axis is parallel to a two-fold rotation axis or perpendicular to a reflection plane, cannot be polarized perpendicular to the sagittal plane.

We now turn to the case of a surface acoustic wave polarized in the sagittal plane. Although the satisfaction of (2.19) decouples the sagittally polarized waves from those polarized perpendicular to the sagittal plane, the solution of (2.17) and the boundary conditions (2.21) is still sufficiently complicated that in general a computer has to be employed. However, if the crystal is oriented with respect to the coordinate axes in such a way that

$$C_{1113} \equiv c_{15} = 0 \ , \quad C_{3313} \equiv c_{35} = 0 \ , \tag{2.28}$$

in addition to the conditions given by (2.19), the frequency and displacement field of the associated surface acoustic wave can be obtained analytically [2.2].

The conditions (2.28) are satisfied if either the x_1 or x_3 axis is along a two-fold rotation axis or is perpendicular to a reflection plane.

In this case the solvability condition for (2.17) is the biquadratic equation

$$c_{33}c_{55}\beta^4 - \left[c_{33}(c_{11} - \varrho v^2) + c_{55}(c_{55} - \varrho v^2) - (c_{13} + c_{55})^2\right]\beta^2$$
$$+ (c_{11} - \varrho v^2)(c_{55} - \varrho v^2) = 0 \ . \tag{2.29}$$

We denote by $\beta_1(v)$ and $\beta_2(v)$ the two solutions of this equation for which $\mathrm{Re}\,\beta_r(v) > 0$. The general solution of the equations of motion of the medium is thus a linear combination of two partial waves:

$$u_1(\boldsymbol{x}, t) = \left[K_1 e^{-k\beta_1(v)x_3} + K_2 e^{-k\beta_2(v)x_3}\right] e^{ikx_1 - ivkt} \ , \tag{2.30a}$$

$$u_3(\boldsymbol{x}, t) = \left[p_1(v) K_1 e^{-k\beta_1(v)x_3} + p_2(v) K_2 e^{-k\beta_2(v)x_3} \right] e^{ikx_1 - ivkt} , \qquad (2.30b)$$

where

$$p_r(v) = \mathrm{i} \frac{c_{11} - \varrho v^2 - c_{55}\beta_r^2(v)}{(c_{13} + c_{55})\beta_r(v)} , \qquad r = 1, 2 . \qquad (2.31)$$

The new amplitudes K_1 and K_2 are obtained by substituting the solution (2.30) into the boundary conditions (2.21), which now take the form

$$\left[c_{55}\frac{\partial u_1}{\partial x_3} + c_{55}\frac{\partial u_3}{\partial x_1} \right]_{x_3=0} = 0 , \qquad (2.32a)$$

$$\left[c_{13}\frac{\partial u_1}{\partial x_1} + c_{33}\frac{\partial u_3}{\partial x_3} \right]_{x_3=0} = 0 . \qquad (2.32b)$$

The result is the pair of equations

$$(\beta_1 - \mathrm{i}p_1)K_1 + (\beta_2 - \mathrm{i}p_2)K_2 = 0 , \qquad (2.33a)$$

$$(c_{13} + \mathrm{i}c_{33}p_1\beta_1)K_1 + (c_{13} + \mathrm{i}c_{33}p_2\beta_2)K_2 = 0 , \qquad (2.33b)$$

for which the solvability condition is

$$(p_1 - p_2)(c_{13} + c_{33}\beta_1\beta_2) + \mathrm{i}(\beta_1 - \beta_2)(c_{13} - c_{33}p_1p_2) = 0 . \qquad (2.34)$$

The repeated use of (2.31) reduces this equation to

$$\beta_1\beta_2 = \frac{\varrho v^2(c_{11} - \varrho v^2)}{c_{33}(c_{11} - \varrho v^2) - c_{13}^2} . \qquad (2.35)$$

If we square this equation we obtain finally the equation determining the speed of surface acoustic waves polarized in the sagittal plane:

$$c_{33}c_{55}\varrho^2 v^4(c_{11} - \varrho v^2) = (c_{55} - \varrho v^2)\left[c_{33}(c_{11} - \varrho v^2) - c_{13}^2\right]^2 . \qquad (2.36)$$

The physical solution v_R of this cubic equation in v^2 is the one for which v is smallest. Indeed, it is a general result that the speed of a surface acoustic wave is lower than the speed of the slowest bulk wave propagating in the same direction. Thus, a surface acoustic wave propagates in a stop band for the normal modes of the corresponding bulk crystal.

From (2.30) and (2.33a) we find that the displacement field in the surface acoustic wave is given by

$$u_1(\boldsymbol{x}, t) = K_1 \left[e^{-k\beta_1(v)x_3} - \frac{\beta_1(v) - \mathrm{i}p_1(v)}{\beta_2(v) - \mathrm{i}p_2(v)} e^{-k\beta_2(v)x_3} \right] e^{ik(x_1 - vt)} , \qquad (2.37a)$$

$$u_3(\boldsymbol{x}, t) = K_1 p_1(v) \left[e^{-k\beta_1(v)x_3} - \frac{p_2(v)}{p_1(v)} \frac{\beta_1(v) - \mathrm{i}p_1(v)}{\beta_2(v) - \mathrm{i}p_2(v)} e^{-k\beta_2(v)x_3} \right]$$
$$\times e^{ik(x_1 - vt)} , \qquad (2.37b)$$

12

where $v = v_R$. The physical displacement components are of course given by the real parts of the right-hand sides of (2.37).

An important special case of the preceding discussion is provided by an isotropic elastic medium. For such a medium

$$c_{33} = c_{11} = \varrho v_l^2 \qquad c_{55} = \varrho v_t^2 , \qquad c_{13} = \varrho(v_l^2 - v_t^2) , \tag{2.38}$$

for any direction of propagation of the wave, on any surface, where v_l and $v_t(< v_l)$ are the speeds of sound for longitudinal and transverse waves in the medium. Equation (2.36) in this case takes the form

$$\xi^6 - 8\xi^4 + 8\left(3 - 2\frac{v_t^2}{v_l^2}\right)\xi^2 - 16\left(1 - \frac{v_t^2}{v_l^2}\right) = 0 , \tag{2.39}$$

where $\xi = v_R/v_t$.

Of the three solutions of this equation for ξ^2 only one, the smallest, satifies the condition $\xi < 1$, that is the condition for the existence of a localized wave. An approximate expression for this root, that is in error by less than 0.5%, is [2.4]

$$\xi = \frac{0.87 + 1.12\sigma}{1 + \sigma} , \tag{2.40a}$$

where

$$\sigma = \frac{1 - 2v_t^2/v_l^2}{2(1 - v_t^2/v_l^2)} \tag{2.40b}$$

is called *Poisson's ratio*, and decreases from $1/2$ to -1 as v_t^2/v_l^2 increases from 0 to $3/4$. An isotropic elastic medium is stable only for $0 < v_t^2/v_l^2 < 3/4$ [2.5]. For actual materials, σ ranges from 0 to $1/2$.

The components of the displacement field are given by

$$u_1(\boldsymbol{x}, t) = K\left[e^{-k\beta_l x_3} - (\beta_l\beta_t)^{1/2}e^{-k\beta_t x_3}\right]\cos[k(x_1 - v_R t)] , \tag{2.41a}$$

$$u_3(\boldsymbol{x}, t) = -K\beta_l\left[e^{-k\beta_l x_3} - \frac{e^{-k\beta_t x_3}}{(\beta_l\beta_t)^{1/2}}\right]\sin[k(x_1 - v_R t)] , \tag{2.41b}$$

where

$$\beta_l = \left(1 - \frac{v_R^2}{v_l^2}\right)^{1/2} , \qquad \beta_t = \left(1 - \frac{v_R^2}{v_t^2}\right)^{1/2} . \tag{2.42}$$

We see, from (2.41), that the particle displacements in a Rayleigh wave execute ellipses in the sagittal plane. Near the surface the Rayleigh wave motion is retrograde, and it reverses its sense at depths greater than approximately one-fifth of a wavelength. The major axes of the ellipses are normal to the surface, and the aspect ratio varies with depth. The components of the displacement field as a function of the distance into the medium from the surface are shown in Fig. 2.1.

13

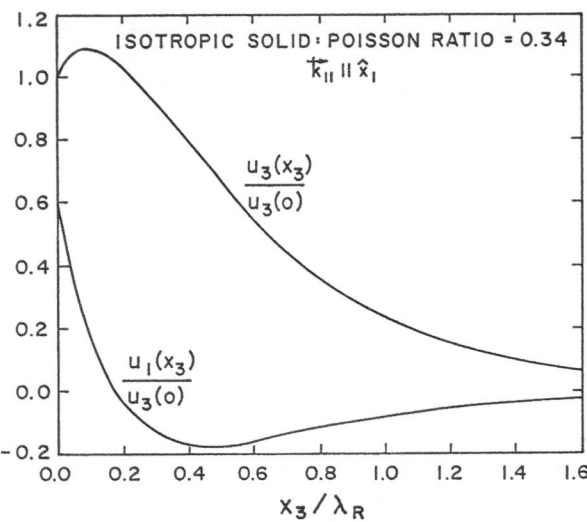

Fig. 2.1. Normalized displacement components in a Rayleigh wave on an isotropic medium as functions of the distance into the medium from the surface [2.6]

In the case of propagation in an arbitrary direction on an anisotropic surface, i.e. in a direction on a surface for which the conditions (2.19, 28) are not satisfied, we are forced to use (2.11, 13, 15, 16) to obtain the dispersion relation for surface acoustic waves and the associated displacement field. The resulting surface acoustic waves have the following general properties.

1. The surface acoustic wave generally has three nonzero displacement components and is the superposition of three partial waves.
2. The partial waves are characterized by the corresponding decay constants $\{\beta_j(v)\}$. Three different situations are possible.
   (a) All three $\{\beta_j(v)\}$ are real and positive. The resulting surface acoustic wave is called an *ordinary Rayleigh wave*.
   (b) It can happen that the sixth-degree equation for the $\{\beta_j(v)\}$ degenerates into a cubic equation in β^2 with real coefficients. Thus not only is β^* a solution if β is, but so is $-\beta$. It is then possible for one of the $\{\beta_j(v)\}$ to be real and the other two to be complex conjugates of each other, with $\mathrm{Re}\{\beta_j(v)\}$ positive in each case. The amplitudes of the latter two waves are such that the corresponding contribution to the displacement field has the form $\exp(-k\beta_R(v)x_3)\sin[k(\beta_I(v)x_3 + \delta)]$. The decay of the amplitude of this part of the displacement field thus has a more complicated dependence on x_3 than the simple exponential decay characteristic of Rayleigh waves on isotropic media. The resulting surface acoustic wave is called a *generalzed Rayleigh wave* [2.18].
   (c) One or more of the $\beta_j(v)$ is complex in such a way that the corresponding partial wave radiates energy into the interior of the solid from

the surface. By conservation of energy the wave is therefore attenuated in the direction of propagation, and k becomes complex. Such surface acoustic waves are called *pseudosurface* or *leaky surface acoustic waves*. They will be discussed in greater detail below.

3. The group velocity vector for a surface acoustic wave is not parallel to the direction of propagation in general.

Of the three types of surface acoustic waves that can exist on a surface of an anisotropic medium a particularly interesting one is the pseudosurface, or leaky surface, wave. This is a wave that depends on the presence of a surface for its existence, and whose phase velocity is greater than that of the slowest bulk transverse wave. The pseudosurface wave is attenuated as it propagates along the surface of a solid. As a consequence k acquires an imaginary part whose sign is such that $\text{Im}\{kx_1\}$ is positive when $\text{Re}\{kx_1\}$ is positive so that the wave is attenuated in the direction of propagation.

In many cases the attenuation is small enough so that the leaky surface acoustic wave is readily observable. Such waves on the surfaces of semi-infinite anisotropic elastic media were discovered by *Engan* et al. [2.9] in 1967.

Leaky surface acoustic waves on anisotropic media were subsequently studied extensively, both theoretically and experimentally, by several authors. It is now known that the presence of anisotropy is not necessary for the existence of leaky surface acoustic waves. In fact, the earliest example of a leaky surface acoustic wave was a wave at the plane interface between an isotropic elastic medium and a liquid [2.10–13]. In this case, the wave is localized in the elastic medium, but as it propagates along the interface it radiates energy into the liquid, and is attenuated in the direction of propagation as a result. Other types of leaky surface acoustic waves at an interface in which attenuation takes place due to the radiation of energy into the medium adjacent to the one in which the wave is localized are Lamb waves in plates immersed in a liquid [2.14, 15], waves at the boundary of a liquid with a solid layer [2.16], and waves at the plane interface between two solid media [2.17, 18].

A more interesting type of leaky surface acoustic wave is the class of waves at a solid–vacuum interface for which energy is radiated not into an adjacent medium but back into the same medium in which the wave is localized. Such waves are thus intrinsic properties of the media supporting them. We have already seen an example of this in the case of leaky surface acoustic waves on anisotropic elastic media. An earlier example is the case of Raleigh waves propagating circumferentially around a cylindrical isotropic solid surface that is concave toward the vacuum [2.19–21]. Yet another example is provided by a semi-infinite isotropic elastic medium in contact with a thin liquid layer [2.22-24]. Leaky surface acoustic waves are known to exist on the planar, stress-free surface of a semi-infinite, isotropic elastic medium [2.25, 26], and their existence has led us to the recent discovery of a new type of leaky surface acoustic wave on anisotropic media by *Camley* and *Nizzoli* [2.27]. Their results for the phase velocities of the several waves that can propagate on the (100) surface of a cubic crystal as

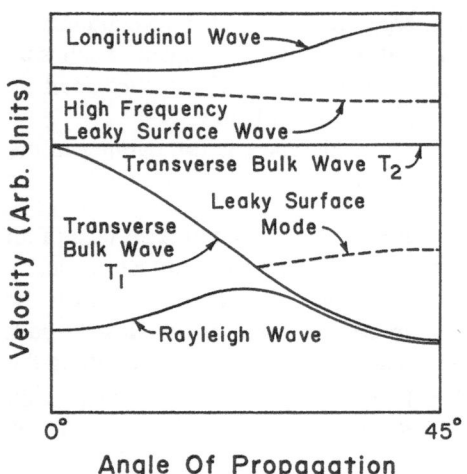

Fig. 2.2. Schematic illustration of the various bulk and surface acoustic modes that propagate on a (001) surface of a cubic crystal. The angle of propagation is the angle from the [100] direction [2.27]

functions of the angle the direction of propagation makes with the [001] direction are depicted schematically in Fig. 2.2. The leaky surface acoustic wave that has a phase velocity between the phase velocities of the slower and faster bulk transverse acoustic waves, which are denoted by T_1 and T_2, respectively, is one that was discovered earlier by *Rollins* et al. [2.28]. The new surface acoustic wave found by *Camley* and *Nizzoli* [2.27] is a high frequency wave whose phase velocity is above that of the T_2 transverse acoustic wave, but below that of the bulk longitudinal wave (L) propagating along the surface. This wave is the analogue of the leaky surface acoustic wave discussed by *Glass* and *Maradudin* [2.26], and like that one is strongly attenuated. This can be understood as follows. As one approaches the isotropic limit, the phase velocity curves labeled T_1 and T_2 collapse into one curve with a phase velocity c_t. The low frequency leaky surface acoustic wave is essentially squeezed out of existence, since it lies between T_1 and T_2. However, the phase velocity of the high frequency leaky surface acoustic wave remains between T_1 and L, and so remains in the isotropic limit. The nature of this mode is that it is a superposition of the longitudinal wave that is localized to the surface and two transverse waves that radiate energy into the bulk and whose amplitudes grow exponentially with increasing distance into the bulk of the solid.

2.1.2 Piezoelectric Media

Deforming a piezoelectric medium produces an electric field in it; applying an electric field to it deforms it. Acoustic surface waves can exist on such media. Piezoelectricity causes only second-order effects on the dominantly mechanical nature of acoustic surface waves. However, the elastic deformation of the medium in such waves is accompanied by an electric field. It is this electric field that makes piezoelectric surface waves interesting from a technological standpoint,

16

for it represents the way in which a surface wave is usually coupled to external electrical circuits, to other surface waves, or to charge carriers in semiconductors.

The equation of motion of a piezoelectric medium are still given by (2.1), except that the stress tensor now possesses an electrical contribution, in addition to the usual mechanical contribution,

$$T_{\alpha\beta} = \sum_{\mu\nu} C_{\alpha\beta\mu\nu} \frac{\partial u_\mu}{\partial x_\nu} - \sum_\nu e_{\mu\alpha\beta} E_\mu , \tag{2.43}$$

where E is the macroscopic field in the medium, the $\{C_{\alpha\beta\mu\nu}\}$ are the elastic moduli (at constant macroscopic field), and $e_{\mu\alpha\beta}$ is the piezoelectric tensor. The latter is symmetric in the second pair of indices. The equations of motion have to be supplemented by the constitutive relation connecting the electric displacement D with the displacement gradients and the macroscopic electric field,

$$D_\alpha = \sum_{\mu\nu} e_{\alpha\mu\nu} \frac{\partial u_\mu}{\partial x_\nu} + \sum_\beta \varepsilon_{\alpha\beta} E_\beta , \tag{2.44}$$

where $\varepsilon_{\alpha\beta}$ is the dielectric tensor of the medium (at constant strain). It is symmetric in the indices α and β. We note that (2.43, 44) are written in the MKS system of units.

Since the disturbances of interest to us propagate with sonic rather than electromagnetic speeds, we can use the quasi-static, or electrostatic, approximation and write the electric field as the gradient of a scalar potential,

$$E_\alpha = \frac{\partial\varphi}{\partial x_\alpha} . \tag{2.45}$$

Equations (2.75, 76) then take the forms

$$T_{\alpha\beta} = \sum_{\mu\nu} C_{\alpha\beta\mu\nu} \frac{\partial u_\mu}{\partial x_\nu} + \sum_\mu e_{\mu\alpha\beta} \frac{\partial\varphi}{\partial x_\mu} , \tag{2.46}$$

$$D_\alpha = \sum_{\mu\nu} e_{\alpha\mu\nu} \frac{\partial u_\mu}{\partial x_\nu} - \sum_\beta \varepsilon_{\alpha\beta} \frac{\partial\varphi}{\partial x_\beta} , \tag{2.47}$$

and the equations of motion become

$$\varrho \frac{\partial^2}{\partial t^2} u_\alpha = \sum_{\beta\mu\nu} C_{\alpha\beta\mu\nu} \frac{\partial^2 u_\mu}{\partial x_\beta \partial x_\mu} + \sum_{\beta\mu} e_{\mu\alpha\beta} \frac{\partial^2\varphi}{\partial x_\beta \partial x_\mu} . \tag{2.48}$$

A closed system of equations is obtained when we include the Maxwell equation

$$\nabla \cdot D = \sum_{\alpha\mu\nu} e_{\alpha\mu\nu} \frac{\partial^2 u_\mu}{\partial x_\alpha \partial x_\nu} - \sum_{\alpha\beta} \varepsilon_{\alpha\beta} \frac{\partial^2\varphi}{\partial x_\alpha \partial x_\beta} = 0 . \tag{2.49}$$

In what follows we use (2.48, 49) as the basis for a discussion of surface acoustic waves on a semi-infinite piezoelectric medium that occupies the region

17

$x_3 > 0$ and is in contact with vacuum in the region $x_3 < 0$. We first formulate the problem of obtaining their dispersion curves in some generality, and then specialize the general treatment to the particular case of a shear horizontal surface acoustic wave that has no counterpart for an elastic medium.

a) **Piezoelectric Surface Acoustic Waves.** In seeking solutions of (2.48, 49) that describe surface waves, we assume forms for the displacement amplitudes and for the potential φ that describe a wave propagating in the x_1 direction that decays exponentially in the x_3 direction, and is independent of the coordinate x_2:

$$u_\beta(\mathbf{x}, t) = A_\beta e^{ikx_1 - \alpha x_3 - i\omega t}, \quad \beta = 1, 2, 3, \tag{2.50a}$$

$$\varphi(\mathbf{x}, t) = A_4 e^{ikx_1 - \alpha x_3 - i\omega t}. \tag{2.50b}$$

When we substitute these expressions into (2.48, 49), we obtain the following determinantal equation for the amplitudes $\{A_\beta\}$:

$$\begin{pmatrix} \Gamma_{11} - \varrho v^2 & \Gamma_{12} & \Gamma_{13} & \Gamma_{14} \\ \Gamma_{12} & \Gamma_{22} - \varrho v^2 & \Gamma_{23} & \Gamma_{24} \\ \Gamma_{13} & \Gamma_{23} & \Gamma_{33} - \varrho v^2 & \Gamma_{34} \\ \Gamma_{14} & \Gamma_{24} & \Gamma_{34} & \Gamma_{44} \end{pmatrix} \begin{pmatrix} A_1 \\ A_2 \\ A_3 \\ A_4 \end{pmatrix} = 0, \tag{2.51}$$

where

$$\Gamma_{11} = c_{11} + 2c_{15}i\beta - c_{55}\beta^2,$$

$$\Gamma_{22} = c_{66} + 2c_{46}i\beta - c_{44}\beta^2,$$

$$\Gamma_{33} = c_{55} + 2c_{35}i\beta - c_{33}\beta^2, \tag{2.52}$$

$$\Gamma_{12} = c_{16} - (c_{14} + c_{56})i\beta - c_{45}\beta^2,$$

$$\Gamma_{13} = c_{15} + (c_{13} + c_{55})i\beta - c_{55}\beta^2,$$

$$\Gamma_{23} = c_{56} + (c_{36} + c_{45})i\beta - c_{34}\beta^2,$$

$$\Gamma_{14} = e_{11} + (e_{15} + e_{31})i\beta - e_{35}\beta^2,$$

$$\Gamma_{24} = e_{16} + (e_{14} + e_{36})i\beta - e_{34}\beta^2,$$

$$\Gamma_{34} = e_{15} + e_{13} + e_{35})i\beta - e_{33}\beta^2,$$

$$\Gamma_{44} = -[\varepsilon_{11} + 2\varepsilon_{13}i\beta - \varepsilon_{33}\beta^2].$$

In writing these equations we have set

$$\beta = \alpha/k, \quad v = \omega/k, \tag{2.53}$$

and have used the Voigt, or contracted, notation in writing the elastic moduli and the elements of the piezoelectric tensor.

In order that (2.51) have a nontrivial solution the determinant of the coefficients must be equated to zero. This leads to an eighth-degree equation for β. For real values of v only four of these roots have a positive real part and so can describe a wave localized at the surface. We denote these by β_j. They are functions of v.

We superpose these four solutions to satisfy the boundary conditions at the surface $x_3 = 0$:

$$u_\alpha(\boldsymbol{x}, t) = \sum_{j=1}^{4} K_j C_\alpha^{(j)} e^{ikx_1 - k\beta_j x_3 - ikvt} , \quad \alpha = 1, 2, 3 , \quad x_3 > 0 \qquad (2.54a)$$

$$\varphi(\boldsymbol{x}, t) = \sum_{j=1}^{4} K_j C_4^{(j)} e^{ikx_1 - k\beta_j x_3 - ikvt} , \quad x_3 > 0 . \qquad (2.54b)$$

In the vacuum region $x_3 < 0$ only the scalar potential is nonzero. It satisfies Laplace's equation,

$$\left(\frac{\partial^2}{\partial x_1^2} + \frac{\partial^2}{\partial x_3^2} \right) \hat{\varphi} = 0 . \qquad (2.55)$$

The solution of this equation that vanishes as $x_3 \to -\infty$ is

$$\hat{\varphi}(\boldsymbol{x}, t) = K_5 e^{ikx_1 + |k|x_3 - i\omega t} \quad x_3 < 0 , \qquad (2.56)$$

where K_5 is the potential in the medium evaluated at the surface $x_3 = 0$. According to (2.54b) this is given by

$$K_5 = \sum_{j=1}^{4} K_j C_4^{(j)} . \qquad (2.57)$$

Satisfaction of (2.57) ensures the continuity of the tangential components of the electric field across the surface $x_3 = 0$.

The boundary conditions that must be satisfied at the surface $x_3 = 0$ are now four in number, which equals the number of unknowns in the problem, viz. K_1, K_2, K_3, K_4. These are the vanishing of the stresses acting on the surface,

$$T_{\alpha 3} = 0 \quad \alpha = 1, 2, 3 , \qquad (2.58)$$

and the continuity of the normal component of the electric displacement,

$$D_3 = -\varepsilon_0 \frac{\partial}{\partial x_3} \hat{\varphi} , \qquad (2.59)$$

where ε_0 is the permittivity of free space.

When the solutions given by (2.54a) are substituted into these homogeneous boundary conditions and the stress components and D_3 are obtained from (2.46, 47), a set of four homogeneous linear equations for the coefficients is obtained. These equations can be written in the form

$$\sum_j M_{ij} K_j = 0 , \qquad (2.60)$$

19

where the elements of the matrix M are given by

$$M_{1j} = \sum_{\mu}(C_{13\mu1} + iC_{13\mu3}\beta_j)C_{\mu}^{(j)} + (e_{113} + ie_{313}\beta_j)C_4^{(j)} \, , \tag{2.61a}$$

$$M_{2j} = \sum_{\mu}(C_{23\mu1} + iC_{23\mu3}\beta_j)C_{\mu}^{(j)} + (e_{123} + ie_{323}\beta_j)C_4^{(j)} \, , \tag{2.61b}$$

$$M_{3j} = \sum_{\mu}(C_{33\mu1} + iC_{33\mu3}\beta_j)C_{\mu}^{(j)} + (e_{133} + ie_{333}\beta_j)C_4^{(j)} \, , \tag{2.61c}$$

$$M_{4j} = \sum_{\mu}(e_{3\mu1} - ie_{3\mu3}\beta_j)C_{\mu}^{(j)} - (\varepsilon_{31} + i\varepsilon_{33}\beta_j + i\varepsilon_0 sgn k)C_4^{(j)} \, . \tag{2.61d}$$

The indexing of the rows in the matrix follows the order of the four boundary conditions given by (2.58, 59).

The condition that the set of equations (2.60) have a nontrivial solution is that the determinant of the matrix M vanish,

$$|M_{ij}| = 0 \, . \tag{2.62}$$

The resulting equation determines the value of v, the speed of surface piezoelectric waves.

Equation (2.62) has to be solved numerically, in general. Even without such a solution, however, we see from the form of (2.60, 61) that piezoelectric surface acoustic waves, like Rayleigh waves, are nondispersive, because of the absence of a characteristic length in the problem. Also, like Rayleigh waves on anisotropic media, the particle motion in a piezoelectric surface acoustic wave is not confined to the sagittal plane: although nothing varies with x_2, there is in general a nonvanishing 2 component of the displacement field.

Numerical solutions of (2.62) have been carried out by several authors [2.29–31] for several piezoelectric surfaces. It is found that the displacement field in a piezoelectric surface acoustic wave is little changed by the piezoelectricity from what it would be for the corresponding purely elastic medium ($e_{\mu\alpha\beta} \equiv 0$, $\varepsilon_{\alpha\beta} \equiv 0$). The new feature is the electric field associated with the surface acoustic wave that is localized to the surface. In Fig. 2.3 we have plotted the results of *Tseng* and *White* [2.29] for the normalized piezoelectric field components versus distance from the surface, both into the medium and into the vacuum outside it, for a surface acoustic wave propagating on the basal plane of CdS. Both the piezoelectric field and the particle displacement vector are elliptically polarized in the sagittal plane. The variation of the electric field outside the crystal with distance from the surface has been measured [2.29]. It is found to decrease exponentially and to have a decay length that is approximately equal to the wavelength of the wave parallel to the surface devided by 2π. Both of these experimental results are in agreement with the predictions of theory [2.29].

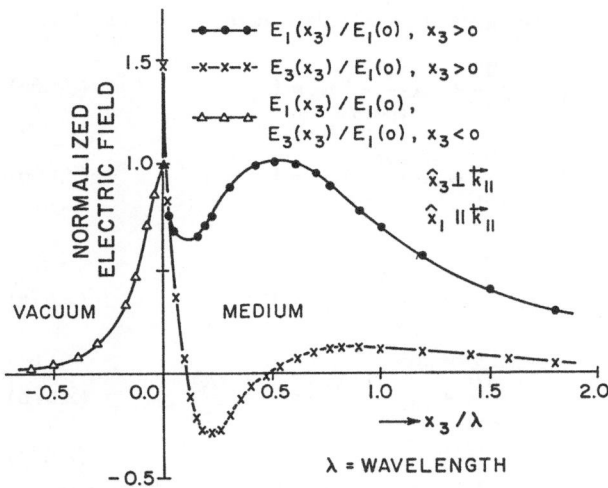

Fig. 2.3. Normalized electric field components in a piezoelectric surface acoustic wave as functions of the distance from the surface [2.29]

b) Bleustein–Gulyaev Surface Acoustic Waves. The equations of motion and the associated boundary conditions, governing the propagation of surface acoustic waves on a piezoelectric medium simplify greatly for propagation of such a wave along a symmetric direction on a surface that is itself a plane of symmetry for the medium. Moreover, under such conditions a new type of surface acoustic wave becomes possible that has no counterpart for purely elastic media. This is a shear horizontal surface acoustic wave discovered theoretically by *Bleustein* [2.32] and *Gulyaev* [2.33]. In what follows we outline the theory of these waves.

Let us return to (2.51). We see from them that if

$$\Gamma_{12} = \Gamma_{14} = \Gamma_{23} = \Gamma_{34} \equiv 0 \tag{2.63}$$

they separate into two uncoupled pairs of equations:

$$(\Gamma_{11} - \varrho v^2)A_1 + \Gamma_{13}A_3 = 0 , \tag{2.64a}$$

$$\Gamma_{13}A_1 + (\Gamma_{33} - \varrho v^2)A_3 = 0 \tag{2.64a}$$

and

$$(\Gamma_{22} - \varrho v^2)A_2 + \Gamma_{24}A_4 = 0 , \tag{2.65a}$$

$$\Gamma_{24}A_2 + \Gamma_{44}A_4 = 0 . \tag{2.65b}$$

It follows from (2.52) that the satisfaction of (2.63) requires that all elements of the elastic modulus tensor $C_{\alpha\beta\mu\nu}$ in which the subscript 2 appears once must vanish, and all elements of the piezoelectric tensor $e_{\mu\alpha\beta}$ in which the subscripts are all 1 and/or 3 must also vanish. These two conditions are satisfied if the x_2 axis is a two-fold rotation axis. In what follows we will assume this is the case.

21

The boundary conditions at the surface $x_3 = 0$ also decouple in this case:

$$\left(c_{15}\frac{\partial}{\partial x_1} + c_{55}\frac{\partial}{\partial x_3}\right)u_1 + \left(c_{55}\frac{\partial}{\partial x_1} + c_{35}\frac{\partial}{\partial x_3}\right)u_3 = 0 , \tag{2.66a}$$

$$\left(c_{13}\frac{\partial}{\partial x_1} + c_{35}\frac{\partial}{\partial x_3}\right)u_1 + \left(c_{35}\frac{\partial}{\partial x_1} + c_{33}\frac{\partial}{\partial x_3}\right)u_3 = 0 , \tag{2.66b}$$

and

$$\left(c_{46}\frac{\partial}{\partial x_1} + c_{44}\frac{\partial}{\partial x_3}\right)u_2 + \left(\varepsilon_{14}\frac{\partial}{\partial x_3} + \varepsilon_{34}\frac{\partial}{\partial x_1}\right)\varphi = 0 , \tag{2.67a}$$

$$\left(e_{36}\frac{\partial}{\partial x_1} + e_{34}\frac{\partial}{\partial x_3}\right)u_2 - \left(\varepsilon_{31}\frac{\partial}{\partial x_1} + \varepsilon_{33}\frac{\partial}{\partial x_3}\right)\varphi = -\varepsilon_0\frac{\partial\hat{\varphi}}{\partial x_3} , \tag{2.67b}$$

$$\varphi = \hat{\varphi} . \tag{2.67c}$$

If solutions to (2.64, 66) exist, they correspond to a surface acoustic wave polarized in the sagittal plane, from which piezoelectric effects are absent. We will not consider them further here, since they are equivalent to the similarly polarized surface acoustic waves that propagate on a purely elastic medium.

The solution of (2.65, 67), if it exists, describes a surface acoustic wave polarized perpendicular to the sagittal plane. To determine the general properties of such a wave let us assume for simplicity that the symmetry of our system is so high that

$$c_{46} = 0 \quad c_{66} = c_{44} , \tag{2.68a}$$

$$e_{14} = e_{36} = 0 , \quad e_{34} = e_{16} , \tag{2.68b}$$

$$\varepsilon_{13} = 0 , \quad \varepsilon_{33} = \varepsilon_{11} . \tag{2.68c}$$

These conditions will be satisfied, for example, for a crystal, such as CdS, belonging to the crystal class C_{6v}, whose axis of six-fold symmetry is parallel to the x_2 direction. In this case (2.65) take the form

$$\begin{pmatrix} c_{44}(1-\beta^2) - \varrho v^2 & e_{16}(1-\beta^2) \\ e_{16}(1-\beta^2) & -\varepsilon_{11}(1-\beta^2) \end{pmatrix}\begin{pmatrix} A_2 \\ A_4 \end{pmatrix} = 0 . \tag{2.69}$$

From the solvability condition for this system of equations, and (2.50), we obtain the result that inside the medium

$$u_2(x_1 x_3; t) = e^{ik(x_1 - vt)}\frac{\varepsilon_{11}}{e_{16}}A_4^{(2)}e^{-k\beta x_3} , \tag{2.70}$$

$$\varphi(x_1 x_3; t) = e^{ik(x_1 - vt)}[A_4^{(1)}e^{-kx_3} + A_4^{(2)}e^{-k\beta x_3}] . \tag{2.71}$$

In these expressions

$$\beta = \left(1 - \frac{\varrho v^2}{\bar{c}_{44}}\right)^{1/2} \tag{2.72}$$

is real and positive, where

$$\bar{c}_{44} = c_{44} + \frac{e_{16}^2}{\varepsilon_{11}} \tag{2.73}$$

is a piezoelectrically stiffened elastic modulus. The satisfaction of the condition that β be real and positive requires that

$$v^2 < \frac{\bar{c}_{44}}{\varrho} . \tag{2.74}$$

The scalar potential in the vacuum is

$$\hat{\varphi}(x_1 x_3; t) = B e^{ik(x_1 - vt)} e^{kx_3} . \tag{2.75}$$

The boundary conditions (2.67) at $x_3 = 0$ now take the form

$$c_{44} \frac{\partial}{\partial x_3} u_2 + e_{16} \frac{\partial}{\partial x_3} \varphi = 0 , \tag{2.76a}$$

$$e_{16} \frac{\partial}{\partial x_3} u_2 - \varepsilon_{11} \frac{\partial}{\partial x_3} \varphi = -\varepsilon_0 \frac{\partial \hat{\varphi}}{\partial x_3} , \tag{2.76b}$$

$$\varphi = \hat{\varphi} . \tag{2.76c}$$

When the solutions given by (2.70, 71, 75) are substituted into (2.76), the solvability condition for the resulting set of homogeneous equations for $A_4^{(1)}$, $A_4^{(2)}$, and B yields the phase velocity for Bleustein–Gulyaev surface acoustic waves,

$$v^2 = \frac{\bar{c}_{44}}{\varrho} \left[1 - \frac{e_{16}^4}{\bar{c}_{44}^2 \varepsilon_{11}^2 (1 + \varepsilon_{11}/\varepsilon_0)^2} \right] . \tag{2.77}$$

The corresponding value of β is

$$\beta = \frac{e_{16}^2}{\varepsilon_{11} \bar{c}_{44}} \frac{1}{1 + \varepsilon_{11}/\varepsilon_0} . \tag{2.78}$$

The continuity of φ and D_3 across the plane $x_3 = 0$ are not the only electrical boundary conditions we can assume in this problem. An alternative set of boundary conditions are the short circuit boundary conditions. These apply when the surface $x_3 = 0$ is slightly metallized so that the mechanical boundary conditions are not affected but the electrical boundary condition now requires that $\hat{\varphi}$ vanish identically. It follows that φ must vanish at $x_3 = 0$, and so must D_3:

$$\varphi = 0 , \tag{2.79a}$$

$$\varepsilon_{16} \frac{\partial}{\partial x_3} u_2 - \varepsilon_{11} \frac{\partial}{\partial x_3} \varphi = 0 . \tag{2.79b}$$

When (2.70, 71) are substituted into (2.76a, 79), we find that

$$v^2 = \frac{\bar{c}_{44}}{\varrho} \left[1 - \frac{e_{16}^4}{\bar{c}_{44}^2 \varepsilon_{11}^2} \right] . \qquad (2.80)$$

The corresponding value of β is

$$\beta = \frac{e_{16}^2}{\varepsilon_{11} \bar{c}_{44}} . \qquad (2.81)$$

From a comparison of the expressions for β given by (2.78, 81) we see that Bleustein–Gulayaev waves on a metallized surface are more useful in device applications than such waves on a surface with free electrical boundary conditions, because β in the former case is larger by a factor $\varepsilon_0 + \varepsilon_{11} \sim 10$ than β in the latter case. This means that the fields, both mechanical and electrical, and hence the energy carried by the wave, are more tightly bound to the surface in the case of a metallized surface [2.34].

We emphasize that a shear horizontal surface acoustic wave does not exist on non-piezoelectric substrates. That such a wave exists on a piezoelectric substrate is due directly to the piezoelectricity of the medium. Indeed the reason that the condition (2.74) is satisfied by the solutions given by (2.77, 80) is the presence of a nonzero piezoelectric constant e_{16}. For CdS $v^2 = 0.9999875 \, \bar{c}_{44}/\varrho$, for the free boundary case.

The first experimental observation of Bleustein–Gulyaev surface acoustic waves was by *Maerfeld* et al. [2.35]. Subsequently, *Koerber* and *Vogel* [2.36] described the criteria for the existence of these waves in crystal classes other than C_{6v}, and they have now been studied theoretically in rhombic and monoclinic crystals [2.37–39].

2.2 Generation and Detection of Surface Acoustic Waves

Surface waves on solids can be excited by externally applied sources, both mechanical and electrical. As discussed in the preceeding sections, these waves are characterized by spatially periodic displacements of the surface that oscillate at the frequency ω. The most common way to excite surface waves, therfore, is to apply forces [2.4, 6, 40], fields, charges [2.4, 40], etc. along the surface with temporal and spatial periodicity identical to that of the surface waves. This distributed coupling approach relies on a coherent (wave vector matched) interaction between the amplitudes of the external and surface wave fields. In the case of non piezoelectric media, excitation requires mechanical surface sources in the form of forces, periodic temperature gradients, etc. If the material is piezoelectric, surface charges and electric fields can also be used to generate surface waves, and in fact the operating principles of the interdigital transducer are based on this case.

The detection of surface waves is the reverse process to generation. Surface wave fields propagating along the surface induce forces, charges, etc. inside the same external sources which were used for generating surface waves. For example, the electric fields associated with surface waves on piezoelectric media induce charges on electrodes placed on the surface. Externally excited surface waves can also be detected by light scattering [2.41]. The strains created by the surface acoustic wave modulate the refractive index inside the material via the electro-optic effect. Surface acoustic waves also cause a "rippling" of the surface that leads to diffraction-grating-like scattering of light incident onto that surface. Both energy and wave vector (parallel to the surface) are conserved in the light–sound interaction so that the frequency of the scattered light is shifted from that of the incident by the acoustic frequency, and the direction of the scattered light is also different from that of the incident light.

Surface and bulk acoustic waves are always present at all frequencies, and traveling in all possible directions on every solid in the form of acoustic "noise". At room temperature, all of the possible acoustic normal modes are excited with an average energy of $k_B T/2$, where k_B is Boltzmann's constant and T is the temperature in Kelvin. Since they are generated by statistical fluctuations, their phases are uncorrelated and their amplitudes interfere destructively when they are picked up by the "external sources" discussed above (which are sensitive to the wave amplitudes). They can, however, be detected by Brillouin scattering, in which the intensity (and not the amplitude) of the light scattered from multiple surface waves is sensed by virtue of the square-law nature of optical detectors [2.42–44].

In the remainder of this section we shall discuss these techniques for generating and detecting surface waves.

2.2.1 Electrical Generation and Detection in Piezoelectric Media

Electrical signals oscillating at megahertz to multigigahertz frequencies can be converted directly into surface waves on a piezoelectric medium by interdigital transducers [2.4, 40]. They consist of interdigitated metal fingers deposited on the piezoelectric substrate, with every alternate finger connected electrically to a common electrical terminal, as shown in Fig. 2.4. This periodic structure has characteristic spatial periodicities given by the Fourier analysis of the surface potential fields created by applying a voltage to the grid structure, that is [2.4, 40]

$$V(x) \propto - \sum_n \frac{V_0}{2n+1} \sin \left[\frac{(2n+1)\pi x}{L} \right] . \tag{2.82}$$

Coupling to the $(2n+1)$th spatial harmonic of the transducer can generate surface waves with wave vector $(2n+1)\pi/L$. The charges induced on the transducer fingers can be considered to drive the electric fields associated with a surface wave in piezoelectric media. When the oscillating voltage is applied, each finger pair emits surface waves in both directions because of the bidirectional symmetry of each finger pair. Constructive interference occurs between these wavelets when the transducer is excited at a frequency given by $\omega = (2n+1)\pi v_s/L$, where

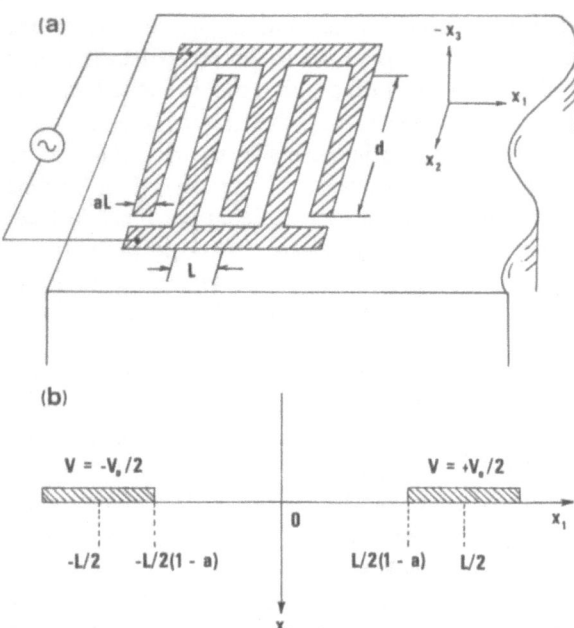

Fig. 2.4a,b. Typical interdigital transducer geometry

v_s is the surface wave velocity. In practice, tranducers have a finite number of fingers, say N, so that the surface waves are emitted over a range of frequencies. For frequencies ω near those required for the synchronous excitation of a surface wave by the $(2n+1)$th spatial harmonic, ω_{2n+1}, the relative surface wave power generated is proportional to

$$\frac{\sin^2 \Delta}{\Delta^2} \; ; \quad \Delta = N\pi \frac{\omega - \omega_{2n+1}}{\omega_{2n+1}} \; . \tag{2.83}$$

A sketch of this response function, which is valid for the weak coupling case, is shown in Fig. 2.5. (For strong coupling, distortions in the response function are produced via multireflections inside the transducer, and by feedback of the generated field when it becomes strong [2.45].) If the electrical impedance of the transducer is matched properly to that of the signal source, almost 100% conversion of electrical to acoustic power occurs and surface waves are generated simultaneously in both directions orthogonal to the fingers. It is not necessary that the substrate medium be piezoelectric. For example, piezoelectric thin films can be used to cover transducers deposited on non-piezoelectric media [2.46].

The actual generation efficiency is determined by many factors such as the material piezoelectricity, the propagation direction and plane, the number of finger pairs, the electrical impedance of the source, etc. The applied electric field-surface wave coupling strength can be estimated from the relative change in the surface wave velocity $\Delta v_s/v_s$ which occurs when the propagation surface is metallized, i.e. metallization changes the electrical boundary conditions at the

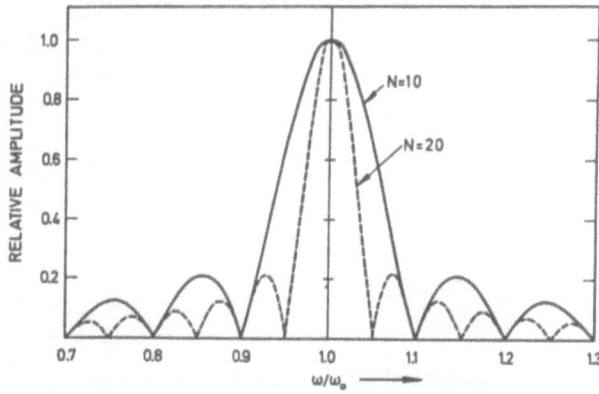

Fig. 2.5. Frequency response of an interdigital transducer for the generation of surface waves on a piezoelectric material with ω_0 the transducer fundamental frequency [2.44]

surface with the result that the tangential surface electric fields are zero. In order to identify promising materials and propagation surfaces, computer techniques have been developed to calculate these velocity differences and extensive compilations are now available [2.47].

When a surface wave is incident onto a transducer that is connected to an electrical circuit, charges are induced by the surface wave on the transducer fingers and hence current flows in the electrical circuit. In practice, some of the energy is lost owing to the simultaneous generation of bulk waves [2.48]. Charges can also be induced onto the tip of a metallic probe placed close to the surface which allows the surface wave to be probed along the propagation surface. (By electrically mixing the probe signal with a reference signal from the oscillator which excited the surface wave, both the phase and magnitude of the surface wave can be measured at any point along the surface [2.49].)

2.2.2 Generation on Non-piezoelectric Media

Before the invention of the interdigital transducer, many different techniques were developed for exciting surface waves by the application of spatially and temporally periodic forces to the propagation surface [2.4, 6, 40]. The most popular approach is shown in Fig. 2.6. Shear or longitudinal waves of velocity v_b are launched via thin-film plate transducers into a bulk material that contacts the surface wave propagation surface at an angle θ to the surface. The wave vectors parallel to the surface must be matched for efficient surface wave generation, which implies that $\cos \theta = v_s/v_b$. Surface wave generation results from the interface stresses and strains produced by the incident bulk wave when it reflects off the interface. Conversely, when a surface wave is incident onto this interface, bulk waves are generated that propagate back to the thin-film transducer for detection.

27

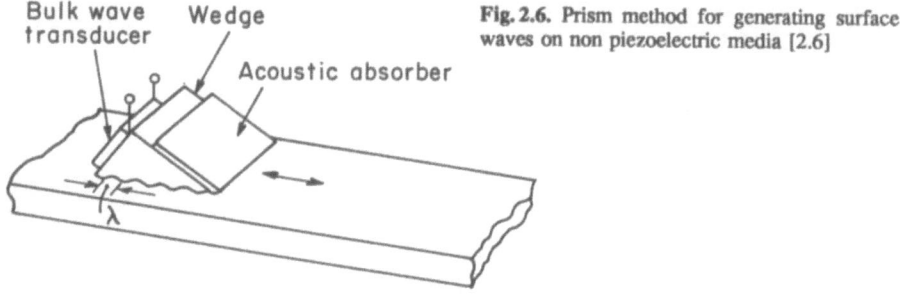

Bulk wave
transducer

Wedge

Acoustic absorber

Fig. 2.6. Prism method for generating surface waves on non piezoelectric media [2.6]

2.2.3 Acousto-optic Detection of Generated Surface Acoustic Waves

When light from, for example, a laser is incident onto a surface on which a surface wave is present, light is scattered both on reflection and transmission [2.41]. For acoustic frequencies of hundreds of megahertz, or lower, light is scattered into multiple diffraction orders (the Raman–Nath limit), with the diffraction angles for the $\pm m$th orders given by [2.41] $\sin\theta_{\pm m} = \sin\theta \pm mk/k_0$, where k is the surface wave vector, k_0 is the incident optic wave vector, and θ is the angle of incidence measured from the normal to the surface. Note that the surface wave vector, and hence the acoustic velocity, can be obtained from the deflection angle [2.41]. For the reflected light, scattering occurs solely from the acoustically produced surface corrugation. The transmitted, deflected light contains contributions from both the corrugation scattering mechanism and the elasto-optic effect, in which the surface wave modulates the refractive index of the medium near the surface. Interference occurs between fields scattered by these two mechanisms and a detailed analysis is required in order to obtain the surface wave power.

The scattered light can be detected and analyzed in many different ways, depending on the amount of acoustic information required. Simply by isolating a single diffraction order in space at a detector, the surface power can be measured. Alternatively, in the presence of stray light of intensity comparable to that of the diffracted light, the generated surface wave can be modulated and phase sensitive detection used to isolate the acoustic signal [2.50]. A Fabry–Pérot interferometer can also be used to isolate any given diffraction order by virtue of its frequency shift [2.41]. The most sophisticated acousto-optic surface wave probes involve heterodyne techniques in which one of the diffraction orders is superimposed on a reference beam (or the undiffracted order) at the detector [2.41, 51]. The most advanced version of this technique also includes automatic compensation for changes in the surface reflectivity and hence can be used to probe acoustic fields inside interdigitated structures, at material interfaces, etc.

2.2.4 Brillouin Scattering from Surface Waves

The acousto-optic interaction can also be used to study acoustic "noise" in and on a sample. The light is frequency shifted by the acoustic frequencies from that of the incident light and is scattered in all directions. A Fabry–Pérot interferometer is used to filter out different frequency components of the scattered light, and the range of frequencies reaching the detector is determined by the bandwidth of the interferometer. The Brillouin signal obtained from thermally excited surface phonons is very weak and the stray scattering from surface imperfections etc. is typically 6–10 orders of magnitude more intense than the signal of interest [2.42, 43]. In order to detect a small signal at one frequency in the presence of a strong signal at a different frequency (stray laser light in this case), the light must be passed through the same interferometer several times (multipassed). In many cases, it is also necessary to use two interferometers swept synchronously in tandem to increase the frequency range available before the Brillouin spectrum repeats itself. Two multipass interferometers swept in tandem, coupled to a computer based stabilization and data-acquisition system represent the state of the art instrumentation for Brillouin scattering from surfaces [2.24–43].

This technique has been used to study surface phonons with frequencies of tens of gigahertz on single surfaces [2.42, 52–55] and guided by thin films [2.53, 56–58]. The scattering geometry defines only the acoustic wave vector in the plane of the surface, in contrast to Brillouin scattering from bulk media in which the scattering geometry defines the three-dimensional acoustic wave vector. Essentially three scattering geometries have been used, as summarized in Fig. 2.7 [2.47]. When either a film or a substrate is optically opaque, scattering from bulk phonons is reduced and the spectrum is dominated by surface phonon features. This is also the case when guided waves are used (Fig. 2.7c) since the optical fields are confined to the vicinity of the film only. From spectra such as the one shown in Fig. 2.8 it is possible to obtain the dispersion relations for surface phonons, and in some cases the attenuation.

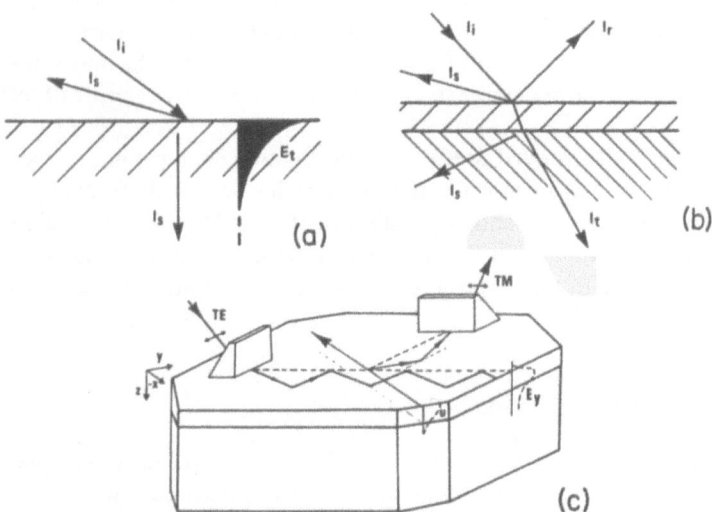

Fig. 2.7a–c. The different scattering geometries used to date in Brillouin scattering from surface waves [2.44]

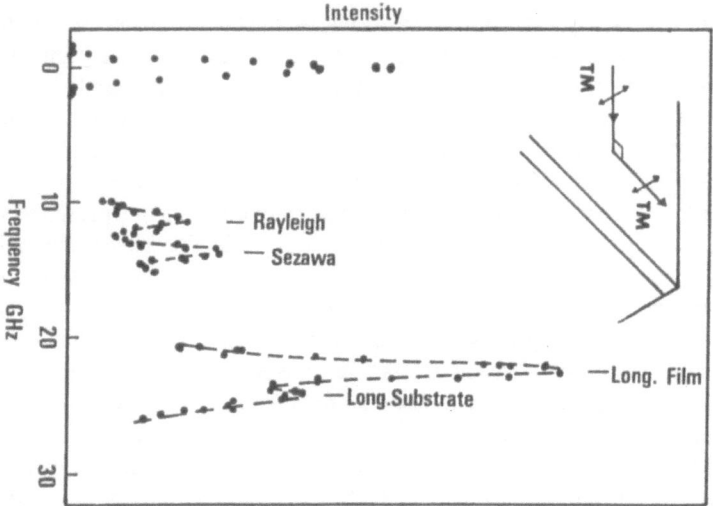

Fig. 2.8. Brillouin spectrum obtained from a thin glass film probed by optical guided waves [2.56]

2.3 Some Applications of Surface Acoustic Waves

Surface acoustic waves have found many applications over the years, starting with early work on seismic waves and proceeding to the present applications to electronics, nondestructive testing, and to the investigation of physical phenomena at surfaces. The most important device applications have been in the areas of signal processing, and it is the development of a "surface wave technology" [2.59, 60] that stimulated the recent, extensive investigations of surface waves at up to gigahertz frequencies. Here our goal is to give some insight into a few applications and devices. More detail can be found in reviews by *Gerard* [2.59] and *Ash* [2.60].

Surface waves have also proved useful for investigating changes in the wave propagation environment and in turn the physics that has led to those changes. Very small changes in the velocity and attenuation can be used to monitor changes in either the surface condition or in the material properties within one acoustic wavelength of the surface. In this section we shall review a representative set of such experiments.

2.3.1 Applications to Electronic Signal Processing

There are several key characteristics that have made surface acoustic waves particularly useful for processing of electronic signals. The interdigital transducer provides a convenient and efficient means of converting electrical signals to acoustic waves, and vice versa. Fairly standard lithographic techniques can be used to generate interdigitated patterns for surface wave generation at frequencies of 10 MHz to 1 GHz, a frequency range in which sophisticated signal processing operations can be put to good use in many systems. Wave trains hundreds of wavelengths long can be fitted onto a crystal surface millimeters long and processed. This is not possible with electromagnetic waves because the speed of sound is orders of magnitude smaller than that of electromagnetic waves. Furthermore, since the waves propagate along the surface, they can be sampled, manipulated, modulated, etc., at any point along their propagation path with surface structures. This characteristic is crucial for signal processing applications.

The properties of the interdigital transducer can be used to manipulate various frequency components of an electrical signal selectively. Typically, transmission through a surface wave device alters the frequency content of an electronic signal applied to an input transducer and subsequently retrieved from the output transducer. It has proved possible to design a throughput device for television receivers which will isolate, for a given channel, the chroma and picture signals from the sound carrier signal for that channel, and from signal components in adjacent channels [2.61]. This is accomplished by a pair of transducers on a surface. In fact, such filters are already in widespread use in many existing models.

Conversely, since the time response is the Fourier transform of the frequency response, interdigital transducers are also used to contour the temporal content of a signal, for example for pulse compression filters in radar systems. The detection sensitivity of a radar system is proportional to the total transmitted energy, and the time (and hence distance) resolution capability is determined by the shortest duration temporal feature that can be identified uniquely. A short electrical pulse is applied to a transducer with a linear taper [2.60] in its spatial periodicity so that the resulting acoustic wave has a linear frequency chirp in time, as shown in Fig. 2.9. This signal is then converted into an electrical signal

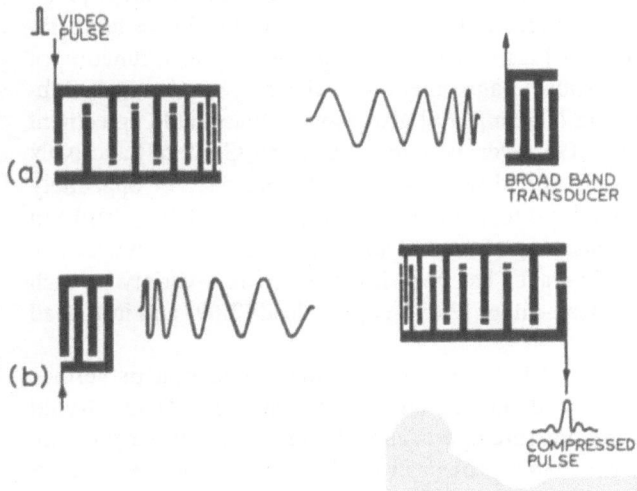

Fig. 2.9a,b. Signal processing used in high power radar systems [2.60]

by a broadband transducer, amplified, and transmitted. On reflection back from an object, the reflected signal is applied to the broadband transducer of an identical surface wave delay line. As the frequency chirped wave passes under the chirped receiving transducer, a strong output signal is obtained only when the spatial structure of the pulse train passing under the interdigital fingers exactly matches that of the interdigitated pattern. Hence the output signal is much narrower in time than the pulse train and the precision timing required for distance evaluation is possible [2.60].

2.3.2 Measurements of Intrinsic Material Properties

Both the velocity and attenuation of a surface wave are sensitive to the acoustic properties of materials placed on top of the propagation surface, and to the material properties of the substrate to a depth of approximately an inverse acoustic wave vector. A number of such studies will be reviewed here.

The change in surface wave attenuation has been used to study phase transitions in thin films deposited onto a surface wave propagation surface. A thin film of the appropriate material is deposited onto the propagation surface of either an α-quartz or a $LiNbO_3$ delay line. The complete sample is then investigated in a low temperature dewar and/or subjected to applied magnetic fields.

Surface acoustic wave investigation of the superconducting properties of metal films depends on the coupling of the electric fields associated with piezoelectric surface waves to the electrons in the metal. In the early work of *Akao* [2.62], the additional surface wave attenuation due to the presence of thin Pb and In films was measured as a function of the reduced temperature (T/T_c, where T_c is the critical temperature). The results were found to be in reasonable agreement with the BCS theory. The application of magnetic fields [2.62, 63] indicated a second-order phase transition as a function of a magnetic field applied perpendicular to the film, and a first-order transition obtained with a parallel field (corresponding to an abrupt change in the order parameter). A second-order phase transition due to surface superconductivity was also observed at lower temperatures. *Jain* and *MacKinnon* [2.64] also measured the gap energy as a function of magnetic field, and again reasonable agreement with theory was obtained. Subsequent work on Zn [2.65] and Nb_3Sn [2.66] also showed reasonable agreement with the standard BCS theory. However, new features in Nb_3Ge [2.67] and NbN [2.67] have been tentatively identified with the possible existence of oppositely polarized vortex pairs. Proximity effects involving Cooper pairs have also been studied in multilayer films consisting of normal and superconducting metals. The attenuation of surface acoustic waves has been investigated in a variety of such sandwich samples, including lead–silver [2.69], copper–lead [2.69] and iron–lead [2.70].

Cheeke and *Morisseau* [2.71] have investigated the attenuation of surface waves when liquid He films are deposited onto the surface of $LiNbO_3$. Rapid increases in attenuation as the pressure approaches the saturated vapour pressure (Fig. 2.10) were identified with rapid growth in the liquid film thickness. Peaks

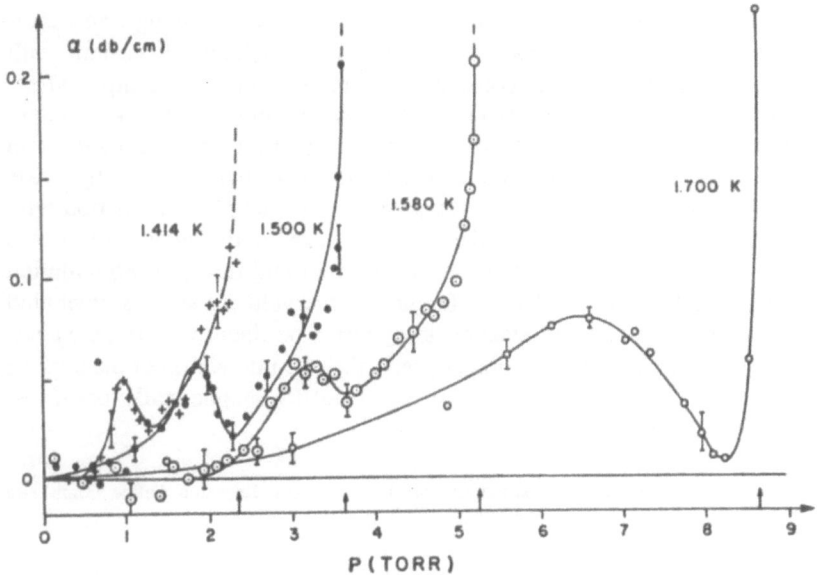

Fig. 2.10. Isothermal curves for the attenuation of surface waves due to a helium film. The arrows at the bottom indicate the saturated vapor pressure for each temperature [2.71]

in the attenuation correspond to the onset of superfluidity in the He film and the subsequent decrease is due to the decoupling of the superfluid from the parallel surface displacement of the surface wave.

Levy et al. [2.72] observed a magnetic field dependent attenuation of surface waves in nickel films deposited on a piezoelectric substrate. The surface wave and magnetization measurements indicated an interaction between the magnetization and the magnetic field induced by the Rayleigh wave. They found that the attenuation depended on the angle between the surface wave propagation wave vector and the axis of easy magnetization of the film. Hysteresis with applied field was interpreted as being due to a gradual rotation of the net magnetization in a metastable equilibrium state and then a sudden jump to a new stable equilibrium state.

Two experiments have been reported on changes in an order parameter at the surface of a material that supported a surface wave. Since the acoustic field samples the material down to a depth equal to the inverse of the acoustic wave vector, usually such experiments are dominated by changes in the parameter of interest in the bulk part of the sample, and specifically surface related effects are difficult to identify unambiguously.

Bjerkan and *Fossheim* [2.73] have investigated structural phase transitions in $SrTiO_3$ and $KMnF_3$. They attempted to make the difference between the sample temperature and the critical temperature, ΔT, small enough that the correlation distance was comparable to the surface wave penetration distance into the sample, making a surface-specific correlation length, if it is different from that of the bulk material, measurable. They were unable to make ΔT small enough and their measurements of the Rayleigh wave velocity were completely consistent with calculations based on bulk elastic constants measured at the same temperature.

Gorodetsky and *Shaft* [2.74] measured the dispersion in surface wave velocity with temperature near two second-order spin reorientation phase transitions in $ErFeO_3$. The correlation length associated with the order parameter in this case was long enough to probe surface-specific properties of the phase transition with surface waves. When compared to the bulk wave case, the main differences were that the surface wave velocity shifts are smaller and the spin reorientation region is narrower by $\simeq 0.7\,K$. The difference in the velocity shift is attributed to additional magnetoelastic and elastic energy terms pertinent to surface waves. The difference in temperature variation suggests that the interaction of the surface wave with the soft bulk spin mode is screened out by magnetically perturbed antiferromagnetic layers.

Acknowledgement. This work was supported in part by the Army Research Office Grant No. DAAG29-85R-0025.

References

2.1   Lord Rayleigh: Proc. Lond. Math. Soc. **17**, 4 (1887)
2.2   D. Royer, J.M. Bonnet, E. Dieulesaint: In *The Mechanical Behavior of Electromagnetic Solid Continua*, ed. by G.A. Maugin (North-Holland, Amsterdam 1984) p.23
2.3   See, for example, K.H. Yen, K.F. Lau, R.S. Kagiwada: Electron. Lett. **15**, 206 (1979) and references therein
2.4   B.A. Auld: *Acoustic Fields and Waves in Solids*, Vol. II (Wiley, New York 1973) p.92
2.5   See, for example, L.D. Landau, E.M. Lifshitz: *Theory of Elasticity* (Pergamon, New York 1959) p.99
2.6   I.A. Viktorov: *Rayleigh and Lamb Waves* (Plenum, New York 1967) Chap. III
2.7   D.C. Gazis, R. Herman, R.F. Wallis: Phys. Rev. **119**, 533 (1960)
2.8   D.A. Tursunov: Akust. Zhur. **13**, 100 (1967) [Soviet Physics – Acoustics **13**, 78 (1967)]
2.9   H. Engan, K.A. Ingebrigtsen, A. Tonning: Appl. Phys. Lett. **10**, 311 (1967)
2.10   E. Strick: Phil. Trans. Roy. Soc. (London) **251**, 488 (1959)
2.11   R.A. Phinney: Bull. Seis. Soc. Am. **51**, 527 (1961)
2.12   L.M. Brekhovskikh: *Waves in Layered Media*, Second Edition (Academic, New York 1980) Chap. I, Sects. 7, 8; Chap. V, Sect. 36
2.13   Ref. [2.6] p.48
2.14   Ref. [2.6] p.118
2.15   Ref. [2.12] Sects. 9–11
2.16   Ref. [2.12] pp.70–74
2.17   G.S. Podlyapol'skii, Y.I. Vasil'ev: Bull. Acad. Sci., USSR, Geophys. Ser. English Translation 859 (1961)
2.18   D.A. Lee, D.M. Corbly: IEEE Trans. on Sonics and Ultrasonics SU-**24**, 206 (1977)
2.19   I.A. Viktorov: Akust. Zhur. **4**, 131 (1958) [Soviet Physics–Acoustics **4**, 131 (1958)]
2.20   B. Rulf: J. Acoust. Soc. Am. **45**, 493 (1969)
2.21   I.A. Viktorov: Akust. Zhur. **7**, 21 (1961) [Soviet Physics-Acoustics **7**, 13 (1961)]
2.22   R.A. Phinney: J. Geophys. Res. **66**, 1445 (1961)
2.23   F. Gilbert: Revs. Geophysics **2**, 123 (1964)
2.24   I.A. Viktorov: Doklady Akad. Nauk SSSR **228**, 579 (1976) [Soviet Physics – Doklady **21**, 272 (1976)]
2.25   J.H. Ansell: Pure Appl. Geophys. **94**, 172 (1972)
2.26   N.E. Glass, A.A. Maradudin: J. Appl. Phys. **54**, 796 (1983)

2.27   R.E. Camley, F. Nizzoli: J. Phys. **C18**, 4795 (1985)
2.28   F.R. Rollins, Jr., T.C. Lim, G.W. Farnell: Appl. Phys. Lett. **12**, 236 (1968)
2.29   C.C. Tseng, R.M. White: J. Appl. Phys. **38**, 4274 (1967)
2.30   C.C. Tseng: J. Appl. Phys. **38**, 4281 (1967)
2.31   J.J. Campbell, W.R. Jones: IEEE Trans Sonics and Ultrasonics **SU-15**, 209 (1968)
2.32   J.L. Bleustein: Appl. Phys. Lett. **13**, 412 (1968)
2.33   Yu.V. Gulyaev: Zhur. Eksper i. Teor. Fiz. Pis. v Red. **9**, 63 (1969) [Soviet Physics – JETP Letters **9**, 37 (1969)]
2.34   R.G. Curtis, M. Redwood: J. Appl. Phys. **46**, 2406 (1975)
2.35   C. Maerfeld, F. Gires, P. Tournois: Appl. Phys. Lett. **18**, 269 (1971)
2.36   G. Koerber, R.F. Vogel: IEEE Trans. Sonics Ultrasonics **SU-19**, 3 (1972)
2.37   C.C. Tseng: Appl. Phys. Lett. **16**, 253 (1970)
2.38   G.G. Kessenikh, V.N. Lyubimov, D.G. Sannikov: Kristallografia **17**, 591 (1972) [Soviet Physics – Crystallography **17**, 512 (1972)]
2.39   G.G. Kessenikh, L.A. Shuvalov: Kristallografia **21**, 5 (1976) [Soviet Physics – Crystallography **21**, 1 (1976)]
2.40   B.A. Auld: *Acoustic Fields and Waves in Solids*, Vol. I (Wiley, New York 1973)
2.41   G.I. Stegeman: IEEE Trans Sonics and Ultrasonics **SU-23**, 33 (1976)
2.42   J.R. Sandercock: In *Light Scattering in Solids*, ed. by M. Cardona, G. Guntherodt (Springer, Berlin, Heidelberg 1982) p.173
2.43   J.G. Dil: Rep. Prog. Phys. **45**, 285 (1982)
2.44   G.I. Stegeman, F. Nizzoli: In *Surface Excitations*, ed. by V.M. Agranovich, R. Loudon (North-Holland, Amsterdam 1984) p.195
2.45   P.R. Emtage: J. Appl. Phys. **43**, 4486 (1972)
2.46   F.S. Hickernell: Proc. IEEE **64**, 631 (1976)
2.47   A.J. Slobodnik, Jr., E.D. Conway: *Microwave Acoustic Handbook*, Vol. 1 (Air Force Cambridge Research Labs, Bedford MA 1970)
2.48   W.S. Goruk, P.J. Vella, G.I. Stegeman: J. Appl. Phys. **50**, 6729 (1979)
2.49   R.C. Williamson: IEEE Trans. Sonics and Ultrasonics **SU-19**, 436 (1972)
2.50   G. Cambon, M. Rouzeyre, G. Simon: Appl. Phys. Lett. **18**, 295 (1971)
2.51   H. Engan: IEEE Trans. Sonics and Ultrasonics **SU-25**, 372 (1978)
2.52   J.R. Sandercock: Solid State Commun. **26**, 547 (1978)
2.53   J.G. Dil, N.C.A. van Hijningen, F. VanDorst, R.M. Aarts: Appl. Optics **20**, 1374 (1981)
2.54   R.T. Harley, P.A. Fleury: J. Phys. **C12**, L863 (1979)
2.55   A. Kueny, M. Grimsditch, K. Miyano, I. Banerjee, C. Falco, I.K. Schuller: Phys. Rev. Lett. **48**, 166 (1982)
2.56   N.L. Rowell, G.I. Stegeman: Phys. Rev. Lett. **41**, 970 (1978)
2.57   V. Bortolani, F. Nizzoli, G. Santoro, A.M. Marvin, J.R. Sandercock: Phys. Rev. Lett. **43**, 224 (1979)
2.58   V. Bortolani, F. Nizzoli, G. Santoro, J.R. Sandercock: Phys. Rev. **B52**, 3442 (1982)
2.59   H.M. Gerard: In *Acoustic Surface Waves*, ed. by A.A. Oliner (Springer, New York, Berlin, Heidelberg 1978) p.61
2.60   E.A. Ash: In *Acoustic Surface Waves*, ed. by A.A. Oliner (Springer, New York, Berlin, Heidelberg 1978) p.97
2.61   A.J. DeVries, R. Adler: Proc. IEEE **64**, 671 (1976)
2.62   F. Akao: Phys. Lett. **30A**, 409 (1969)
2.63   E. Kraetzig, K. Walther, W. Schultz: Phys. Lett. **30A**, 411 (1969)
2.64   M.C. Jain, L. MacKinnon: Phys. Lett. **32A**, 275 (1970)
2.65   W.E. Bailey, B.J. Marshall: Phys. Rev. **B19**, 3467 (1979)
2.66   H.P. Fredericksen, H.L. Salvo, Jr., M. Levy, R.H. Hammond, T.H. Geballe: Phys. Lett. **75A**, 389 (1980)
2.67   H. Salvo, Jr., H.P. Fredericksen, M. Levy: J. Low Temp. Phys. **48**, 189 (1982)
2.68   H.P. Fredericksen, M. Levy, J.R. Gavaler, M. Ashkin: Phys. Rev. **B27**, 3065 (1983)
2.69   E. Kraetzig: Phys. Lett. **33A**, 343 (1970)
2.70   G.N. Bishop: J. Phys. **F5**, 278 (1974)
2.71   J.D.N. Cheeke, P. Morisseau: J. Low. Temp. Phys. **46**, 319 (1982)
2.72   J. Feng, H. Fredericksen, C. Krischer, M. Tachiki, M. Levy: Proc. 1977 IEEE Ultrasonics Symp. 328 (1977)
2.73   L. Bjerkan, K.L. Fossheim: Solid State Comm. **21**, 1147 (1977)
2.74   G. Gorodetsky, S. Shaft: Phys. Rev. **B23**, 6755 (1981)

3. The Green's Function Method in the Surface Lattice Dynamics of Ionic Crystals

G. Benedek and L. Miglio

With 10 Figures

The mathematical technique for studying the dynamics of perturbed systems, known as the Green's function (GF) method, was first applied to solid state problems by *Lifshitz* [3.1] and *Koster* and *Slater* [3.2]. They have stimulated the development of the GF method as a powerful tool in the quantum theory of scattering [3.3–5]. Although the general formulation of the GF method in solid state physics has become a textbook subject [3.6–8], only in recent decades has the method been extensively used in the dynamics of real systems, such as solids with defects, disorder, or boundary surfaces [3.9, 10].

This article is devoted to the GF method in surface lattice dynamics, a subject which is now progressing further because of recent advances in surface phonon spectroscopy [3.11–13].

Besides the classical Lifshitz GF theory [3.1], which has been used in a variety of models with short range interactions [3.14–18], there are several variations of the GF method for surface phonons, applicable to more specific and complex situations. We mention, among others, GF matching [3.19], the continued fraction technique [3.20], the method of generating coefficients [3.21], and the invariant GF method, which we have developed for treating ionic crystal surfaces [3.22–24].

Since there is a substantial equivalence among these formulations [3.23] we shall restrict the present theoretical survey to the invariant method. We want nevertheless to stress a peculiar aspect of this method which makes it applicable to ionic crystals: the reduction of long-range Coulomb interactions, as modified by the surface, to an effective short-range perturbation, which is in turn self-consistently defined through invariance conditions. The finite range of the perturbation subspace is a prerequisite for the standard GF method.

3.1 Outline of the Time-Independent Green's Function Method

Consider the eigenvalue problem for a linear Hermitian operator L_0 in the configuration representation

$$\int d^3r' L_0(r, r')\psi_0(r') - \lambda\psi_0(r) = 0 , \qquad (3.1)$$

Springer Series in Surface Sciences, Vol. 21   **Surface Phonons**
Editors: W. Kress · F.W. de Wette   © Springer-Verlag Berlin, Heidelberg 1991

whose eigenvalues are arranged along the real axis either in a discrete set $\{\lambda_n\}$ or a continuum $\{\lambda_c\}$, or both. The corresponding inhomogeneous equation

$$\int d^3r' L_0(r, r')\psi(r') - \lambda\psi(r) = f(r) , \qquad (3.2)$$

produced by a perturbing field, $f(r)$, admits different types of solutions according to whether λ belongs to or is outside the continuum spectrum $\{\lambda_c\}$.

For $\lambda \in \{\lambda_c\}$ the perturbed solution can be written as a sum of any eigenvector $\psi_0(r)$ of eigenvalue λ and a particular solution of equation (3.2), i.e.

$$\psi(r) = \alpha\psi_0(r) + \int d^3r' G_0(r, r'; \lambda)f(r') , \qquad (3.3)$$

where the Green's function $G_0(r, r'; \lambda)$ fulfills the equation

$$\int dr''[L_0(r, r'') - \lambda\delta(r - r'')]G_0(r'', r'; \lambda) = \delta(r - r') . \qquad (3.4)$$

This is just equation (3.2) for a perturbing field consisting of unitary stimulus $-\delta(r - r')$ at position r'. Thus $G_0(r, r'; \lambda)$ represents the λ-th component of the response of the system at position r to the unitary stimulus at r' [3.26]. The constant α is arbitrary, but clearly has to be zero if $\lambda \notin \{\lambda_c\}$, namely for localized solutions of the perturbed problem.

The important point is that (3.4) is formally solved once we know the complete set of eigenvalues λ_j and eigenvectors $\psi_{0j}(r)$. One immediately sees that the complex functions

$$G_0^{\pm}(r, r'; \lambda) = \sum_{j=\{n,c\}} \frac{\psi_{0j}(r)\psi_{0j}^*(r')}{\lambda_j - \lambda \pm i0^+} \qquad (3.5)$$

are solutions of (3.4) for any λ, by virtue of completeness of the eigenvector set.

The continuous spectrum of L_0 gives rise to a branch cut for G_0 along a portion of the real axis.

Here G_0^+ and G_0^- are different and complex non-Hamiltonian operators, whereas outside the eigenvalue spectrum G_0^+ and G_0^- are equal and Hermitian. In general we can split $G_0^{\pm}$ into their Hermitian and anti-Hermitian parts:

$$G_0^{\pm} = \tfrac{1}{2}(G_0^+ + G_0^-) \pm \tfrac{1}{2}(G_0^+ - G_0^-) ; \qquad (3.6)$$

the anti-Hermitian and Hermitian parts are respectively related to the (Hermitian) *spectral* (Sp) and *dispersive* (Dp) parts of the GF operator

$$\mathrm{Sp}\{G_0^{\pm}\} = \mp\frac{i}{2}(G_0^+ - G_0^-) = \pm\pi \sum_{j=\{n,c\}} \psi_{0j}(r)\psi_{0j}^*(r')\delta(\lambda_j - \lambda) \qquad (3.7)$$

$$\mathrm{Dp}\{G_0^{\pm}\} = \frac{1}{2}(G_0^+ + G_0^-) = \mathrm{P} \sum_{j=\{n,c\}} \psi_{0j}(r)\psi_{0j}^*(r')(\lambda_j - \lambda)^{-1} , \qquad (3.8)$$

where P means that the Cauchy principal part has to be taken in the integration over the continuous part of the spectrum.

Since $G_0^\pm$ are eigenfunctions of the Hilbert transform with eigenvalues $\pm i$, i.e.

$$\frac{1}{\pi} P \int_{-\infty}^{+\infty} \frac{d\lambda'}{\lambda' - \lambda} G_0^\pm(\lambda') = \pm i G_0^\pm(\lambda) , \tag{3.9}$$

their spectral and dispersive parts, as well as their imaginary and real parts fulfill the Kramers–Kronig relations:

$$\frac{1}{\pi} P \int_{-\infty}^{+\infty} \frac{d\lambda'}{\lambda' - \lambda} Sp\{G_0^\pm(\lambda')\} = \pm Dp\{G_0^\pm(\lambda)\} , \tag{3.10a}$$

$$\frac{1}{\pi} P \int_{-\infty}^{+\infty} \frac{d\lambda'}{\lambda' - \lambda} Dp\{G_0^\pm(\lambda')\} = \mp Sp\{G_0^\pm(\lambda)\} , \tag{3.10b}$$

$$\frac{1}{\pi} P \int_{-\infty}^{+\infty} \frac{d\lambda'}{\lambda' - \lambda} Im\{G_0^\pm(\lambda')\} = \pm Re\{G_0^\pm(\lambda)\} , \tag{3.11a}$$

$$\frac{1}{\pi} P \int_{-\infty}^{+\infty} \frac{d\lambda'}{\lambda' - \lambda} Re\{G_0^\pm(\lambda')\} = \mp Im\{G_0^\pm(\lambda)\} . \tag{3.11b}$$

The density of states (DOS) is directly obtained from the spectral part:

$$\varrho_0(\lambda) = \pm \frac{1}{\pi} Tr\{Sp\{G_0(\lambda)\}\} = \pm \frac{1}{\pi} Tr\{Im\{G_0(\lambda)\}\} . \tag{3.12}$$

Note that outside the continuum $G_0^+ = G_0^-$ have poles at $\lambda = \lambda_n$ so that the localized states of the system are found at the singularities of the GF.

The time-dependent GFs, obtained by Fourier-transforming G_0^+ and G_0^-, are seen to represent the response of the system either following or anticipating the stimulus, respectively. Causality implies that G_0^+ is physically meaningful. Since we shall always use causal Green's functions, we shall hereafter drop the superscript +.

The GF formalism is particularly useful also in the homogeneous problem for the operator L whenever it admits the decomposition

$$L = L_0 + \Lambda , \tag{3.13}$$

L_0 being an operator whose eigenvalues and eigenvectors are known, and Λ a well defined localized perturbation. Here we use abstract matrix notation since we do not need to specify the coordinate space where Λ exhibits localization.

In the direct space we deal with

$$f(\boldsymbol{r}) = - \int \Lambda(\boldsymbol{r}, \boldsymbol{r}')\psi(\boldsymbol{r}')d^3r' , \tag{3.14}$$

and (3.3) transforms into its integral form, the Lippmann–Schwinger equation for the perturbed eigenvector

$$\psi(r) = \alpha\psi_0(r) - \iint d^3r' d^3r'' G_0(r, r''; \lambda)\Lambda(r'', r')\psi(r') , \qquad (3.15)$$

where α acts now as a normalization constant.

This equation can be solved in practice provided we dispose of the particular representation in which Λ is localized. Hence we proceed using abstract notation by re-writing equation (3.15) as

$$|\psi\rangle = \alpha|\psi_0\rangle - G_0\Lambda|\psi\rangle , \qquad (3.16)$$

where

$$G_0 = (L_0 - z)^{-1} \qquad (3.17)$$

and

$$z \equiv \lambda + i0^+ . \qquad (3.18)$$

We expect from (3.9) three types of solutions:

i) solutions for λ inside the continuous spectrum $\{\lambda_c\}$ consisting of the distortion of the unperturbed solutions ψ_0;

ii) localized solutions with $\lambda \notin \{\lambda_c\}$, but possibly close to λ_n, as modifications of the unperturbed localized states (if any);

iii) additional localized solutions wiht $\lambda \notin \{\lambda_c\}$, well split from the edges of the continuous spectrum as an effect of the perturbation.

We consider first the localized solutions, $\lambda \notin \{\lambda_c\}$. In this case (3.16) reduces to

$$|\psi\rangle = -G_0\Lambda|\psi\rangle \qquad (3.19)$$

and the new eigenvalues are solutions of the secular equation

$$\det(I + G_0\Lambda) = 0 . \qquad (3.20)$$

By introducing the perturbed GF

$$G = (L - z)^{-1} = (L_0 + \Lambda - z)^{-1} = (I + G_0\Lambda)^{-1}G_0 \qquad (3.21)$$

we see that the solution of (3.20), i.e. the perturbed localized eigenvalues, are poles of G. Note that the perturbed GF is a solution of the Dyson equation

$$G = G_0 - G_0\Lambda G , \qquad (3.22)$$

which is equivalent to (3.16) for the wave function.

When $\lambda \in \{\lambda_c\}$ we deal with (3.16). This can be formally solved to give

$$|\psi\rangle = \alpha(I + G_0\Lambda)^{-1}|\psi_0\rangle \qquad (3.23a)$$

$$= \alpha|\psi_0\rangle - \alpha G_0 T|\psi\rangle , \qquad (3.23b)$$

where the transition matrix

$$T \equiv \Lambda(1 + G_0\Lambda)^{-1} \tag{3.24}$$

has nonzero elements only in the perturbation subspace. T is in turn a solution of the Dyson equation

$$T = \Lambda - \Lambda G_0 T . \tag{3.25}$$

Equation (3.23a) expresses the solution of a perturbative problem as a linear superposition of unperturbed wave functions, whereas (3.23b) represents the solution in the language of scattering theory, i.e., as a sum of incident wave $|\psi_0\rangle$ and diffused wave $-G_0 T|\psi_0\rangle$.

In the first-order Born approximation, i.e., when (3.25) is solved to first order by replacing T with V,

$$|\psi\rangle = \alpha(1 - G_0\Lambda)|\psi_0\rangle . \tag{3.26}$$

Here, however, the peculiar effects of the denominator in (3.24) are lost. Indeed the factor $1 + G_0\Lambda$ may be responsible for resonant enhancement of the scattered wave. In general we speak of *resonance* when the perturbation induces a peak in the spectral part of the perturbed GF (3.21). This occurs for values of $\lambda = \lambda_R$ fulfilling the resonance condition

$$\text{Re}\{\det(1 + G_0\Lambda\} = 0 . \tag{3.27}$$

By expanding around λ_R one sees that a resonance appears as a Lorentzian peak in the perturbed DOS

$$\varrho(\lambda) = \frac{1}{\pi}\text{Tr}\{\text{Im}\{G(\lambda)\}\} . \tag{3.28}$$

Clearly inside the continuum $\text{Im}\{\det(1+G_0\Lambda)\}$ is nonvanishing. If this is a slowly varying function around λ_R the peak half width is approximately given by

$$\frac{\Gamma}{2} = \frac{\text{Im}\{\det[1 + G_0(\lambda_R)\Lambda]\}}{\frac{\partial}{\partial\lambda}(\text{Re}\{\det[1 + G_0(\lambda_R)\Lambda]\})} . \tag{3.29}$$

In order for the peak to be sharp and to produce a significant feature in the perturbed DOS, $\text{Im}\{\det(1 + G_0\Lambda)\}$ has to be small. Under resonance conditions, $G(z)$ and $T(z)$ exhibit a pole in the complex plane at

$$z \cong \lambda_R - i\Gamma/2 . \tag{3.30}$$

Thus $G(z)$ takes a physical meaning in the whole complex plane, in that the complex poles indicate resonances as well as real poles identify localized solutions. Since we are actually working with the causal GF, Γ has to be positive. Therefore, a value of λ_R yielding in (3.29) a negative Γ does not contribute a real resonance but an "anti-resonance", i.e. a depletion region in $\varrho(\lambda)$, with respect to $\varrho_0(\lambda)$, in favor of the intensity which has been concentrated in local and resonant modes.

This can be better understood by considering the change of DOS induced by the perturbation. By remembering that, for any operator $\hat{O}$, $\mathrm{Tr}\{\ln \hat{O}\} = \ln \det \hat{O}$, we can rewrite expressions (3.12) and (3.28) as

$$\varrho_0(\lambda) = \frac{1}{\pi} \frac{\partial}{\partial \lambda} \mathrm{Im}\{\ln \det G_0(\lambda)\} , \tag{3.31}$$

$$\varrho(\lambda) = \frac{1}{\pi} \frac{\partial}{\partial \lambda} \mathrm{Im}\{\ln \det G(\lambda)\} . \tag{3.32}$$

Hence $\Delta\varrho \equiv \varrho - \varrho_0$ is given by

$$\Delta\varrho(\lambda) = -\frac{1}{\pi} \frac{\partial}{\partial \lambda} \mathrm{Im}\{\ln \det[I + G_0(\lambda)\Lambda]\} . \tag{3.33}$$

The change of DOS at a resonance can be related with a little algebra to the half width:

$$\Delta\varrho(\lambda_R) = 2/\pi\Gamma . \tag{3.34}$$

Thus a positive Γ implies a resonance enhancement; a negative Γ a depletion. The larger $\Delta\varrho(\lambda_R)$, the smaller Γ. For a local mode, $\Delta\varrho(\lambda_R) = +\infty$ and $\Gamma = 0^+$.

Expression (3.33) offers a straightforward way of calculating the spectral modifications directly from the resonant denominator $\det(I + G_0\Lambda)$. Indeed such a determinant is identically equal to the same determinant evaluated in the perturbation subspace. In the latter, only the elements of G_0 in the Λ subspace (in the representation where it exhibits localization), i.e., the elements of the *projected* GF g_0, are needed. However, $\det G_0 \neq \det g_0$, so that distinction has to be made between total and projected DOS according to whether whole-space or projected GFs are used in (3.31) and (3.32). For projected DOS the trace is intended over the perturbation subspace in (3.12) and (3.28). As general concepts, spectral localization and localization in space of perturbed waves are strictly related, which means that the sharper the resonance, the larger its amplitude in the perturbation subspace. While a resonant state tends to a plane wave behavior at large distances from the perturbation region, a localized state cannot propagate at all in the unperturbed region of the system and therefore it must decay exponentially out of the perturbation subspace.

3.2 The Green's Function Method in Surface Dynamics

We consider a crystal lattice constituted by N unit cells at positions $\boldsymbol{r}_l (l = 1, 2, \ldots, N)$ with s atoms per unit cell located at $\boldsymbol{r}_\kappa (\kappa = 1, 2, \ldots, s)$ with respect to the conventional cell center. Thus the equilibrium atomic positions are

$$\boldsymbol{r}_{l\kappa} = \boldsymbol{r}_l + \boldsymbol{r}_\kappa . \tag{3.35}$$

Then we write the secular equation for lattice vibrations in the harmonic approximation as

$$\sum_{\beta} \left[\phi_{\alpha\beta}(l\kappa, l'\kappa') - \omega^2 M_{\kappa} \delta_{ll'} \delta_{\kappa\kappa'} \delta_{\alpha\beta} \right] u_{\beta}(l'\kappa') = 0 , \tag{3.36}$$

where ϕ is the force constant matrix, M_{κ} the k-th atom mass, ω the phonon angular frequency and $u(l'\kappa')$ the atomic displacement vector from the equilibrium position. The invariance of the crystal potential energy with respect to infinitesimal rigid translations and rotations implies the following conditions for the tensor $\phi(l\kappa, l'\kappa')$:

$$\mathcal{T}\phi \equiv \sum_{l\kappa} \phi(l\kappa, l'\kappa') = 0 , \quad \forall l'\kappa' , \tag{3.37}$$

$$\mathcal{R}\phi \equiv \sum_{l\kappa} r_{l\kappa} \times \phi(\lambda\kappa, l'\kappa') = 0 , \quad \forall l'\kappa' , \tag{3.38}$$

respectively.

The linear operator $\mathcal{T}$ and $\mathcal{R}$ in the $(l\kappa\alpha)$ space, respectively determining translational invariance (TI) and rotation invariance (RI) conditions, are given by the $3 \times 3NS$ rectangular matrices

$$\langle \alpha | \mathcal{T} | l\kappa\beta \rangle = \delta_{\alpha\beta} , \tag{3.39}$$

$$\langle \alpha | \mathcal{R} | l\kappa\beta \rangle = (\delta \times r_{l\kappa})_{\alpha\beta} , \tag{3.40}$$

where δ is the unit tensor in the Cartesian subspace of components $\delta_{\alpha\beta}$.

For a three-dimensional periodic lattice, l represents a set of three integer numbers (l_1, l_2, l_3) such that $r_l = l_1 a_1 + l_2 a_2 + l_3 a_3$, $\{a_j\}$ being any set of direct lattice vectors. For a semi-infinite lattice with a single surface (or a slab with two parallel surfaces) one can always choose a_1 and a_2 in the surface plane, so that the pairs $(l_1, l_2) = L$ label the translations along the surface, and l_3 labels the atomic layers parallel to the surface.

We consider an infinitely thick slab originated by perturbing an infinite lattice with three-dimensional cyclic boundary conditions [3.22, 23]. The free surfaces are constructed by cutting the infinite lattice along an ideal plane Σ, as shown in Fig. 3.1. The resultant perturbation of the cyclic lattice force constants $\phi_{0\alpha\beta}(l\kappa, l', \kappa') = \phi_{0\alpha\beta}(L - L'; l_3 - l'_3, \kappa\kappa')$ is then described by setting to zero all the interatomic force constants crossing the plane Σ; in addition it contains all force constant changes near the two surfaces produced by the elastic relaxation. Therefore we define the perturbation matrix as

$$\Lambda_{\alpha\beta}(L - L'; l_3\kappa l'_3\kappa) = \phi_{\alpha\beta}(L - L'; l_3\kappa l'_3\kappa) - \phi_{0\alpha\beta}(L - L'; l_3\kappa l'_3\kappa) , \tag{3.41}$$

where both ϕ, the slab force constant matrix, and ϕ_0 are assumed to satisfy TI and RI conditions. Thus, Λ also fulfills the TI condition and, in the absence of elastic relaxation, the RI condition [3.29]. In (3.41) it is intended that the slab force constant elements connecting atoms across Σ are zero. In practice, Λ works as a perturbation only when its nonzero elements are restricted within a small perturbation subspace σ (Fig. 3.1); for example, if $l_3 = 1$ and $l_3 = N_L$ denote

43

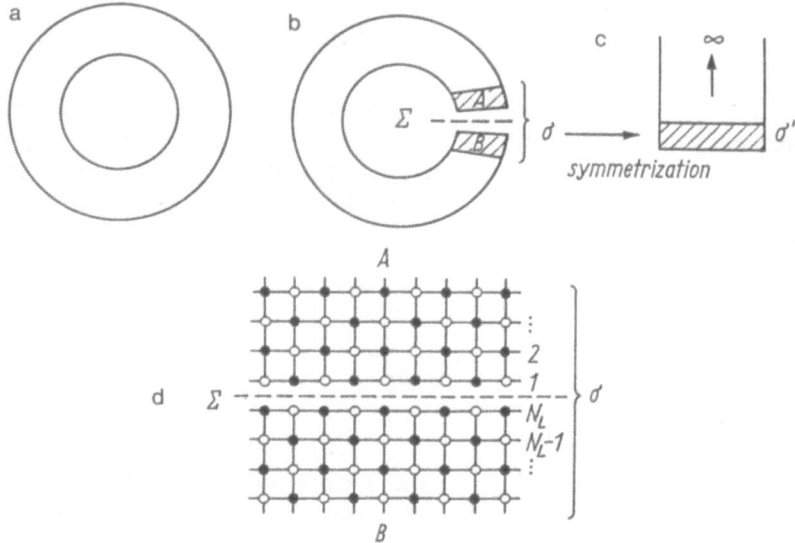

Fig. 3.1. Representation of an ideal lattice with cyclic boundary conditions (a), and of the same lattice after two free surfaces are created by a cut along the plane Σ (b); σ denotes the perturbation subspace. In the representation of symmetrized coordinates we obtain a semi-infinite lattice with a perturbation subspace σ', restricted to a single surface (c). For an ionic lattic (NaCl-like) the inversion symmetry with respect to Σ is recovered after a shear translation of the surface B with respect to A (d)

the two surface layers, the subspace σ includes $l_3 = 1, 2, \ldots N_p$ and $N_l, N_L - 1, \ldots, N_l - N_p + 1$, where N_p, the number of atomic layers on each side of the slab involved in the cutting procedure, is "small". This requirement seems to rule out the applicability of the perturbation method to lattices with long-range forces, like ionic crystals. However, we shall see that in the limit $N_L \rightarrow \infty$ it is possible to work with a small perturbation matrix also for ionic crystals.

In abstract notation (3.36) for stationary vibrational states reads

$$(\phi_0 + \Lambda - M\omega^2)u = 0 , \tag{3.42}$$

where M is the mass matrix. We assume that (3.42) has already been Fourier-transformed with respect to the coordinate along which translational symmetry is maintained, i.e. we work in the surface wave vector representation.

The eigenvalue spectrum of $D_0 = M^{-1/2}\phi_0 M^{-1/2}$, the dynamical matrix of the infinite lattice, is represented for each wave vector K by a set of $3s$ continuous bands. According to the analysis made in Sect. 3.1 we first consider ω^2 outside the spectrum $\{\omega_0^2\}$ of D_0; the values of ω^2 for which

$$\det \left[I + (\phi_0 - M\omega^2)^{-1}\Lambda \right] = 0 \tag{3.43}$$

at any K give the dispersion relations of the surface modes; they are localized at the surface since no frequency outside $\{\omega_0^2\}$ can propagate inside the lattice.

When ω^2 falls into $\{\omega_0^2\}$ we have a wave propagating in the bulk which is distorted by the surface through scattering processes.

Resonant states, i.e., enhanced scattering waves, occur when

$$\text{Re}\{\det(I + g_0 \Lambda)\} = 0 , \tag{3.44}$$

where

$$g_0 = (\phi_0 - Mz)^{-1} \text{ projected onto } \sigma \tag{3.45}$$

is the unperturbed projected GF matrix, and

$$z = \omega^2 + 2i\omega 0^+ . \tag{3.46}$$

For an infinite lattice the perturbative effects of the free surface on the continuous frequency bands are understood in terms of unperturbed and perturbed projected phonon densities

$$\varrho_0(K,\omega) = \frac{2\omega}{\pi} \text{Tr}\{\text{Im}\{g_0(K,\omega^2)\}\} , \tag{3.47}$$

$$\varrho(K,\omega) = \frac{2\omega}{\pi} \text{Tr}\{\text{Im}\{g(K,\omega^2)\}\} , \tag{3.48}$$

respectively, where Tr denotes trace in the subspace σ of the perturbation, and

$$g = (\phi - Mz)^{-1} \text{ projected onto } \sigma \tag{3.49}$$
$$= (I + g_0 \Lambda)^{-1} g_0 \tag{3.50}$$

is the perturbed projected GF matrix [3.30].

For K along the symmetry directions of the surface Brillouin zone all modes have a polarization which is either in the plane defined by the directions of the K vector and the normal to the surface (saggital ($\perp$) polarization) or transverse and parallel to the surface (parallel ($\parallel$) polarization). This yields the factorization

$$\{\omega_0^2\} = \{\omega_0^2\}_\perp \cup \{\omega_0^2\}_\parallel \tag{3.51}$$

and a corresponding block diagonalization of all the above matrices. Having in mind (3.50), we see that the following cases for the values ω^2 fulfilling (3.44) occur:

1) Surface modes: ω^2 is outside $\{\omega_0^2\}$. The imaginary part of g_0 is infinitesimal and $\varrho(K,\omega)$ exhibits a δ-peak.
2) Pseudosurface modes: ω^2 falls into a band of $\{\omega_0^2\}$, but the displacement vector u is orthogonal to all vectors u_0 of that band; again we have a local mode, crossing a transparent bulk band. This occurs, e.g. when ω^2 belongs to $\{\omega_0^2\}_\perp$ but is outside $\{\omega_0^2\}_\parallel$ (or vice versa). Clearly, pseudosurface modes exist only along symmetry directions.
3) Surface resonances: ω^2 falls into a band of $\{\omega_0^2\}$ whose u_0 are not orthogonal to u. In this case we have a resonance: $\varrho(K,\omega)$ displays a Lorentzian-

shaped peak, whose width is proportional to $\text{Im}\{g_0\}$ and is finite. When deviating from a symmetry direction all pseudosurface modes transform into resonances. However, surface resonances may exist also along symmetry directions in addition to local modes.

3.3 The Intrinsic Perturbation for a Semi-Infinite Lattice

For a slab with identical surfaces we can always find a symmetry transformation $\mathcal{S}$ producing a simultaneous block diagonalization of the Green's function and perturbation matrices, namely

$$\mathcal{S} g_0 \mathcal{S}^{-1} = \begin{vmatrix} g_0 & 0 \\ 0 & g_- \end{vmatrix}, \quad \mathcal{S} \Lambda \mathcal{S}^{-1} = \begin{vmatrix} \Lambda_+ & 0 \\ 0 & \Lambda_- \end{vmatrix}, \tag{3.52}$$

and therefore

$$\mathcal{S} g \mathcal{S}^{-1} = \begin{vmatrix} \tilde{g}_+ & 0 \\ 0 & \tilde{g}_- \end{vmatrix}, \tag{3.53}$$

where

$$\tilde{g}_\pm = (1 + g_\pm \Lambda_\pm)^{-1} g_\pm . \tag{3.54}$$

For example, in many cases the lattice exhibits a mirror symmetry with respect to the plane Σ; sometimes, the mirror symmetry is recovered by a shear translation of the surface B with respect to A, as for the (001) surface of the NaCl lattice (Fig. 3.1c). In these cases, by re-labeling the layers of A and B sides by $l_3 = 1/2$, $3/2 \ldots N_p - 1/2, \ldots$ and $l_3 = -1/2, -3/2, \ldots, -N_p + 1/2, \ldots$, respectively, it is possible to replace the set of coordinates $(l_3 \kappa \alpha)$ with a new set of symmetrized coordinates $(|l_3|, \kappa \alpha, p)$ defined by the transformation $\mathcal{S}$,

$$\mathcal{S} |l_3 \kappa \alpha\rangle = \||l_3|, \kappa \alpha, p\rangle = \frac{1}{\sqrt{2}} (|l_3, \kappa \alpha\rangle + p \, \text{sgn}(\alpha)| - l_3, \kappa \alpha\rangle) , \tag{3.55}$$

where $p = \pm 1$ is the parity index, and $\text{sgn}(\alpha) = 1$ for $\alpha = z$ (the surface orientation), and $= -1$ for $\alpha = x, y$. This enables us to work in a reduced $3sN_p$ dimensional subspace σ'.

When $N_L \to \infty$, a slab with identical surfaces becomes equivalent to a semi-infinite lattice with a single surface, provided that the displacement correlation between the two slab surfaces tends to zero all over the spectral region. This means that all the modes of the slab become degenerate in pairs. Therefore in the subspace σ' we have

$$\tilde{g}_+ = \tilde{g}_- \equiv \tilde{g} ; \tag{3.56}$$

$\tilde{g}$ is interpreted as the perturbed projected Green's function for the single surface of the semi-infinite lattice. Equation (3.54) can be written as

$$\tilde{g} = (I + \bar{g}\bar{\Lambda})^{-1}\bar{g} ,$$ (3.57)

where

$$\bar{g}^{-1} = \tfrac{1}{2}(g_{+}^{-1} + g_{-}^{-1})$$ (3.58)

and

$$\bar{\Lambda} = \tfrac{1}{2}(\Lambda_{+} + \Lambda_{-})$$ (3.59)

are the inverse of the unperturbed projected Green's function for the semi-infinite lattice and the pertaining perturbation, respectively. Notice that all the inversions are performed in the subspace σ'. The resonance condition (3.44) in the subspace σ' reads

$$\mathrm{Re}\{\det(I + \bar{g}\bar{\Lambda})\} = 0 .$$ (3.60)

Apart from dimensional reduction, the symmetrization (3.55) has the advantage that the symmetrized perturbation $\bar{\Lambda}$ takes a very simple form. Indeed, for the case of symmetry mentioned above, we have

$$\Lambda(\boldsymbol{k}; l_3\kappa, l_3'\kappa'))\mathrm{sgn}(\alpha)\mathrm{sgn}(\beta)\Lambda_{\alpha\beta}(\boldsymbol{K}; -l_3\kappa, -l_3'\kappa') ,$$ (3.61)

since the same property holds for ϕ_0 and ϕ [3.31]. The symmetrized components are readily obtained by means of (3.48):

$$\Lambda(\boldsymbol{K}; |l_3|\kappa p; |l_3'|\kappa' p')$$
$$= \delta_{pp'} \{\Lambda_{\alpha\beta}(\boldsymbol{K}; |l_3|\kappa, |l_3'|\kappa') + \mathrm{sgn}(\alpha)p\Lambda_{\alpha\beta}(\boldsymbol{K}; |l_3|\kappa, -|l_3'|\kappa')\} ,$$ (3.62)

and according to (3.59)

$$\bar{\Lambda}_{\alpha\beta}(\boldsymbol{K}; l_3\kappa, l_3'\kappa') = \Lambda_{\alpha\beta}(\boldsymbol{K}; l_3\kappa, l_3'\kappa') , \quad (l_3, l_3' > 0) .$$ (3.63)

Thus the semi-infinite lattice perturbation $\bar{\Lambda}$ contains only the elements of Λ which connect pairs of atoms on one side with respect to the cutting plane Σ. These elements represent the reaction of the lattice in the region A to the bond cutting as required by conservation of total momentum and total angular momentum. The two conservation laws are reflected in the force constant matrix through the TI and RI conditions, respectively. Therefore the matrix elements of $\bar{\Lambda}$ contain exclusively *what is implied by* TI *and* RI *conditions*. More precisely the diagonal part Λ^{T} of $\bar{\Lambda}$ (the Einstein part) comes from TI condition, so that we can write

$$\bar{\Lambda} = \Lambda^{\mathrm{T}} + \Lambda^{\mathrm{R}} .$$ (3.64)

The procedure for obtaining Λ^{T} and Λ^{R} is based on the identity

$$\Lambda_p = \bar{\Lambda} - \tfrac{1}{2}(g_p^{-1} - g_{-p}^{-1}) , \quad (p = \pm) ,$$ (3.65)

directly obtainable from (3.56). By applying to both numbers the blocks $\mathcal{T}_p$ and $\mathcal{R}_p$ of the symmetrized TI and RI operators $\mathcal{T}\mathcal{S}^{-1}$ and $\mathcal{R}\mathcal{S}^{-1}$, respectively, and noting that $\mathcal{T}_p \Lambda_p = \mathcal{R}_p \Lambda_p = 0$, we have

$$\mathcal{T}_p \bar{\Lambda} = \tfrac{1}{2} \mathcal{T}_p (g_p^{-1} - g_{-p}^{-1}) \,, \tag{3.66}$$

$$\mathcal{R}_p \bar{\Lambda} = \tfrac{1}{2} \mathcal{R}_p (g_p^{-1} - g_{-p}^{-1}) \,. \tag{3.67}$$

This is a set of linear inhomogeneous equations having the elements of $\bar{\Lambda}$ as unknown. One can prove that, whatever the extension of the *intrinsic* surface perturbation, the set (3.66, 67) contains as many nontrivial independent equations as the number of independent elements of $\bar{\Lambda}$ are needed.

We see that the perturbation matrix $\bar{\Lambda}$ turns out to be entirely self-consistently defined in terms of elements of inverse unperturbed GFs, i.e. in terms of the intrinsic bulk dynamics. This is an interesting illustration of the general statement that all that happens at the free surface of a solid, such as the change of force constants, atomic equilibrium positions, or electronic structure, is due to intrinsic properties of the bulk Hamiltonian manifesting themselves through the symmetry breaking.

In the GF method applied to surface problems the bulk dynamical structure is the basic ingredient, whereas the perturbation of an intrinsic surface, induced by the symmetry breaking, turns out to be fully described through the TI and RI conditions. This procedure has been shown [3.25] to be equivalent to a more general theory of interface dynamics based on the Green's function matching when applied to free surfaces.

The elements of the unperturbed bulk GF matrix contain all the information we need about both bulk dynamics and intrinsic surface perturbation. Thus the GF method, in the form into which it has developed, is just a formal procedure to transfer to the surface all of what we know, *a priori*, in the ideal lattice with cyclic boundary conditions.

Another interesting property of $\bar{\Lambda}$ is its sharp localization. The application of $\mathcal{T}_p$ and $\mathcal{R}_p$ on the right-hand member of (3.66) and (3.67) implies a sum over the atom index κ. In ionic crystals the sum over positive and negative ions yields a cancelation of the long-range Coulomb contributions and the elements of (3.63) fall off rapidly for increasing l_3 and/or l'_3. Thus $\bar{\Lambda}$ can be truncated and restricted to a very small number of layers. Although truncation would prevent a complete separation of the two sides A and B, the free-surface behavior is always recovered through TI and RI conditions, which ensure the existence and correct behavior of the Rayleigh wave in the long-wave limit.

Let us consider an application of these concepts to the (001) surface of alkali halides with rocksalt structure. The Einstein part of $\bar{\Lambda}$ has the form

$$\Lambda_{\alpha\beta}^{\mathrm{T}}(\boldsymbol{K}; l_3\kappa, l'_3\kappa') = \delta_{l_3 l'_3} \delta_{\kappa\kappa'} \delta_{\alpha\beta} f_{\kappa\alpha}(l_3) \tag{3.68}$$

with $l_3 > 0$, and the force constants

$$f_{\kappa\alpha}(l_3) \equiv \sum_{\boldsymbol{L}, \kappa', l'_3 > 0} \phi_{0_{\alpha\alpha}}(\boldsymbol{L}; l_3\kappa, -l'_3\kappa') \tag{3.69}$$

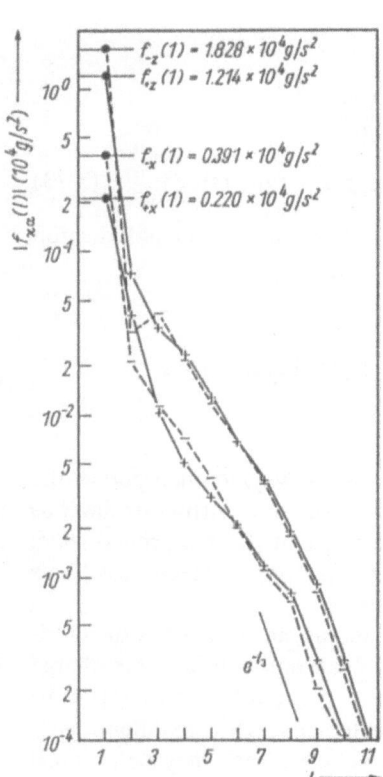

$f_z(1) = 1.828 \times 10^4 g/s^2$
$f_z(1) = 1.214 \times 10^4 g/s^2$
$f_x(1) = 0.391 \times 10^4 g/s^2$
$f_x(1) = 0.220 \times 10^4 g/s^2$

e^{-l_3}

Fig. 3.2. Spatial behavior along the direction perpendicular to the surface of the double Fourier transformed force constants, calculated using the breathing shell model and 0 K input data for the infinite lattice

are obtained from the bulk dynamical model. A calculation performed for NaCl(001) (Fig. 3.2) [3.22] shows for $f_{\kappa\alpha}(l_3)$ a steep exponential decay for increasing l_3, to such an extent that $\bar{A}$ can be restricted to the first surface layer ($l_3 = 1/2$). After truncation we have to replace the force constants (3.69) with those self-consistently obtained from the TI condition:

$$f_{\kappa\alpha} = \text{sgn}(\alpha) \sum_{\kappa'} \left[(\kappa\alpha|g_-^{-1} - g_+^{-1}|\kappa'\alpha)/2 \right]_{K=0} . \qquad (3.70)$$

Then from the RI condition we obtain, for $l_3 = l_3' = 1/2$,

$$\Lambda_{\alpha\beta}^{R}(\boldsymbol{K}; \kappa\kappa') = i(1 - \delta_{\kappa\kappa'})\varepsilon_{\alpha\beta}(\boldsymbol{K})(f_{\kappa\alpha}\delta_{\beta z} + f_{\kappa'\beta}\delta_{\alpha z}) , \qquad (3.71)$$

where

$$\varepsilon_{\alpha\beta}(\boldsymbol{K}) \equiv (\delta_{\alpha x}\delta_{\beta z} - \delta_{\alpha z}\delta_{\beta x}) \sin(r_0 K_x) + (\delta_{\alpha y}\delta_{\beta z} - \delta_{\alpha z}\delta_{\beta y}) \sin(r_0 K_y) \quad (3.72)$$

and r_0 is the interionic distance.

The elements of the unperturbed GF matrices for $l_3 = l_3' = 1/2$ are found to be

49

$$(\kappa\alpha|g_p(\mathbf{K}\omega^2)|\kappa'\beta)$$

$$= \frac{1}{\pi(M_\kappa M_{\kappa'})^{1/}} \int_0^\pi d\xi \sum_j \frac{e_\alpha(\kappa|qj)e_\beta^*(\kappa'|qj)}{\omega_{qj}^2 - \omega^2 - 2i\omega 0^+}$$

$$\times [(1 + \mathrm{sgn}\alpha\mathrm{sgn}\beta)(1 + p\mathrm{sgn}\alpha\cos\xi) + ip(\mathrm{sgn}\alpha - \mathrm{sgn}\beta)\sin\xi], \qquad (3.73)$$

where ω_{qj} and $e(k|qj)$ are bulk phonon angular frequencies and polarization vectors of wave vector $q \equiv (\mathbf{K}, \xi/r_0)$ and branch index j.

3.4 The Electronic Contribution to Surface Dynamics in the Framework of Shell Models

Shell models [3.110, 32–34] are perhaps the simplest way to incorporate the effects of electron dynamics in the lattice dynamical matrix. Virtual excitations of the electron system modify the interionic force constants for the ground state. This happens through a change in the shapes of the external electron orbitals surrounding ion cores, during their vibrational motion.

The electronic charge density can be represented as superpositions of multipoles centered at convenient symmetry points in each unit cell, and the charge density modulation is described as an oscillation of the multipoles [3.35]. Thus the multipoles work as additional internal coordinates of scalar (monopole), vector (dipole), 2nd-rank tensor (quadrupole), etc., character. They may be grouped in a single pseudovector v having as many components as there are such new degrees of freedom. The latter correspond to the different virtual electronic transitions we shall consider in our description of the electron–phonon interaction. In principle such charge-density deformations can also include contributions from free-electron excitations and collective oscillations of electrons, i.e., plasmons, so that there is no real objection in applying such extended versions of shell models to metals. It is just a matter of choosing appropriate centers for rapidly converging multipole expansions, not necessarily located at the ion core positions. Such re-organization of part of the charge density into multipoles, i.e. shells centered at symmetric interstitial sites, leads to the concepts of "pseudoatoms" or "quasi-particles" recently introduced in the literature [3.36–38].

A successful application of the quasi-particle model to the bulk [3.35] and surface [3.38] dynamics of noble metals has been given recently and accounted for in the Chap. 8 of this book.

The shell-model dynamical equations for the bulk are written as

$$\begin{cases} M\omega^2 u = Ru + Qv \\ m\omega^2 v = Q^+u + Hv \end{cases}, \qquad (3.74)$$

where u is the core displacement vector, R the rigid-ion force constant matrix, Q the force-constant matrix connecting core displacements to multipolar coordinates, and H the coupling between multipolar coordinates. Of course, matrices

R, Q and H may all incorporate Coulomb interactions. The elements of the matrix m represent the inertial properties of the shells against deformations and are therefore of the order of the electron mass. Here they take the meaning of effective masses depending on the band structure and on the specific virtual transition involved. In the adiabatic approximation we simply set $m = 0$. If not, we note that the equation

$$m\omega^2 v = H v \qquad (3.75)$$

just represents the dynamics of electrons in a rigid lattice ($u = 0$) seen from the point of view of multipole coordinates. By eliminating the electronic coordinates in (3.74), we have a single dynamical equation for cores,

$$M\omega^2 u = [R - Q(H - m\omega^2)^{-1}Q^+]u , \qquad (3.76)$$

and the eigenvalues are given by the roots of the equation

$$\det[R - Q(H - m\omega^2)^{-1}Q^+ - M\omega^2] = 0 . \qquad (3.77)$$

We note that the electron dynamics participates through its Green's function $(H - m\omega^2)^{-1}$ given in the multipole representation. Since phonon frequencies ω are much smaller than those of electron excitations, the diagonal elements of $H - m\omega^2$ are positive and the second term in the determinant (3.77) is effectively subtracted from T, yielding a softening of the phonon frequencies with respect to rigid-ion model predictions ($Q = 0$). If there are electronic transitions of frequencies as small as phonon frequencies, as is the case in metals or narrow-gap semiconductors, the corresponding poles in the electron Green's functions contribute resonant electron–phonon coupling and a nonperturbative softening of phonon frequencies in certain spectral regions (Kohn or interband anomalies).

In surface problems the elimination of internal coordinates v should occur after the perturbation and symmetrization have been applied, in order to treat rigorously the perturbation of electron dymanics and electron–phonon interaction. Only by working in the extended space of nuclei and electron coordinates does the semi-infinite perturbation $\bar{\Lambda}$, as defined through the TI and RI conditions, attain self-consistency. This is particularly important in those crystals where large anomalies driven by the electron–phonon interaction appear in the phonon dispersion, e.g. in charge-density-wave (CDW) materials, mixed-valence crystals, superconductors and ferroelectrics.

The procedure has been discussed in a previous review article [3.39] for the shell model, where the number of shell coordinates is equal to that of the nuclear coordinates, but can be easily extended to the present multipole picture. We write the dynamical equations in the adiabatic approximation, using lower-case letters for the surface perturbation matrices, as

$$\begin{cases} M\omega^2 u = (R + r)u + (Q + q)v \\ 0 = (Q^+ + q^+)u + (H + h)v \end{cases} . \qquad (3.78)$$

After eliminating the multipolar coordinates v, we obtain a single equation in u which we solve as usual in the perturbation subspace. The eigenfrequencies are the solutions of the determinantal equation

$$\det[I + (R - QH^{-1}Q^+ - M\omega^2)^{-1}\Lambda] = 0 , \tag{3.79}$$

where the perturbation matrix is defined as

$$\begin{aligned}
\Lambda = {} & r - qH^{-1}Q^+ - QH^{-1}q^+ + QH^{-1}t_h H^{-1}Q^+ \\
& - qH^{-1}q^+ + qH^{-1}t_h H^{-1}Q^+ + QH^{-1}t_h H^{-1}q^+ \\
& + qH^{-1}t_h H^{-1}q^+ ,
\end{aligned} \tag{3.80}$$

and where

$$t_h = h(I + H^{-1}h)^{-1} \tag{3.81}$$

is the transition matrix associated with the electronic perturbation. The first row of (3.80) contains terms which are first order in the perturbations r, q or t_h, the second and the third rows terms which are quadratic and cubic in the perturbations, respectively. However, t_h contains the electronic perturbation h to all orders. The symmetrization of the perturbation matrix again has the effect of cutting off long-range Coulomb terms owing to charge-neutrality cancellation and therefore only the short-range part of the electron response, represented by H^{-1}, plays an important role and has to be retained. The inclusion of nonadiabatic effects is obtained by replacing H^{-1} with the electronic Green's function $(H - m\omega^2)^{-1}$.

In the above expression for the surface perturbation, all terms incorporate what amounts to the purely geometrical perturbation, namely the cut of the interatomic bonds implied by the creation of a free surface, as well as the local changes of force constants (due to elastic relaxation), of electronic states (in the matrix h) and electron–phonon interaction (due to electronic relaxation) in the surface region.

Despite its short-range character, the perturbation matrix, when all the required electronic coordinates are included, can admittedly be rather large in size, so that much of the advantage in using the GF method could be in question. Thus in Green's function calculations so far performed, a shortcut has been used consisting in the elimination of the electronic coordinates v before applying the surface perturbation. In this way the surface perturbation consists in cutting or modifying a number of effective interatomic force constants, which contain a hidden dependence on the electronic parameters. Clearly the surface changes in the electronic structure and electron–phonon interaction (e.g., the change of polarizability for surface ions) can hardly be taken into account in such a picture. We may, however, exploit the short-range nature of Λ and the relationship connecting the surface perturbation to the inverse Green's functions and define the effective surface perturbation in the nuclear coordinate subspace for the semi-infinite lattice by the *ansatz*

$$\overline{\Lambda^*} = \overline{\Lambda} + \Delta_\alpha(g_+^{-1} - g_-^{-1})/2 , \tag{3.82}$$

where the symbol Δ_α denotes the difference between the Green's function expressions calculated for perturbed and unperturbed values of the electronic and electron–phonon interaction parameters in the perturbation subspace.

3.5 Surface Phonon Polaritons

Electromagnetic surface modes associated with optical phonons, i.e., surface phonon polaritons, can be obtained in principle from the Green's function formalism. In the absence of retardation, the macroscopic field associated with bulk longitudinal optical phonons works as an external perturbation and yields the LO-TO splitting as a consequence of the symmetry break.

In a semi-finite ionic solid, the frequency of the surface phonon polariton appears automatically in the spectral part of the total Green's function G, once the part concerning the macroscopic field is added to the surface perturbation matrix. To prove this, we regard the "unperturbed" Green's function for the semi-infinite lattice G as already containing the microscopic surface modes, but not yet the part of the surface perturbation $\bar{\Lambda}_M$ bearing the macroscopic field effects. The elements of this matrix are obtained from (3.59) by noting that odd symmetry modes contribute the same macroscopic field as in the cyclic lattice (as if no surface cut would be made), whereas even symmetry modes do not contribute any macroscopic field owing to the exact cancellation of the contributions from the opposite slab surfaces. Thus

$$\Lambda_{M+} = 0 , \quad L_{M-} = 4\pi Z^2/\mu v_c , \tag{3.83}$$

where Z is the ionic effective charge, μ the cell reduced mass, v_c the cell volume, and

$$\bar{\Lambda}_M = (2\pi/\mu v_c)Z^2 . \tag{3.84}$$

The perturbed Green's function $\tilde{G}$, including both macroscopic and microscopic perturbations, is then given by

$$\tilde{G} = \bar{G}(I + \bar{\Lambda}_M \bar{G})^{-1} . \tag{3.85}$$

Remembering the relationship between Green's function and dielectric susceptibility [3.40],

$$\chi = (Z^2/\mu v_c)\bar{G} , \tag{3.86}$$

we can express $\tilde{G}$ as

$$\tilde{G} = \bar{G}(1 + 2\pi\chi)^{-1} , \tag{3.87}$$

and expressing $\bar{G}$ in terms of χ via (3.86) and introducing the dielectric function $\varepsilon(\omega) = 1 + 4\pi\chi$, we obtain after some algebra, the spectral part of the perturbed Green's function,

$$\text{Sp}\{\tilde{G}\} = -(\mu v_c / \pi Z^2)\text{Im}\{(1 + \varepsilon(\omega))^{-1}\} \, . \tag{3.88}$$

In this derivation we have considered that Z and χ represent possibly anisotropic but still diagonal matrices. Equation (3.88) tells us that the total response of the crystal surface includes, besides the microscopic modes which constitute the spectral structure of G, also the frequencies of the macroscopic electromagnetic modes. The latter are the solution of the equation

$$\varepsilon(\omega) = -1 \, . \tag{3.89}$$

This well-known implicit equation for the surface polariton frequencies is normally obtained by solving Maxwell equations for a semi-infinite medium with the appropriate surface boundary condition [3.41].

It is however interesting to note that $\varepsilon(\omega)$ in the denominator of (3.88) also contains all the information on microscopic modes born by the semi-infinite lattice Green's function. This can be easily seen, starting from a simplified expression for the susceptibility, with only two oscillators: the transverse bulk optical mode of frequency ω_T, and a microscopic surface optical mode of frequency ω_s. The latter may have a spectral weight $w^2 \ll 1$, since it is restricted to a small portion of the lattice, in the surface region. We write

$$\chi = \frac{z^2}{\mu v_c} \left(\frac{1}{\omega_T^2 - \omega^2 - i\gamma} + \frac{w^2}{\omega_s^2 - \omega^2 - i\gamma} \right) . \tag{3.90}$$

Substitution of this expression into (3.88) yields

$$\frac{1}{1 + \varepsilon(\omega)}$$
$$= \frac{(\omega_T^2 - \omega^2)(\omega_s^2 - \omega^2)}{2(\omega_T^2 - \omega^2)(\omega_s^2 - \omega^2) + (4\pi Z^2/\mu v_c)(\omega_s^2 - \omega^2 + w^2\omega_T^2 - w^2\omega^2) - i\Gamma} \, , \tag{3.91}$$

where

$$\Gamma = \gamma \left[\frac{\omega_s^2 - \omega^2}{\omega_T^2 - \omega^2} + w^2 \frac{\omega_T^2 - \omega^2}{\omega_s^2 - \omega^2} \right] . \tag{3.92}$$

The real part of the denominator has two zeros in ω^2, yielding two peaks in the spectral function, given by

$$\omega_\pm^2 = \frac{1}{2} \left\{ \omega_{FK}^2 + \omega_s^2 + \frac{1}{2} w^2 \omega_P^2 \right.$$
$$\left. \pm \sqrt{(\omega_{FK}^2 + \omega_s^2 + \frac{1}{2} w^2 \omega_P^2)^2 - 4\omega_s^2 \omega_{FK}^2 - 2w^2 \omega_P^2 \omega_T^2} \right\} . \tag{3.93}$$

Here

$$\omega_P^2 \equiv 4\pi Z^2/\mu v_c \, ,$$
$$\omega_{FK}^2 \equiv \omega_T^2 + \frac{1}{2}\omega_P^2 \tag{3.94}$$

are the plasma and the Fuchs–Kliewer (unperturbed surface polariton) squared frequencies, respectively. In the limit $w^2 \to = 0^+$, the two zeros are just the Fuchs–Kliewer and the microscopic surface mode squared frequencies. To first order the finite amplitude of the microscopic mode perturbs both ω_{FK}^2 and ω_m^2 as

$$\omega_+^2 \equiv \tilde{\omega}_{FK}^2 \cong \omega_{FK}^2 + \frac{w^2}{2} \frac{\omega_P^4}{\omega_{FK}^2 - \omega_s^2} , \qquad (3.95)$$

$$\omega_-^2 \equiv \tilde{\omega}_s^2 \cong \omega_s^2 - \frac{w^2}{2} \frac{\omega_P^2(\omega_T^2 - \omega_s^2)}{\omega_{FK}^2 - \omega_s^2} . \qquad (3.96)$$

The spectral part of $\tilde{G}$ turns out to be

$$\mathrm{Sp}\, \tilde{G} = \frac{2\pi}{\omega_P^2} \left| \frac{(\omega_T^2 - \omega^2)(\omega_s^2 - \omega^2)}{\tilde{\omega}_{FK}^2 - \tilde{\omega}_s^2} \right| [\delta(\omega^2 - \tilde{\omega}_{FK}^2) + \delta(\omega^2 - \tilde{\omega}_s^2)] . \qquad (3.97)$$

Clearly, for $\omega^2 = \tilde{\omega}_s^2$ the prefactor becomes of the order of w^2 as a consequence of (3.96), whereas the intensity of the Fuchs–Kliewer mode remains of the order of unity. For a very thin slab the two frequencies may be appreciably shifted as a consequence of the mutual perturbation, particularly when the surface optical mode falls close to the Fuchs–Kliewer polariton. This may happen, e.g., in certain alkali halides (see Sect. 3.6) for the shear-vertical S_2 surface mode. On the other hand, the longitudinal and shear-horizontal optical surface modes (the Lucas modes S_4 and S_5, respectively (see Sect. 3.6)) fall slightly below, very close to ω_T, and therefore ω_{FK} is only slightly stiffened, whereas ω_S is practically unaffected.

The very weak interaction between Fuchs–Kliewer modes and the microscopic surface modes fully justifies a surface dynamical calculation without the inclusion of the macroscopic field effects in the surface perturbation, as is done in all calculations reviewed in the next section.

3.6 Surface Vibrations in Alkali Halides

We have reviewed elsewhere [3.39] our GF calculations of surface phonons for the (001) surface of eight alkali halides (LiF, NaF, NaCl, NaI, KCl, KBr, KI and KF), which have been published from 1973 onwards [3.22, 24, 42, 43]. All these calculations were based on the breathing shell model (BSM) [3.44]. Shell model calculations based on direct dynamical matrix diagonalization for a thin slab have been performed since 1970 and are reported in an extended paper by *Chen* et al. [3.45] for several rocksalt structures (LiF, NaF, NaCl, NaI, RbF, RbCl and MgO) and for CsCl. Recently *Kress* et al. [3.46] have extended this study to all rocksalt alkali halides with a selfconsistent calculation of the surface relaxation and its effects on surface dynamics. A review is given by de Wette in Chap. 4 and therefore we restrict ourselves to two examples: GF calculations for

LiF and NaF (001) surfaces, where a direct comparison to He scattering data is possible.

With respect to the direct dynamical matrix diagonalization applied to a thin slab, the GF method has the formal advantage of reducing a big problem to a small one: it enables one to work in the perturbation subspace rather than in the large slab space. In practice, however, the GF method has a drawback in the severe computational difficulties which derive from the singular nature of the surface-projected Green's function [3.23].

The BSM dynamical matrix is directly related to a set of phenomenological input data such as the elastic constants c_{ji}, ionic polarizabilities $\alpha_{\pm}$, net charge Z, bulk transverse optical frequency ω_{TO}, static and high-frequency dielectric constants ε_s and ε_0. In addition to nearest neighbor repulsion the model allows for a repulsive second neighbor ($2n$) interaction only between larger ions. Thus $2n$ repulsion between halogen ions is considered in all crystals, except in KF, where the $2n$ interaction is between potassium ions.

The GFs are calculated for each surface wave vector K over a mesh of 101 equally spaced values of the frequency from zero to the maximum value of the crystal. The numerical integration over the wave vector component normal to the surface from 0 to π/r_0 has been performed on a mesh of 97 points, which is equivalent to a slab calculation with 192 layers.

In such a large value rests the main difference between the GF method and *Chen et al.* slab calculations. Here, unlike in slab calculations, there is no interference between deeply penetrating surface modes belonging to opposite surfaces, such as long-wave Rayleigh modes; moreover, the surface projected bulk bands form a continuum, well separated from local modes. Actually, the main advantage of the GF method is to allow for an accurate determination of the surface-projected phonon densities which are useful quantities in the calculation of surface thermodynamical properties and vibrational response functions.

In order to offer a guideline for understanding the calculated dispersion curves we present a classification of surface phonons according to their character and polarization.

Along symmetry direction $\overline{\Gamma X} : K = (\xi, \xi); \overline{\Gamma X} : K = (\xi, 0)$ the bulk bands and related surface modes have either sagittal ($\perp$) or parallel ($\parallel$) polarization. Sagittal modes can be either quasi-shear vertical (SV) or quasi-longitudinal (SP). Parallel and shear horizontal (SH) polarizations are synonyms. We associate TA_1, LA, TO_1 and LO bulk modes with sagittal bands, TA_2 and TO_2 bands with parallel bands. Often, but not always, the band edges correspond to certain dispersion curves of the bulk along symmetry directions.

With regard to the identification and classification of surface modes the simplest way is to associate each surface mode with the corresponding bulk band which the surface mode comes from. Roughly speaking we expect at least one surface mode from each of the six bulk modes.

In doing that, we follow, and strongly suggest as a convention, the general principle that labels should correspond to well defined spectroscopic entities, with special regard to polarization and (optical or acoustic) character. Therefore we adopted, as far as possible, *Chen* et al. [3.45] notations (S_j), but we were forced to change them wherever they do not fulfill the above requirement. This often occurs for the noncrossing behavior of quasi-shear horizontal and quasi-sagittal branches in nonsymmetry directions.

In Table 3.1 we display the classification of surface modes according to the above concepts.

Table 3.1. Classification and conventional names of surface modes according to their bulk bands of origin and polarization

Bulk band	Surface Mode	Polarization	Conventional name
TA$_1$	S_1	SV	Rayleigh wave
	S_8	SV	crossing mode [3.48]
TA$_2$	S_7	SH	—
LA	S_6	SP	—
TO$_1$	S_4	SP	Sagittal Lucas mode
TO$_2$	S_4	SH	Parallel Lucas mode
LO	S_2	SV	—
	S_3	SV	—

The sagittal mode S_1 corresponds to a Rayleigh wave in the continuum limit ($K \rightarrow 0$). The SH mode S_7 is another acoustic surface mode with ∥ polarization existing along the (110) direction in cubic crystals; it was found by G.P. *Alldredge* [3.47]. S_6 is an acoustic mode peeled off from the LA lower edge; normally it is localized at the zone boundary. The sagittal resonance S_8, crossing the LA band in all directions, may appear in some crystals as a folded prolongation of the Rayleigh wave. Its intensity is appreciable in crystals with nearly equal ionic masses (NaF, KCl) [3.48].

S_4 and S_5 form the pair of Lucas modes (LM) [3.49]. They have an optical character and become degenerate at the zone center ($\bar{\Gamma}$ point). For $K \rightarrow 0$ along x, S_4 and S_5 become linearly polarized along x and y, respectively. S_2 is a microscopic optical mode whose polarization is quasi-SV everywhere, and exactly normal to the surface at $\bar{\Gamma}$ and $\bar{M}$; its existence has been proved by *R.F. Wallis* in one of his pioneering works in surface lattice dynamics [3.50]. At the zone boundary, S_2 turns out to be associated with TO$_1$ whereas another mode, S_3, comes from the edge of LO.

According to *Maradudin* et al. [3.9] and to *De Wette* [3.51], S_2 can be strongly hybridized with the surface phonon polariton of the semi-infinite crystal provided that the dispersion of the optical branches is so effective as to push this mode down in the gap below the LO band. The mechanism has been discussed in the previous section. The calculations for NaCl and MgO slabs in the rigid ion model (Refs. [3.51] and [3.52], respectively) offer an example.

The connection between the polarization of a surface mode and that of the related band modes might appear to be rather complicated at first glance. We give here a simple guideline. Sagittal surface modes are normally eliptically polarized except at $K = (1,0)\pi/r_0$ ($\bar{M}$-point), where all modes have linear polarization for symmetry reasons. We note that at $K = 0$ ($\bar{\Gamma}$-point) longitudinal and transverse bulk modes are polarized respectively along z and x (or y), whereas at the $\bar{M}$ point the bulk modes at the lower edge are polarized along x if longitudinal and along z if transverse (TA$_1$ or TO$_1$). Therefore S_1 is normally z-polarized at $\bar{M}$ and becomes more and more elliptical as $K \rightarrow 0$. The polarization, however, remains elliptical in the continuum limit because of the macroscopic nature of RW. In contrast the microscopic optical modes S_4 and S_2 become linearly polarized as $K \rightarrow 0$, obviously along x and z, respectively, like the corresponding TO$_1$ and LO bands at $\bar{\Gamma}$.

The behavior of optical S_4 and $S_2(S_3)$ modes for K varying across the surface Brillouin zone up to the zone boundary cannot be reduced to the simple scheme shown above because of their mutual hybridization. Labels and polarizations are often interchanged. This is seen, for instance, at $\bar{M}$, where x- and y-polarized surface modes are degenerate in pairs for symmetry reasons, like at $\bar{\Gamma}$: in some cases (e.g. Na halides) the degenerate pair is (S_5, S_2) whereas at $\bar{\Gamma}$ the degenerate pair is (S_4, S_5).

We show the calculated surface phonon dispersion curve of LiF and NaF for which extensive experimental investigation has been performed (Figs.

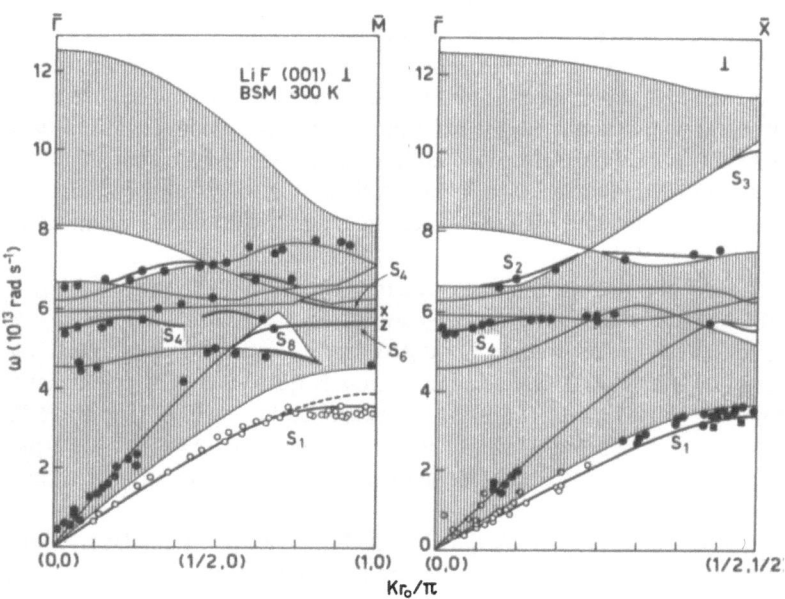

Fig. 3.3. Surface phonon dispersion curves of LiF(001) along symmetry directions for sagittal polarization ($\perp$). Calculations were made with the GF method and BSM room-temperature data. Heavy lines are surface modes; shaded areas correspond to bulk bands projected to the surface. Atom scattering data are displayed: open circles are from [3.54]; black points from [3.55]

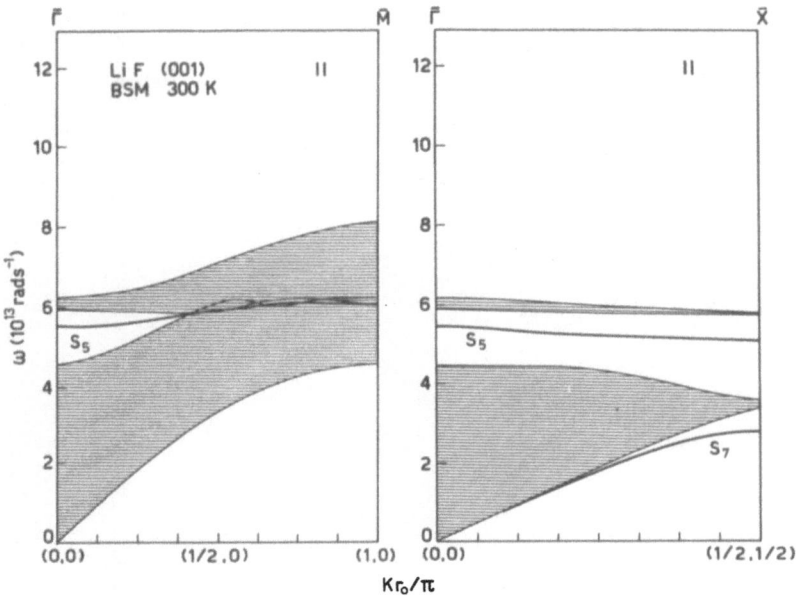

Fig. 3.4. The same as Fig. 3.3 for parallel polarization (∥)

3.3–8). For the sake of clarity, along symmetry directions (Figs. 3.3, 4, 6, 7) we plot sagittal and parallel dispersion curves separately. Along the zone boundary $\overline{MX}$, however, sagittal and parallel components are mixed together and are shown superimposed in Figs. 3.5 and 3.8. Heavy lines are surface modes. Thin lines are band edges of bulk modes. When a dispersion curve enters a band, the surface-localized mode transforms into a resonance. The RW dispersion curve

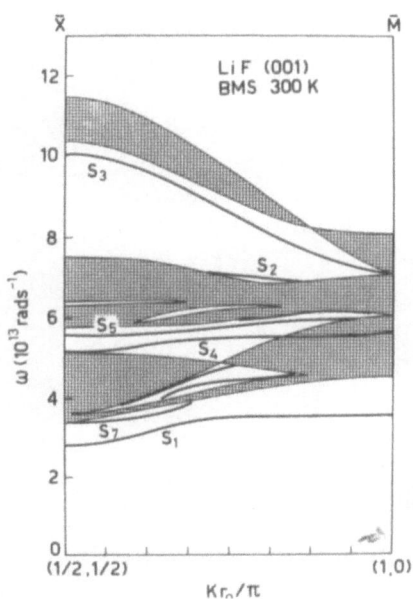

Fig. 3.5. Surface phonon dispersion curves of LiF(001), along the $\overline{MX}$ direction. Here sagittal and parallel components intermix and are displayed together

59

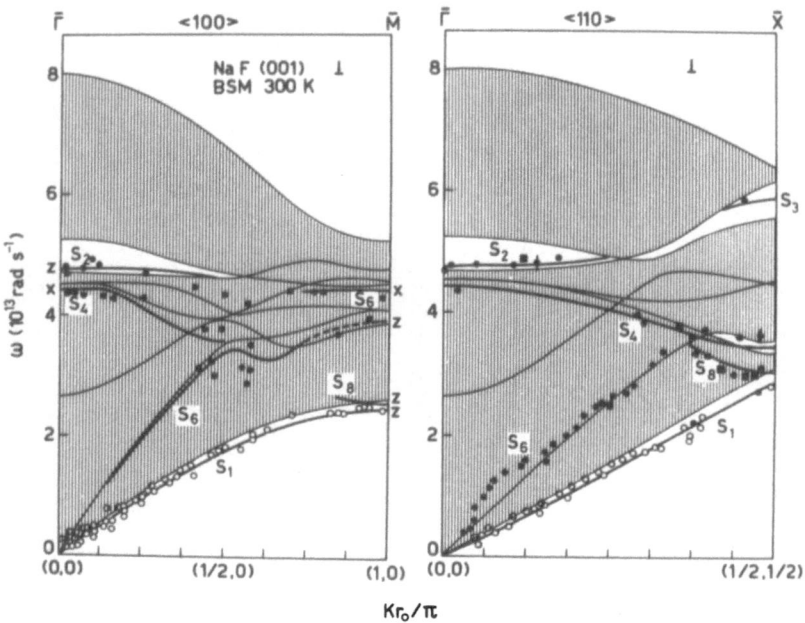

Fig. 3.6. Surface phonon dispersion curves of NaF(001) along symmetry directions for sagittal polarization ($\perp$). Calculations were made with the GF method and BSM room temperature data. Heavy lines are surface modes; shaded areas correspond to bulk bands projected to the surface. Open circles are from [3.58], black dots from [3.48]

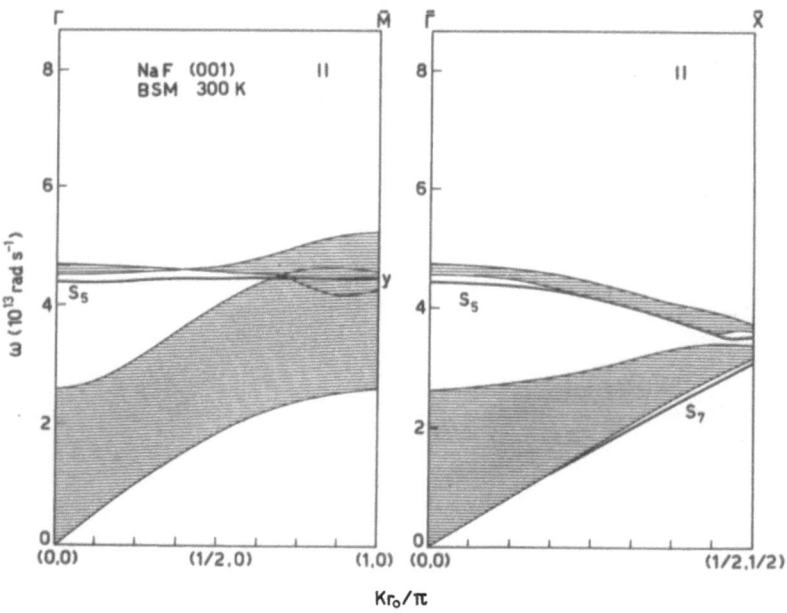

Fig. 3.7. The same as Fig. 3.6 for parallel polarization ($\parallel$)

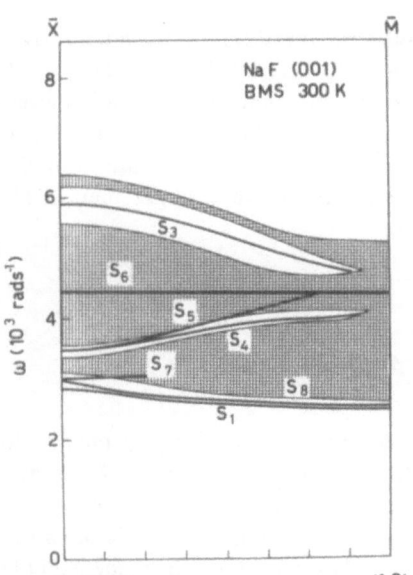

Fig. 3.8. The same as Fig. 3.5 for NaF(001)

(S_1) appearing in Figs. 3.3 and 3.6 is compared to the data obtained from He scattering time-of-flight (TOF) spectra (open circles [3.53, 54] and black squares [3.55]).

The experimental points follow very well the theoretical dispersion curve along the two directions in both crystals. However both slab [3.55] and GF calculations [3.42] do not predict correctly the Rayleigh frequency at the M point (broken line) of LiF. Among the various mechanisms which can lower the zone-boundary frequency [3.56], we considered the polarizability change because the misfit is found when the anions move in the $\bar{M}$ point Rayleigh mode, as in LiF, and is not found when the anions are at rest, as in NaF (Fig. 3.6, open circles).

The problem of relaxation in LiF(001) has been carefully investigated by de Wette, Schroeder and Kress in a shell-model slab calculation, where surface dynamics is consistently based on the surface equilibrium configuration. The change of three-body forces is the most efficient mechanism for shifting the Rayleigh mode frequency at the $\bar{M}$ point [3.57].

In NaF, fluorine ions move mainly in the optical modes. Since anions have usually a larger interaction with He atoms than cations, owing to their large size, optical surface phonons have been detected in NaF(001) (Fig. 3.6, black dots) [3.48, 58]. The experimental points are in good agreement with the theoretical S_2, S_4 and S_6 branches. However, for S_4 and S_6 a simpler hybridization scheme is suggested by experiment: S_4 should stay above and continue up to the zone boundary, while S_6 connects to the z-polarized mode at M.

The experimental evidence of optical modes in NaF(001) has been confirmed recently by electron energy loss spectroscopy (EELS) by Thiry et al. [3.59], who were able to resolve the very weak features of microscopc surface modes from the large tail of the Fuchs–Kliewer mode. There the S_4 mode appears to have a flatter dispersion than in calculated and He scattering results, but the experimental error in EELS measurements is normally much larger than in He scattering experiments.

3.7 Further Developments:
The Study of Surface Phonon Anomalies

The GF method applies well to more complex perturbations, e.g. a change in surface force constants [3.16, 60]. The difficult cases of stepped surfaces and reconstruction have been successfully treated by the method of GF generating coefficients developed by *Armand* and *Masri* [3.61]. More recently, *Goldhammer* et al. have approached the problem of the (7×7) reconstruction of the Si(111) surface and the (8×8) reconstruction of the Ge(111) surface with related phonon instabilities by means of the GF method [3.17]. Many of the surface dynamical phenomena owe their complexity to the role of electron–phonon interaction. He scattering measurements are revealing aspects of surface dynamics that were unexpected on the basis of the underlying bulk dynamics. Many microscopic mechanisms which are forbidden or inhibited in the bulk for symmetry reasons are promoted by a symmetry reduction and manifest themselves at the surface. A typical example is that of surface phonon anomalies induced by electron–phonon coupling, which may possibly occur also in crystals with regular bulk phonon dispersion.

Most of the existing theories of surface lattice dynamics have been conceived for ideal, unrelaxed surfaces. There the change of atomic equilibrium positions at the surface (elastic relaxation) of electron density of states (electronic relaxation), and the consequent change in the electron–phonon interaction are not taken into account, or, at best, not in a self-consistent way.

The case of LiF(001) gave an indication that the change in many-body force constants at the surface may induce important deviations from the ideal surface behavior in the surface phonon spectrum. The role of surface electron states and surface electron phonon interaction becomes apparently dramatic in noble metal surfaces. High resolution He TOF spectra from Ag(111) give evidence, in addition to the Rayleigh mode, of a resonance below the LA bulk edge [3.62]. The downward shift of this resonance with respect to the LA edge is not expected in the ideal-surface GF calculation performed by *Armand* [3.63] by means of the generating coefficients. This behavior is reproduced in a slab calculation performed by *Bortolani* et al. [3.64] with a suitable parametrization of bulk dynamics and a softening of in-plane surface force constants, and has been recently explained on the basis of the quasi-particle model [3.38].

More spectacular surface phonon anomalies have been predicted in superconducting refractory materials as a consequence of the deep anomalies existing in the bulk phonon dispersion. These anomalies have been associated with the electron–phonon coupling which determines the superconducting transition. Interesting questions are whether anomalies do occur in the dispersion of surface phonons; whether surface anomalies are shifted with respect to their bulk counterparts, as a consequence of the difference in the electron–phonon coupling at the surface; and whether all this has any relevance in surface superconductivity.

In a cluster shell model calculation of TiN dispersion curves *Miura* et al. [3.65] have explained the deep anomaly occurring in the LA branch at 2/3 of the zone by a cluster-deformation model. This model is directly related to the microscopic properties and emphasizes the importance of excitations of p-d hybridized states near the Fermi surface for the lattice vibrations in transition metal compounds. The calculated surface dynamics of TiN(001) by the GF method (Fig. 3.9) [3.66] show that the dispersion curves of the Rayleigh wave (S_1) and the resonance S_6 and the optical mode S_2 are anomalous in the (100) direction. The S_6 anomaly remains very close to the bulk anomaly, as a consequence of its prevailing bulk character, while the Rayeigh wave anomaly is shifted back to 1/2 of the zone. This means that the effective range of the electron–phonon interaction, determining the position of anomalies, is changed at the surface. The anomalous behavior of the S_4 branch consists in the large softening towards the zone boundary, in contrast with the optical band edge increase and with the regular behavior of S_2 in large-gap insulators (e.g. NaI [3.39]).

Such an anomaly has been recently found in transition metal carbides by means of high resolution EELS by *Oshima* et al. [3.67, 68]. The comparison with the EELS data for a very similar crystal, TiC [3.67], presented in Fig. 3.10, indicates a very good agreement between theoretical prediction for TiN and experiment for the surface optical branches. Of course, no acoustic anomalies

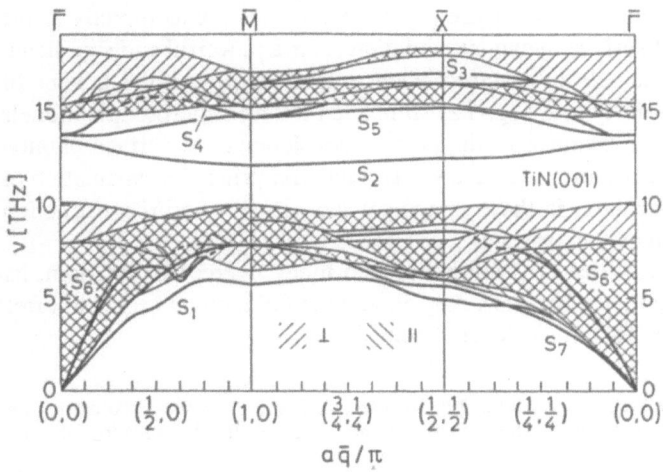

Fig. 3.9. GF calculation of the dispersion curves of surface phonons in TiN(001) (from [3.66])

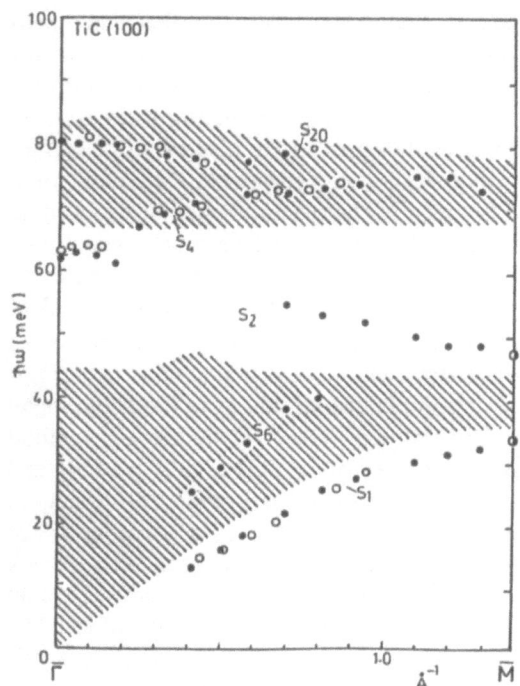

Fig. 3.10. Experimental EELS data for surface phonons in TiC(001) measured by *Oshima* et al. [3.67]. A comparison with the calculated TiN surface phonons is discussed in the text

occur in TiC because of its very low superconducting T_c. On the other hand the S_1 mode in the good superconductor $NbC_{0.96}$(001) of [3.68] appears to be localized below the acoustic edge above 1/2 of the zone and suddenly rises up into the band above 3/4 of the zone, which suggests an anomaly shift with respect to the LA bulk anomaly at 2/3 of the zone, in agreement with the prediction for TiN(001).

Clearly, the interpretation of surface phonon anomalies points directly to the microscopic theory of surface dynamics, and represents a powerful and significant test for the present models of electron–phonon interaction. As the advent of neutron spectroscopy thirty years ago has stimulated the first microscopic models and theories of lattice dynamics, so the present development in surface phonon spectroscopy will trigger more extensive work in the first-principles calculation of surface phonons, particularly in those systems where stability and reconstruction problems have a primary importance.

In this case the difficulty of dealing with too many degrees of freedom, as implied by slab calculations, becomes crucial, and the GF method seems to offer more practicable paths for future developments.

Acknowledgements. The authors are grateful to J.P. Toennies, G. Brusdeylins and J. Skofronick for several useful discussions and to Dr. P. Thiry for the information on his EELS results in NaF prior to publication.

References

3.1   I.M. Lifshitz: Nuovo Cimento Suppl. **3**, 732 (1956) and references quoted therein
3.2   G.F. Koster, I.C. Slater: Phys. Rev. **94**, 1392 (1954); **95**, 1167 (1954)
3.3   J. Callaway: J. Math. Phys. **5**, 783 (1964); Phys. Rev. **154**, 515 (1967)
3.4   R.A. Brown: Phys. Rev. **156**, 889 (1967)
3.5   M.V. Klein: In *Physics of Color Centers*, ed. by W.B. Fowler (Academic, New York 1968)
3.6   E.N. Economu: *Green's Function in Quantum Physics* (Springer, Berlin, Heidelberg 1978)
3.7   S. Doniach, E.H. Sandheimer: *Green's Function for Solid State Phyisicists* (Benjamin, London 1974)
3.8   G. Rickeyzen: *Green's Functions and Condensed Matter* (Academic, New York 1980)
3.9   A.A. Maradudin, E.W. Montroll, G.H. Weiss, I.P. Ipatova: *Theory of Lattice Dynamics in the Harmonic Approximation*, Suppl. 3 to *Solid State Physics* (Academic, New York 1971)
3.10   H. Bilz, D. Strauch, R.K. Wehner: In *Encyclopedia of Physics*, Vol. XXV/2d (Light and Matter Id) (Springer, Berlin, Heidelberg 1984)
3.11   G. Brusdeylins, R.B. Doak, J.P. Toennies: Phys. Rev. **B27**, 3662 (1983)
3.12   S. Lehwald, J.M. Szeftel, H. Ibach, T.S. Rahman, D.L. Mills: Phys. Rev. Lett. **50**, 518 (1981); J.M. Szeftel, S. Lehwald, H. Ibach, T.S. Rahman, J.E. Black, D.L. Mills: Phys. Rev. Lett. **51**, 268 (1983)
3.13   G. Benedek: Physica **127B**, 59 (1984)
3.14   A.A. Maradudin, J. Melngailis: Phys. Rev. **133**, A1188 (1964)
3.15   L. Dobrzynski, G. Leman: J. Phys. (Paris) **30**, 116 (1969)
3.16   S.W. Musser, K.H. Rieder: Phys. Rev. **B2**, 3034 (1970)
3.17   W. Goldammer, W. Ludwig, W. Zierau, C. Falter: Surf. Sci. **141**, 139 (1984); W. Zierau, W. Goldammer, C. Falter, W. Ludwig: Proc. Int. Conf. on Superlattices, Urbana, Ill. (USA) 1984
3.18   J.E. Black, B. Lacks, D.L. Mills: Phys. Rev. **B22**, 1818 (1980)
3.19   F. Garcia Moliner: Ann. Phys. (Paris) **2**, 179 (1977)
3.20   R.E. Allen: Surf. Sci. **76**, 91 (1978)
3.21   G. Armand: Phys. Rev. **B14**, 2218 (1976)
3.22   G. Benedek: Phys. Stat. Sol. **B58**, 661 (1973)
3.23   G. Benedek: Surf. Sci. **61**, 603 (1976)
3.24   G. Benedek, L. Miglio: In *Ab Initio Calculation of Phonon Spectra*, ed. by J.T. Devreese, V.E. van Doren, P.E. van Camp (Plenum, New York 1983) p.215
3.25   G. Platero, V.R. Velasco, F. Garcia Moliner, G. Benedek, L. Miglio: Surf. Sci. **143**, 243 (1984)
3.26   Owing to this special meaning the GF is often defined with opposite sign. Here we keep to the definition given in many classic mathematics texts, e.g., I.G. Petrovsky: *Lectures on Partial Differential Equations* (Interscience, New York 1954) §24. This has been adopted in Bilz, Strauch and Wehner's handbook [3.10] and also in our previous papers
3.27   A. Messiah: *Mécanique Quantique* (Dunod, Paris 1959)
3.28   J.M. Ziman: *Elements of Advanced Quantum Theory* (Cambridge University Press, Cambridge 1969)
3.29   We note that the operator $\mathcal{R}$ is a function of the equilibrium positions. Only when they do not change by cutting the bands across Σ is the operator $\mathcal{R}$ the same for ϕ and ϕ_0. However, the RI condition has been proved to be equivalent to equilibrium condition and therefore it contributes a fictitious force field at the surface, required to equilibrate the surface in the unrelaxed configuration. The relationship between rotational invariance and equilibrium and the effect of elastic relaxation are discussed in [3.24].
3.30   Since the eigenvalue in equations (3.45) and (3.49) is multiplied by the mass matrix, ϱ_0 and ϱ are not merely frequency densities. They are normalized to $1/s\mu$, μ being the unit cell reduced mass
3.31   R.E. Allen, G.P. Alldredge, F.W. de Wette: Phys. Rev. **B4**, 1648, 1661, 1682 (1971); **B2**, 2570 (1970)
3.32   B.G. Dick, Jr., A.W. Overhauser: Phys. Rev. **112**, 90 (1958)
3.33   A.D.B. Woods, W. Chochran, B.N. Brockhouse: Phys. Rev. **119**, 980 (1960)
3.34   W. Cochran: C.R.C. Critical Reviews in Solid State Sciences **2**, 1 (1971)
3.35   This method is developed in C.S. Jayanthi, H. Bilz, W. Kress, G. Benedek: Phys. Rev. Lett. **59**, 795 (1987), in connection with the surface dynamics of noble metals. An application to

the bulk dynamics of these crystals was previously given in C.S. Jayanthi, H. Bilz, W. Kress: In *Phonon Physics*, ed. by. J. Kollar, N. Kroo, N. Menyhard, T. Siklos (World Scientific, Singapore 1985) p.630

3.36   M.S. Daw, M.J. Baskes: Phys. Rev. B **29**, 6443 (1984)

3.37   C. Falter, W. Ludwig, A.A. Maradudin, M. Selmke, W. Zierau: Phys. Rev. B **32**, 6510 (1985); C. Falter, M. Selmke, W. Ludwig, K. Kunc: Phys. Rev. B **32**, 6518 (1985)

3.38   Y.R. Wang, A.W. Overhauser: Phys. Rev. B **35**, 497 (1987); 501 (1987)

3.39   L. Miglio, G. Benedek: In *Structure and Dynamics of Surfaces II*, ed. by W. Schommers, P. von Blanckenhagen (Springer, Berlin, Heidelberg 1987) p.35

3.40   G. Benedek, G.F. Nardelli: Phys. Rev. **155**, 1004 (1967)

3.41   A.D. Boardman (ed.): *Electromagnetic Surface Modes* (Wiley, New York 1982)

3.42   G. Benedek, G.P. Brivio, L. Miglio, V.R. Velasco: Phys. Rev. B **26**, 497 (1982)

3.43   G. Benedek, F. Galimberti: Surf. Sci. **71**, 87 (1978); **118**, 713 (1982)

3.44   U. Schröder: Solid State Commun. **4**, 347 (1966); U. Schröder, V. Nüsslein: Phys. Status Solidi **21**, 309 (1967)

3.45   T.S. Chen, F.W. de Wette, G.P. Alldredge: Phys. Rev. B **15**, 1167 (1977) and references therein

3.46   W. Kress, F.W. de Wette, A.D. Kulkarni, U. Schröder: Phys. Rev. B **35**, 2467 (1987)

3.47   G.P. Alldredge: Phys. Rev. Lett. **41A**, 281 (1972)

3.48   G. Benedek, L. Miglio, G. Brusdeylins, J.G. Skofronick, J.P. Toennies: Phys. Rev. B **35**, 6593 (1987). The sagittal resonance is also found in NaCl [3.39] where the mass difference is appreciable. F.W. de Wette, W. Kress and U. Schröder [Phys. Rev. B **33**, 2835 (1986)] have proposed another mechanism for folding based on the TA-TO hybridization of bulk modes, which may work also in crystals with different masses. Since the pure geometrical condition may not be sufficient for folding to be observed (even in NaF not all modes do show evident folded companions), the hybridization mechanism may be invoked. A discussion is given in Chap. 5 of the present book. It should be noted that in KCl and NaCl the folded branch S_8 apparently extends up to the zone center [3.39], where it takes the character of the so-called structure-induced surface resonance recently discussed by J.A. Stroscio, M. Persson, S.R. Bare, W. Ho: Phys. Rev. Lett. **54**, 1428 (1985), for metal surfaces. Such a resonant mode was actually found in the original GF calculation of 1973 [3.22] and stems from a secondary lower edge of a bulk band associated with a relative minimum of the dispersion relation in the z direction

3.49   A.A. Lucas: J. Chem. Phys. **48**, 3156 (1968)

3.50   R.F. Wallis: Phys. Rev. **116**, 302 (1959)

3.51   F.W. de Wette: In *Lattice Dynamics*, ed. by M. Balkanski (Flammarion Sciences, Paris 1978) p.275

3.52   G. Lakshmi, F.W. de Wette: Phys. Rev. **22**, 5009 (1980)

3.53   G. Brusdeylins, R.B. Doak, J.P. Toennies: Phys. Rev. Lett. **44**, 1417 (1980); **46**, 437 (1981)

3.54   G. Benedek, J.P. Toennies, R.B. Doak: Phys. Rev. B **28**, 7277 (1983)

3.55   G. Bracco, E. Cavanna, A. Gussoni, C. Salvo, R. Tatarek, S. Terreni, F. Tommasini: Vuoto Sci. Tecn. **16** (1986); G. Bracco, M. D'Avanzo, C. Salvo, R. Tatarek, S. Terreni, F. Tommasini: Surface Sci. **189/190**, 684 (1987)

3.56   E.R. Cowley, J.A. Barker: Phys. Rev. B **28**, 3124 (1983)

3.57   F.W. de Wette, W. Kress, U. Schröder: Phys. Rev. B **32**, 4143 (1985)

3.58   G. Brusdeylins, R. Rechsteiner, J.G. Skofronick, J.P. Toennis, G. Benedek, L. Miglio: Phys. Rev. Lett. **54**, 466 (1985)

3.59   J.L: Longueville, P.A. Thiry, J.J. Pireaux, R. Candano: Communicat. to the 10th European Conference on Surface Science, Bologna 5–8 Sept. 1988

3.60   R.E. Black, F.C. Shanes, R.F. Wallis: Surf. Sci. **133**, 199 (1983)

3.61   G. Armand, P. Masri: Surf. Sci. **130**, 89 (1983)

3.62   R.B. Doak, U. Harten, J.P. Toennies: Phys. Rev. Lett. **51**, 587 (1983)

3.63   G. Armand: Solid State Commun. **48**, 261 (1983)

3.64   V. Bortolani, A. Franchini, F. Nizzoli, G. Santoro: Phys. Rev. Lett. **52**, 429 (1984)

3.65   M. Miura, W. Kress, H. Bilz: Z. Phys. B**54**, 103 (1984)

3.66   G. Benedek, M. Miura, W. Kress, H. Bilz: Phys. Rev. Lett. **52**, 1907 (1984)

3.67   C. Oshima, T. Aizawa, M. Wuttig, R. Souda, S. Otani, Y. Ishizawa, H. Ishida, K. Terakura: Phys. Rev. B **36**, 7510 (1987)

3.68   C. Oshima, R. Souda, M. Aono, S. Otani, Y. Ishizawa: Phys. Rev. Lett. **56**, 2401 (1986)

4. Study of Surface Phonons by the Slab Method

F.W. de Wette

With 20 Figures

During the past decade, important progress has been made in the study of vibrations of crystal surfaces as a result of recent advances in two important experimental techniques: low energy light atom He scattering and electron energy loss spectroscopy [EELS]. The need for interpretation of the new experimental data has stimulated further theoretical and computational developments.

In this chapter we review the contributions and results which the so-called *slab method* has made to our understanding of the vibrational and related properties of crystal surfaces. In a strict sense, the term *surface phonons* (appearing in the title of this volume) applies to only part of these vibrational properties, namely to those vibrations of the system which are localized at the surface. But, of course, the slab method is concerned with the vibrational properties of the entire system (i.e. both bulk and surface phonons), all of which, in principle, play a role in surface scattering phenomena and thermodynamic properties of the system. In this connection it is useful to make a distinction between *surface properties* and *surface-excess properties*. The first are structural or dynamic properties which show differences for surface and bulk particles (e.g. the mean-square amplitudes of vibration), whereas the second are system properties which are modified by the presence of surfaces (e.g. the surface-excess specific heat). A full understanding of the influence of surfaces on the physical properties of crystals requires a knowledge of the dynamical properties of the entire system. In this context the main theme of this volume, *surface phonons*, has to be understoood in the broad sense of including *all* vibrations near a surface, i.e. surface-localized vibrations themselves as well as manifestations at the surface of bulk vibrations.

The recent emphasis in surface dynamics on the interpretation and understanding of experimental data has created the need for microscopic theories or crystal models which describe correctly, not only the structural features (bulk and surface), but also the dynamics. The development of microscopic models on which to base realistic dynamical calculations for both the bulk and the surface is an extremely difficult task, which, however, is beginnning to show progress. Nevertheless, for the immediate future we will in many cases continue to have to rely on interaction *models* designed to represent specific materials or classes of materials. In the context of using models there is for surface studies an additional difficulty caused by our basic lack of knowledge of how particle interactions are modified at the surface. Aside from the obvious fact of lower particle coordination at the surface, one reason why surface interactions can differ from those in the bulk is that the latter, because of bulk symmetry, may implicitly contain a number of simplifications, cancellations or two-body parametrizations of many-body interactions, which cannot be represented in the same way at the surface. In fact, the existence of lower symmetry at the surface can provide a stringent test for interaction models, in particular in regimes close to a structural or dynamic instability; in some cases the presence of a surface may trigger the transition to a new state.

The problem of wether and how to modify particle interaction near a surface is, of course, strongly dependent on the type of binding of the crystal. In this sense the noble gas crystals and the simple ionic crystals offer a significant contrast to covalent crystals and metals. In these cases of closed-shell interactions one may have some confidence that the semi-empirical pair interaction models, fitted to bulk properties, retain their validity in the neighborhood of surfaces, so that surface relaxation and surface-induced changes in the dynamical matrix may be obtained in a fairly straightforward way. In the case of covalent crystals and metals the situation is much more complicated. Despite important progress in the development of self-consistent electron theories of covalent and metallic surfaces – in a number of cases surface structure and relaxation have been obtained from microscopic calculations – the microscopic theories for surface dynamics are much less developed. Thus, for the immediate future there will be a trend to continue to rely on Born–von Kármán force constant parametrizations, including ad hoc modifications of certain force constants near the surface, in order to obtain the best possible representation of experimental results.

The Slab Method: During the past twenty years, the slab method has proved to be a powerful and convenient method to study the dynamics of crystals with surfaces and thin films. Much of our basic understanding, systematic discussion and nomenclature of surface vibrations has been developed by slab calculations. After originally being introduced in 1965 [4.1], the first systematic application of the slab method to specific systems, namely noble gas solids [4.2, 3] and ionic crystals [4.4] appeared in 1969. A very general and systematic study, introducing much of the description and classifications of surface vibrations as it is used today, followed in 1971 [4.5, 6].

In the slab method we consider a crystal bounded by two planar surface and containing a finite number of parallel crystal planes, extending to infinity in the directions parallel to the surface. The latter condition means in practice that we apply periodic boundary conditions in the parallel directions, enclosing within one period of these boundary conditions $\tilde{N}$ two-dimensional (2D) surface unit cells. The two principal approaches to determining the dynamics of such a system are *molecular dynamics* (MD) and *lattice dynamics* (LD); in a sense the latter can be considered a linearized form of the former. In this article we will concentrate on the LD approach.

In LD the vibrational states of this model crystal, including surface states, can be found by a straightforward diagonalization of the dynamical matrix (DM) for a representative set of 2D wave vectors. The main limitation of this method is that, because of the finite thickness of the slab, it does not correctly reproduce the deeply penetrating surface modes of a macroscopic crystal in the extreme long wavelength limit.

Until recently a related practical limitation of the slab method concerned the statistics of certain calculated quantities. Because of the limited number of layers in a slab, there are only a correspondingly limited mumber of vibrational frequencies for each 2D wave vector. This could lead to unsatisfactory statistics in calculated quantities like the "surface projected density of states", which is needed in surface scattering cross-section calculations. However, this limitation has been overcome by the use of supercomputers, which allow calculations on very thick slabs. We note in this connection that, except for providing a better representation of the deeply penetrating modes at long wavelengths, the only important effect of increasing the thickness of the slab is to provide more bulk mode frequencies for each wave vector. For most wave vectors, the positions of the surface mode frequencies as well as the limits of the bulk bands (ranges of bulk mode frequencies) do not change significantly beyond a thickness of about ten layers. An advantage of the slab method is, of course, that it gives a true description for very thin films of finite extent. Thus, it is the method of choice for the study of effects of finite thickness and size on a variety of physical properties.

The material of this chapter is partly of a tutorial nature in that we review the general principles and results of surface dynamics, while introducing the slab method and discussing specific results. In Sect. 4.1 we develop the basic formalism of slab dynamics and discuss general properties of slab vibrational modes, following closely the work of *Allen* et al. [4.5, 6] (cf. also the review by de *Wette* and *Alldredge* [4.7]). This material is generally applicable, i.e. also when surface reconstruction or adsorption are present, as long as 2D periodicity is maintained. No assumptions are necessary concerning the effective interactions between the particles, the number of particles per unit cell, the crystal structure, or the surface orientation. Unless stated otherwise, there are no assumptions or approximations beyond 2D periodicity, and the usual adiabatic and quasi-harmonic approximations.

Sections 4.2 and 4.3 deal with computational considerations and interaction models, respectively. In Sect.4.4 we present selected results for noble gas crystals, ionic crystals and metals. We conclude, in Sect. 4.5, with a brief overview of temperature dependent surface properties such as vibrational correlation functions and thermodynamic quantities, which follow from a knowledge of the vibrational properties of the system.

4.1 Formalism

4.1.1 Slab Dynamics

We consider crystal slabs with two parallel surfaces, consisting of a finite number of crystalline planes perpendicular to the z-direction, and of infinite extension in the x and y directions. The unit cell of this crystal extends through the entire thickness of the slab and is specified by the basis vectors a_1, a_2, a_3, of which a_1 and a_2 are parallel to the surface and a_3 usually perpendicular to the surface. The cross section of a crystal plane with the slab unit cell forms the 2D unit cell in that plane, and it may have different contents for different planes. It should be stressed that from the point of view of symmetry operations, a slab (whatever its thickness) is a 2D crystal (in three dimensions), because it is generated by applying the operations of a 2D translation group to the unit cell.

The instantaneous position of a particle (atom or ion) is given by

$$r(l\kappa) = r_0(l\kappa) + u(l\kappa) , \tag{4.1}$$

where $r_0(l\kappa)$ gives the mean position of the particle and $u(l\kappa)$ the time dependent displacement. The set of integers $l = (l_1, l_2, l_3)$ and the index κ specify a particular particle: l_3 labels a crystalline plane parallel to the surface, l_1 and l_2 specify the points in the 2D lattice which spans a plane; κ distinguishes different particles in the 2D unit cell in the place l_3.

It is useful to introduce 2D vectors with only x and y components and indicate these with the subscript $\|$: if $r = (x, y, z)$ and $l = (l_1, l_2, l_3)$, then $r_\| = (x, y)$ and $l_\| = (l_1, l_2)$. The position of the 2D lattice point $l_\|$ is represented by

$$r_{0\|}(l_\|) = l_1 a_1 + l_2 a_2 . \tag{4.2}$$

As a result of the presence of the surfaces, there is no translational symmetry in the z direction. Therefore, there are only 2D reciprocal lattice vectors, associated with the 2D periodic structure, given by

$$G_\| = n_1 b_1 + n_2 b_2 , \tag{4.3}$$

where n_1 and n_2 are integers and where the 2D reciprocal basis vectors b_1 and b_2 are defined by

$$b_1 = 2\pi \frac{a_2 \times (a_1 \times a_2)}{|a_1 \times a_2|^2} , \quad b_2 = 2\pi \frac{a_1 \times (a_2 \times a_1)}{|a_2 \times a_1|^2} . \tag{4.4}$$

In this notation (4.1) can be written as

$$r(l\kappa) = r_{0\|}(l_\|) + r_0(l_3\kappa) + u(l\kappa) . \tag{4.5}$$

Here $r_0(l_3\kappa)$ is the "basis vector" that gives the mean position of a particle (of type κ in layer l_3) within the slab unit cell associated with $l_\|$; its projection on the xy plane is $r_{0\|}(l_3\kappa)$.

We assume that the total potential energy ϕ of the system is a function of the particle positions and can be expanded in a Taylor series

$$\phi - \phi_0 = \sum_{l\kappa\alpha} \phi_\alpha(l\kappa) u_\alpha(l\kappa) + \frac{1}{2} \sum_{l\kappa\alpha} \sum_{l'\kappa'\beta} \phi_{\alpha\beta}(l\kappa; l'\kappa') u_\alpha(l\kappa) u_\beta(l'\kappa') + \dots , \tag{4.6}$$

where

$$\phi_\alpha(l\kappa) = \left(\frac{\partial\phi}{\partial u_\alpha(l\kappa)}\right)_0 ,$$

$$\phi_{\alpha\beta}(l\kappa; l'\kappa') = \left(\frac{\partial^2\phi}{\partial u_\alpha(l\kappa)\partial u_\beta(l'\kappa')}\right)_0 . \tag{4.7}$$

A number of relations between the coefficients $\phi_{\alpha\beta}(l\kappa; l'\kappa')$ follow from the behavior of ϕ and the force on an atom under rigid body translations and rotations (cf. [4.8]). One of these relations leads to an expression for the "self-interaction (force) constants",

$$\phi_{\alpha\beta}(l\kappa; l\kappa) = \phi_{\beta\alpha}(l\kappa; l\kappa) = -\sum_{l'\kappa'}{}' \phi_{\alpha\beta}(l\kappa; l'\kappa') . \tag{4.8}$$

The zero subscripts in (4.7) indicate that these quantities are evaluated with all particles at their mean positions and the prime on the summation in (4.8) means that the term $l'\kappa' = l\kappa$ is omitted.

In the quasi-harmonic approximation (QHA), one neglects all terms except the second one on the right-hand side of (4.6). In this approximation the equations of motion are

$$M_\kappa(l_3)\frac{d^2}{dt^2}u_\alpha(l\kappa) = -\sum_{l'\kappa'\beta'} \phi_{\alpha\beta}(l\kappa; l'\kappa')u_\beta(l'\kappa') . \tag{4.9}$$

We note that in the QHA, the force constants $\phi_{\alpha\beta}(l\kappa; l'\kappa')$ should, in principle, be evaluated at the true mean positions of the particles, i.e. with surface and bulk thermal expansion, surface relaxation and reconstruction taken into account.

Because of the 2D translational invariance of the crystal, the force constants $\phi_{\alpha\beta}$ depend only on the difference of $l'_\parallel$ and $l_\parallel$:

$$\phi_{\alpha\beta}(l\kappa; l'\kappa') = \phi_{\alpha\beta}(l_3\kappa; l'_3\kappa'; l'_\parallel - l_\parallel) . \tag{4.10}$$

This 2D translational property of the force constants implies that the normal mode solutions to (4.9) have the form of 2D Bloch functions:

$$u_\alpha(l\kappa) = [\bar{N}M_\kappa(l_3)]^{-1/2}Q_0\xi_\alpha(l_3\kappa) \exp[iq_\parallel \cdot r_{0\parallel}(l_\parallel)$$
$$+ q_\parallel \cdot r_{0\parallel}(l_3\kappa) - \omega t)] . \tag{4.11}$$

where $\bar{N}$ is the number of 2D lattice points ($\bar{N} \to \infty$), $M_\kappa(l_3)$ the mass of the particle κ in the plane l_3, Q_0 the amplitude of vibration, $\xi_\alpha(l_3\kappa)$ a polarization vector component which is normalized to unity (see below); $q_\parallel$ is a 2D *wave vector* in the xy plane and ω is the vibrational frequency. Substitution of (4.11) into (4.9) leads to the eigenvalue equation

$$\sum_{l'_3\kappa'\beta} D_{\alpha\beta}(l_3\kappa; l'_3\kappa'; q_\parallel)\xi_\beta(l'_3\kappa'; q_\parallel p) = \omega^2(q_\parallel p)\xi_\alpha(l_3\kappa; q_\parallel p) , \tag{4.12}$$

where the elements of the dynamical matrix (DM) are defined by

$$D_{\alpha\beta}(l_3\kappa; l_3'\kappa'; \boldsymbol{q}_\parallel) = [M_\kappa(l_3)M_{\kappa'}(l_3')]^{-1/2} \sum_{l'} \phi_{\alpha\beta}(l_3\kappa; l_3'\kappa'; \boldsymbol{l}_\parallel' - \boldsymbol{l}_\parallel)$$

$$\times \exp\{i\boldsymbol{q}_\parallel \cdot [\boldsymbol{r}_{0\parallel}(\boldsymbol{l}_\parallel' - \boldsymbol{l}_\parallel) + \boldsymbol{r}_{0\parallel}(l_3'\kappa') - \boldsymbol{r}_{0\parallel}(l_3\kappa)]\} . \tag{4.13}$$

The *polarization index* p distinguishes the different modes corresponding to a particular $\boldsymbol{q}_\parallel$. If there are $s(l_3)$ particles in the 2D unit cell in layer l_3 [i.e. $\kappa = 1, 2, \ldots, s(l_3)$], then $p = 1, 2, \ldots 3\mathcal{N}$, where $\mathcal{N} = \sum_{l_3} s(l_3)$; i.e., for a given 2D wave vector $\boldsymbol{q}_\parallel$ there are as many frequencies (vibrational modes) as there are degrees of freedom in the slab unit cell. The polarization vectors $\xi_\alpha(l_3\kappa; \boldsymbol{q}_\parallel p)$ have $3\mathcal{N}$ components, namely three for each particle $l_3\kappa$ in the unit cell: they indicate how the particle $(l_3\kappa)$ vibrates in the particular mode $\boldsymbol{q}_\parallel p$. Since the DM is Hermitian, the polarization vectors can be chosen to satisfy the usual orthonormality condition

$$\sum_{l_3\kappa\alpha} \xi_\alpha(l_3\kappa; \boldsymbol{q}_\parallel p)\xi_\alpha^*(l_3\kappa; \boldsymbol{q}_\parallel p') = \delta_{pp'} , \tag{4.14}$$

which implies that the polarization vectors of a given mode are normalized to unity over the whole thickness of the slab. This enables us to determine for a given mode, by inspection of the variation of the ξ_α's for that mode over the thickness of the slab, whether it is a *bulk* mode, a *surface* mode or a *mixed* mode (resonance). These points will be discussed in greater detail below.

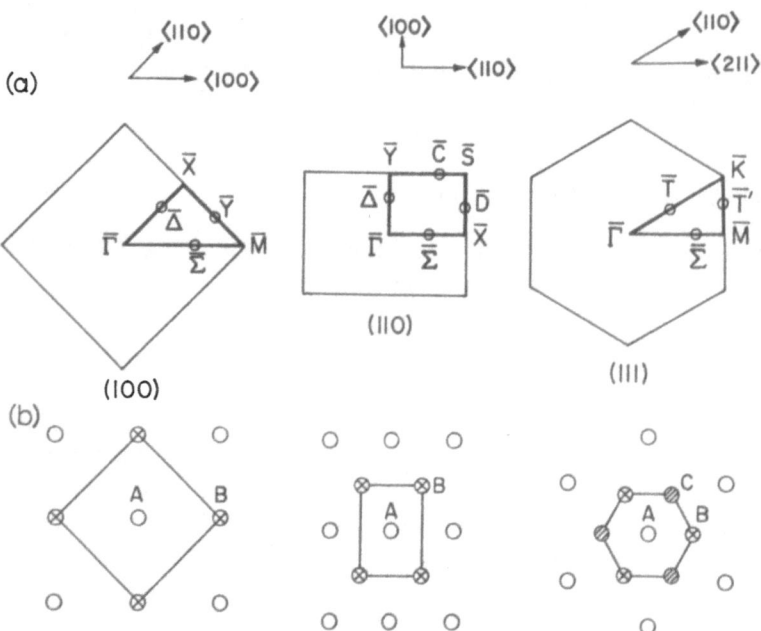

Fig. 4.1. (a) Surface Brillouin Zones (SBZ) for the fcc (100), (110), and (111) surfaces; an irreducible element is outlined in each case. (b) Surface unit cells generating (a); projections of lattice points of the stacking sequence $ABAB\ldots$ and $ABCABC$ are shown

The physically distinct solutions (4.11) of the equations of motion correspond to 2D wave vectors $q_\parallel$ lying in the first Brillouin zone (BZ) of the 2D reciprocal lattice (defined by b_1, b_2), associated with the 2D lattice of the slab planes (defined by a_1, a_2); this BZ is often referred to as the *surface* BZ (SBZ). In Fig. 4.1, we display the SBZ's and the surface unit cells (SUC) for the (100), (110) and (111) surfaces of the fcc lattice. The distinct values of $\omega(q_\parallel p)$ correspond to values of $q_\parallel$ lying in the irreducible element, which is that part of the SBZ which can be mapped into the rest of the SBZ by the 2D symmetry operations of the slab. The irreducible element is indicated by heavy lines and depicted with the notation for labeling the symmetry points and lines.

4.1.2 Use of Symmetry

The DM (4.13) is in general a complex Hermitian matrix of dimension p, where p is the total number of degrees of freedom in the slab unit cell ($p = 3\mathcal{N}$). Symmetry should be used as much as possible to reduce the computational requirements of storage ($p(p+1)/2$ words for the real part, $p(p-1)/2$ for the imaginary) and time (diagonalization time is proportional to the cube of the dimension). However, the amount of crystal symmetry in surface problems available to simplify and organize the solutions of the dynamical equation is far less than in bulk problems. Here we mention a few of the important symmetry reductions and simplifications; for a more detailed discussion, we refer to [4.2, 5, 6].

a) Simplifications Holding for General $q_\parallel$ in the SBZ. For an arbitrary crystal, a particle moves in an elliptical path about its mean position if the crystal is vibrating in a single normal mode. This ellipse may have any orientation in general, but if the crystal has "axial-inversion symmetry" then one axis of the ellipse is normal to the surface. (This condition holds, for example, in the case of an fcc crystal with one atom per unit cell which has a (100) or (110) surface.) For such crystals, the original dynamical matrix can always be reduced to a real symmetric matrix of the same size. If the crystal has finite thickness and a three-dimensional center of inversion, then the eigenvalue problem can be reduced still further; the final result is a pair of eigenvalue equations, each involving a real symmetric matrix which is (approximately) one-quarter the size of the original matrix. The surface modes will always occur in nearly degenerate pairs, if the thickness is large compared to the penetration depth of the modes.

These reductions can be useful in numerical calculations, since the required computer time and memory are greatly reduced.

b) Simplifications Holding for $q_\parallel$ on High Symmetry Lines and Points in the SBZ. When the *sagittal plane* (SP) (defined by $q_\parallel$ and the surface normal) coincides with a reflection plane of the slab, the DM can be reduced into two blocks, one containing two-thirds of the modes, of which the polarizations are ellipses in the sagittal plane; these will be labeled SP (sagittal plane) modes. The other block contains the remaining one-third of the modes, which are linearly polarized normal to the sagittal plane; these will be labeled SH (shear horizontal) modes.

Finally, we mention that there is no general symmetry-related restriction concerning the dependence or amplitude or polarization of the vibrational mode on the distance from the surface. In general, the polarization for a given mode can change from one layer to the next. For example, the particles in one layer may vibrate perpendicular to the surface while those in an adjacent layer vibrate parallel to the surface. In general, the amplitude of a surface mode need not decrease monotonically, or even regularly, with distance from the surface, and it certainly need not decrease exponentially.

4.1.3 Slab Vibrational Modes and Their Dispersion Curves

At this point it is instructive to consider and discuss some examples of slab vibrational modes and the corresponding dispersion curves. First of all, as we have seen, there exist for each 2D wave vector $q_\parallel$ as many vibrational modes as there are degrees of freedom in the slab unit cell. This is illustrated in Fig. 4.2 where the dispersion curves for a monatomic fcc slab with (111) surfaces are plotted for wave vectors $q_\parallel$ along the high symmetry lines $\overline{\Gamma M}$, $\overline{MK}$ and $\overline{K\Gamma}$ which form the boundary for the irreducible SBZ (Fig. 4.1a, 3rd picture), for increasing numbers of layers in the slab (taken from [4.6]). The number of dispersion curves ($p = 3N_3$ in this case) increases with the number of layers N_3. It is seen that most of these curves bundle together in bands (bulk bands) which get denser as N_3 increases; in addition there are identifiable individual dispersion curves, which represent, as we will see, surface localized modes. Their location does not change with N_3 beyond a certain slab thickness.

To illustrate the connection between slab and bulk dispersion curves, we consider the case of a 15-layer slab of LiF with (001) surfaces (LiF(001)). Since the unit cell contains two particles per layer (LiF; $s = 2$) it contains $3sN_3 = 3 \times 2 \times 15 = 90$ degrees of freedom, and there are 90 slab dispersion curves. These are shown for q values along $\overline{\Gamma M}$ (cf. Fig. 4.1a, 1st picture) in the right-hand side of Fig. 4.3.

The connection of these slab dispersion curves with the bulk dispersion curves can now be seen as follows: in the bulk there exists translational symmetry in the x, y, and z directions and consequently q_x, q_y and q_z are appropriate quantum numbers of the Bloch functions; however, for the slab the translational symmetry in the z direction is absent, and q_z is no longer a proper quantum number. Another way of expressing this is by saying that the slab vibrational modes can be traveling waves in the xy directions, but only standing waves in the z direction (satisfying the free surface boundary conditions). In the left side of Fig. 4.3 we have depicted the bulk dispersion curves of LiF in the q_z direction ($q_x = q_y = 0$) of the bulk BZ. Since, for the slab, q_z is no longer a proper quantum number, the frequencies of the corresponding bulk modes are all associated with $q_\parallel = 0$ for

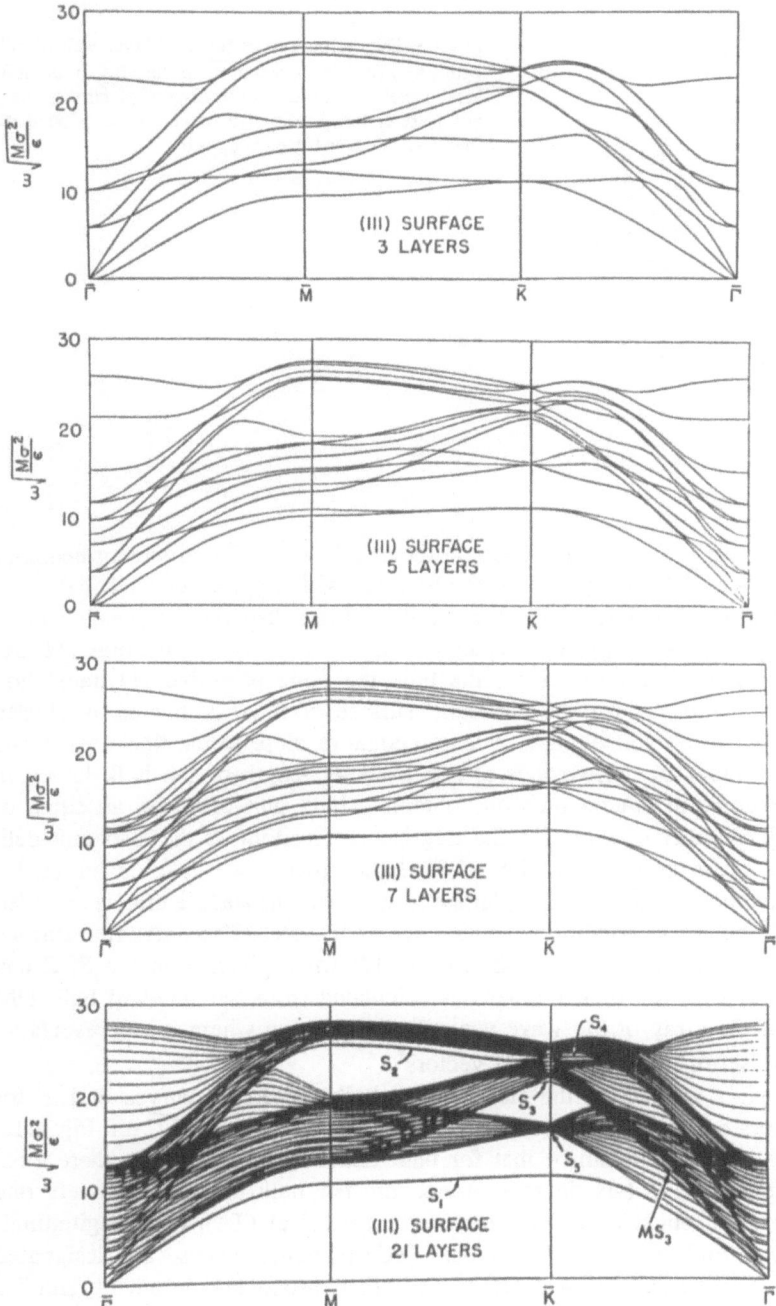

Fig. 4.2. ω vs. $q_{\parallel}$ dispersion curves for 3-, 5-, 7- and 21-layer monatomic fcc slabs with (111) surfaces for $q_{\parallel}$ along the boundary of the irreducible SBZ (cf. Fig. 4.1). There are 9, 15, 21, 63 dispersion curves, respectively. M is the particle mass and ε and σ are the Lennard-Jones potential parameters

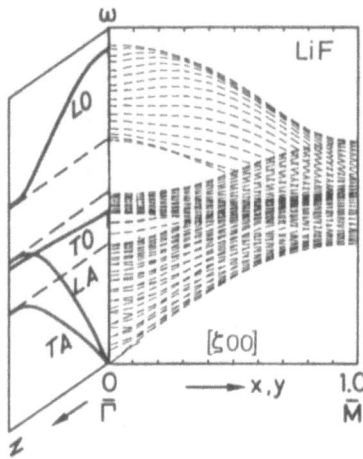

Fig. 4.3. Dispersion curves for a 15-layer slab of LiF with (001) surfaces, along $\overline{\Gamma M}$ of the SBZ. In the left-hand panel are the LiF bulk dispersion curves along the q_z direction. In the slab-adapted description these frequencies occur for $q_x = q_y = 0$

the slab, i.e. they are projected back on the ω axis at $\overline{\Gamma}$. The argument becomes more complicated for the points of the SBZ for which $q_{\|x}$ and/or $q_{\|y} \neq 0$.

To illustrate this connection between bulk and slab dispersion curves in a more precise way, *Allen* et al. [4.6] introduced the notion of *slab-adapted* bulk spectra. In this procedure one chooses for the bulk the same primitive 2D translation vectors a_1 and a_2 as for the particular surface of interest, but a_3 is chosen perpendicular to a_1 and a_2, and of length equal or close to the thickness of the free slab (so that a_3 corresponds to a periodicity length of the bulk lattice in the z direction); in other words one obtains a bulk unit cell with an equal or comparable number of particles and degrees of freedom as the slab unit cell. Thus for a given free-surface slab of N_3 lattice planes we have constructed a comparison *bulk* slab of $N_3'(\simeq N_3)$ lattice planes, with the surface effects removed by periodic boundary conditions applied across its faces. The $3sN_3$ free-surface frequencies calculated for the slab from (4.12) for a given q in the SBZ can be compared with the bulk frequencies calculated from the standard bulk DM equation at the array of 3D wave vectors $q = q_\| + q_z$, where q_z represents an appropriate set of z direction wave vectors[1].

In Fig. 4.4a are shown the slab-adapted bulk dispersion curves of LiF for wave vectors q along the high symmetry lines $\overline{\Gamma X}$, $\overline{XM}$ and $\overline{M\Gamma}$ (cf. Fig. 4.1a, 1st picture). It is well known that for bulk diatomic ionic crystals there exist six dispersion curves (six degrees of freedom per bulk unit cell), namely one longitudinal-acoustical (LA), two transverse-acoustical (TA), one longitudinal-optical (LO), and two transverse-optical (TO) branches. (These are designated in the left-hand side of Fig. 4.3. Note that in this particular case the TA and TO branches are doubly degenerate.) The 15-layer slab-adapted bulk has 90 disper-

---

[1] We have here glossed over the details of the slab-adapted bulk description. For a precise definition we refer to [4.5, 7].

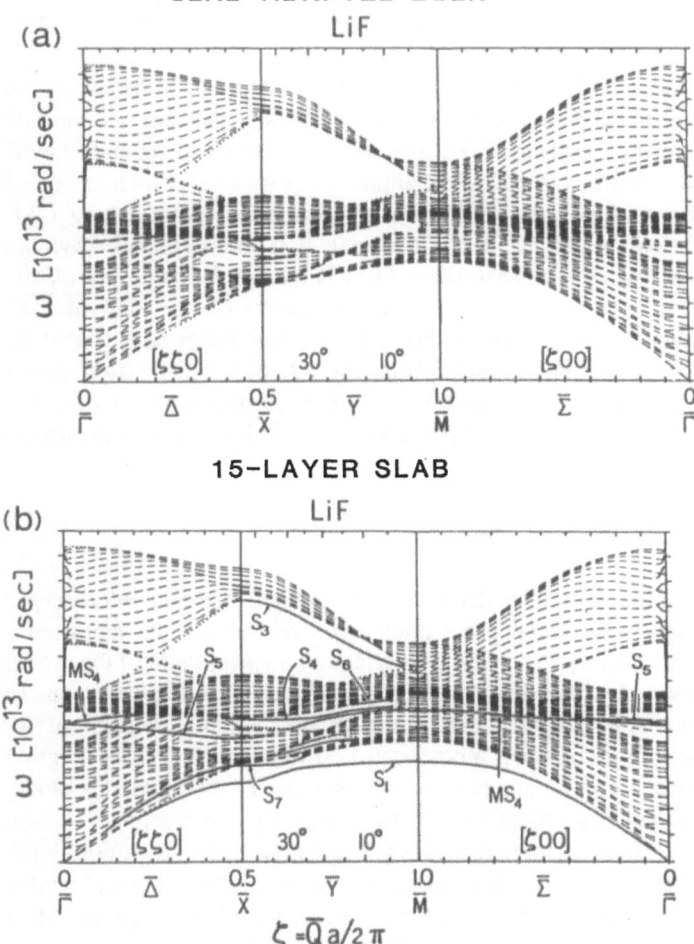

SLAB-ADAPTED BULK

(a)

LiF

ω [10^{13} rad/sec]

[$\zeta\zeta$0] 30° 10° [ζ00]

0    $\bar{\Delta}$ 0.5 $\bar{Y}$ 1.0 $\bar{\Sigma}$ 0
$\bar{\Gamma}$ $\bar{X}$ $\bar{M}$ $\bar{\Gamma}$

15-LAYER SLAB

(b)

LiF

ω [10^{13} rad/sec]

MS$_4$ S$_5$ S$_3$ S$_4$ S$_6$ S$_5$

S$_7$ S$_1$ MS$_4$

[$\zeta\zeta$0] 30° 10° [ζ00]

0    $\bar{\Delta}$ 0.5 $\bar{Y}$ 1.0 $\bar{\Sigma}$ 0
$\bar{\Gamma}$ $\bar{X}$ $\bar{M}$ $\bar{\Gamma}$

$\zeta = \bar{Q} a/2\pi$

Fig. 4.4. (a) Slab-adapted bulk dispersion curves for LiF(001) for $q_{\parallel}$ along the boundary of the irreducible SBZ (the periodicity in the z direction is taken to be 15 layers). (b) Dispersion curves of an unrelaxed 15-layer LiF(001) slab with free surfaces. Surface modes are indicated by solid lines

sion curves which are arranged in the six densely populated *bulk (sub)bands*, carrying the same designations as the bulk dispersion curves that give rise to them, namely LA, TA$_{1,2}$, LO, and TO$_{1,2}$. In Fig. 4.4b are plotted the dispersion curves of an unrelaxed 15-layer LiF(001) slab with free surfaces. Comparing with (a), we see that the bulk bands are unchanged, but that there are individual dispersion curves below the bulk bands, and in the gaps between them. These dispersion curves represent surface-localized vibrations or surface modes; this designation indicates that, in these modes, particles at or near the surface have finite vibrational amplitudes, while particles in the interior of the slab (bulk particles) are at rest. In addition to these isolated surface modes there exist resonances inside the bulk bands which correspond to modes that have both surface-localized and bulk-like components; these resonances are sometimes called "mixed modes" or "pseudosurface modes".

In principle, when a surface is created, each bulk band can have associated with it, over a part of the BZ, one or more surface modes of the same vibrational character but with lower frequencies; i.e. the surface modes are "peeled down" from the bulk band. As an example, we mention S_1 (Fig. 4.4b) which is peeled down from the bottom edge of the TA bulk bands. For wave vectors along $\overline{\Gamma M}$, S_1 has SP polarization (mainly $\perp$ to the surface); it is the so-called Rayleigh surface wave which was first described by Lord Rayleigh for isotropic elastic media. In the particular case of LiF, S_1 has SH character on the interval $\overline{\Gamma X}$, while S_7 is the Rayleigh mode (SP) on that interval. The point here is that S_1 and S_7 derive their vibrational character from the TA bulk bands from which they are peeled down. Another example is S_3, which is peeled down from the LO bulk band into the LO–TO gap on the interval $\overline{XM}$; it is an optical surface mode.

4.1.4 Macroscopic and Microscopic Surface Modes

Vibrational surface modes can be divided into *macroscopic* and *microscopic* modes. For the *macroscopic* modes the attenuation of the particle amplitudes away from the surface is in some way proportional to the wavelength of the mode; therefore, for long wavelengths these modes extend over considerable distances into the crystal. Since at long wavelengths the atomistic crystal structure (but not its anisotropy) is unimportant, such modes can be found (at long wavelengths) in elastic and dielectric continuum theory. An example is the Rayleigh mode, mentioned above, which was identified by Lord Rayleigh in elasticity theory. The *microscopic* modes are characterized by the fact that their penetration depth into the crystal extends over only a few interplanar distances for *all* wavelengths, although the details of the penetration pattern depend on the wavelength. Most surface modes that are encountered in practice are microscopic modes.

There exists, in fact, a third class of surface modes which is intermediate between the macroscopic and the microscopic. This class does not exist in the continuum limit of lattice dynamics (the "long-wavelength limit"), but for certain propagation directions it does exist at arbitrarily long wavelengths ("long-wavelength regime") and the penetration distance of these modes scale, not as the wavelength, but as the square of the wavelength [4.9].

4.1.5 Attenuation Curves

In principle, microscopic surface modes and macroscopic surface modes (at large wave vectors) can be identified by examining the squares of the polarization vectors of a given mode $q_{\parallel}p$, for all the particles in the unit cell. Using the abbreviation $\xi(m, \kappa)$ for $\xi(l_3\kappa; q_{\parallel}p)$, where m represents the layer index (increasing away from the surface), we examine the so-called *attenuation curve*

$$|\xi(m,\kappa)|^2 = |\xi_x(m,\kappa)|^2 + |\xi_y(m,\kappa)|^2 + |\xi_z(m,\kappa)|^2 = M_\kappa |u(m,\kappa)|^2 \qquad (4.15)$$

as a function of m.

When a dispersion curve is found outside a bulk band or inside a gap, such a curve can be identified immediately as belonging to a surface mode, without the need to examine the attenuation curve . However, if a surface mode is surrounded by modes of its own symmetry type of polarization, one must examine the attenuation curve for identification of its surface-localized character. In such a case, the surface mode will, to some extent, interact with the surrounding bulk modes, and become, in effect, a mixed mode, pseudosurface mode or surface resonance (e.g. MS_4 in Fig. 4.4b). In the case that a surface mode is surrounded by bulk modes of a different symmetry type, no interaction with the surrounding bulk modes can occur; hence it exists as a pure surface mode (e.g. S_5 and S_7 in Fig. 4.4b).

The simplest kind of attenuation curve of a surface mode is that exhibiting an exponential decay with distance from the surface. Although the above mentioned surface mode S_5 of LiF exhibits this kind of behavior, for most surface modes the attenuation curves are considerably more complex. As examples, in Fig. 4.5, we present the attenuation curves for the fcc (111) surface modes appearing in the bottom part of Fig. 4.2. Notice that the vertical scale in Fig. 4.5 is logarithmic and that all of the modes S_i decrease by more than an order of magnitude between the surface and the center of the crystal slab.

The mode S_1 is typical of the "generalized Rayleigh waves" found in continuum theory, in that it lies below the bulk bands and persists into the long-wavelength limit and shows an approximately exponential decay in amplitude with increasing distance from the surface (approximate straight line in Fig. 4.5).

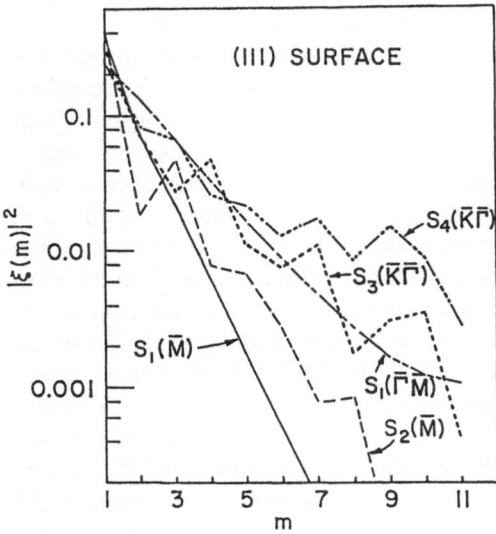

Fig. 4.5. Surface mode attenuation curves: $|\xi(m)|^2$ vs. layer index m for surface modes shown in Fig. 4.2. Here $m = 1$ for a surface plane and $m = 11$ for the center layer. The vertical scale is logarithmic. $S_1(\overline{\Gamma M})$ corresponds to the midpoint of the S_1 branch between $\overline{\Gamma}$ and $\overline{M}$. $S_3(\overline{K\Gamma})$ and $S_4(\overline{K\Gamma})$ correspond to these branches just off the point $\overline{K}$, and $S_1(\overline{M})$ and $S_2(\overline{M})$ correspond to the $\overline{M}$ point. (After [4.6])

None of these statements is true for the "gap modes" S_2, S_3 and S_4; their amplitudes decay in a very complicated fashion away from the surface, yet they all have their largest amplitude in the surface layer. In the surface layer, S_1 is primarily a "shear-vertical" (SV) mode, i.e., it is primarily associated with vibrations normal to the surface. S_3 and MS_3 are primarily "shear-horizontal" (SH), i.e., associated with vibrations perpendicular to $q_\parallel$ and parallel to the surface. S_2 and S_4 are primarily "longitudinal" modes, i.e., vibrations primarily polarized in the direction of $q_\parallel$. S_5 (cf. Fig. 4.2) exists as a surface mode only in a narrow region around $\bar{K}$ just below the TA bulk band; it has the same polarization (SV) as S_1, but largest amplitude in the second layer: one says that S_1 and S_5 belong to the same *family* of modes.

4.1.6 Systematic Features of Surface Modes

Despite the complexity of surface mode spectra, their etiology and general features can be understood in the context of a simple phenomenological scheme. Referring again to Fig. 4.4a we saw that for a diatomic crystal *without surfaces* there exist six intertwining bulk bands. The presence of a surface introduces a perturbation consisting of two parts: a large "first-order" perturbation due simply to the truncation of the crystal, and a "second-order" perturbation due to the changes in the force constants near the surface. The strength of the total perturbation depends on the location of $q_\parallel$ in the BZ. If the perturbation is strong enough it will peel one or more surface modes off a given bulk band, and since the perturbation entails a softening of the total forces acting on the surface particles, the surface modes are peeled off the bottom of the bulk band.

Ordinarily, the total perturbation first peels off (from a given bulk band) a mode primarily localized in the first layer. If strong enough, the perturbation then peels off a mode primarily localized in the second layer; and so on. As just mentioned, in Fig. 4.2 the modes S_1 and S_5 belong to such a family.

When a mode is peeled off one of four things will happen. (a) It may fall under all of the bulk bands, in which case it will necessarily be a surface mode. (b) It may fall into a gap inside the bulk bands, in which case, again, it will necessarily be a surface mode. (c) Along a symmetry line associated with a reflection plane, it may fall into a region occupied only by a sub-band of bulk modes to which it is automatically orthogonal by virtue of symmetry. In this case, once more, it will necessarily be a surface mode. (d) It may fall into a region occupied by bulk modes to which it is not automatically orthogonal. In this case it will not be able to survive as a pure surface mode and will be a mixed mode or surface resonance instead. To illustrate the effect of weakening of the forces, as for instance resulting from relaxation, we show in Fig. 4.6 the dispersion curves for an 21-layer fcc(100) slab in which the particles interact through a simple Lennard–Jones pair interaction. Two cases are shown: (a) without and (b) with static relaxation taken into account. Notice the overall lowering of the surface mode dispersion curves in (b), resulting from the weakening of the forces. At $\overline{M}$

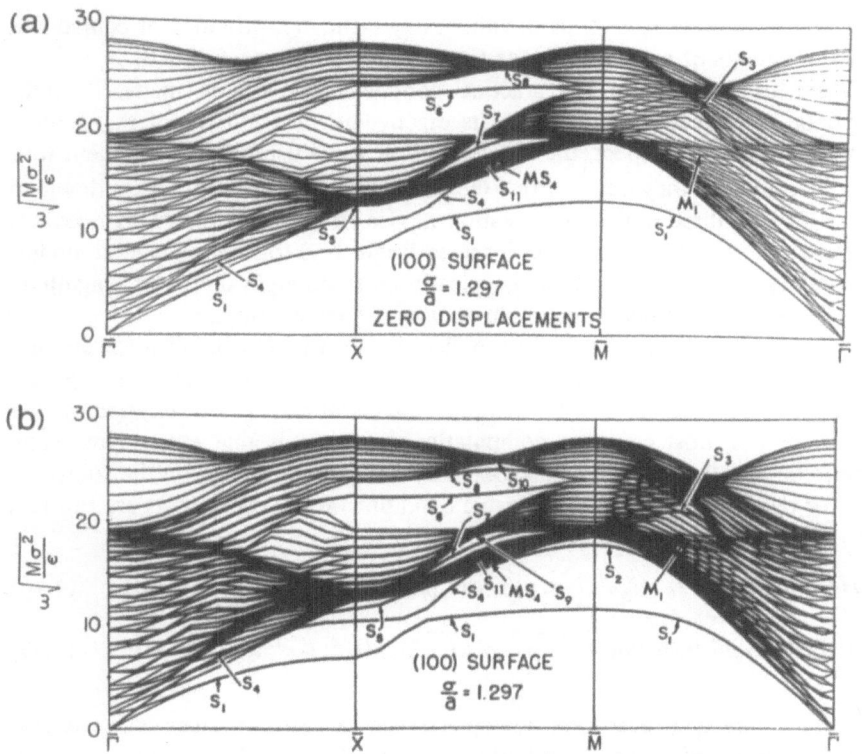

(a)

(b)

Fig. 4.6. Dispersion curves $\omega(q_{\parallel}p)$ for a 21-layer fcc(100) slab of an LJ crystal (**a**) without and (**b**) with surface relaxations taken into account. Surface (S_i) and pseudosurface (MS_i or M_i) mode branches are labeled according to an arbitrary scheme. (After [4.6])

a new mode S_2 has peeled off from the bottom of the TA bulk band; it is the second layer mode belonging to S_1. Similarly, on $\overline{XM}$ S_{10} is peeled off from the bottleneck in the LA bulk band; it is the third layer mode of the family S_6, S_8, S_{10}.

4.2 Computational Considerations

4.2.1 Diagonalization Techniques

Computational applications and practice have developed to such an extent in recent years that it is hardly necessary to comment on matrix diagonalization techniques required to obtain eigenvalues and eigenvectors of the DM equation (4.12). The IMSL or EISPACK libraries are generally available on most computer systems. The routines from these libraries can be applied to the diagonalization of the matrices at hand without any difficulty, provided of course that the requirements of memory storage (in-core or virtual) are fulfilled.

There are, however, a few practical comments that can be made. The DM should be positive definite because its eigenvalues are squares of the phonon frequencies. In this respect, the eigenvalues at $q_{||} = 0$ provide a stringent test, since in principle they range from the maximum of the bulk, ω_{max}^2, down to $\omega^2 = 0$ for the three uniform-translation modes. If the computed DM does not contain any errors, the eigenvalue (squared frequency) for the translation modes, while rarely precisely zero, should typically be some eight orders of magnitude smaller than the computed maximum squared frequency (ω_{max}^2).

Occasionally, when a newly coded slab program is diagonalized for the first time one obtains negative values for ω^2, most frequently for wave vectors at and near $q_{||} = 0$. The reasons can be two-fold: *computational* and/or *physical*.

One of the most common computational reasons is that errors have been made in coding the DM elements. One may be able to track down the source of the error by checking whether the usual conditions for DM elements are fulfilled [4.8], namely

Hermiticity: $D_{\beta\alpha}(l_3'\kappa'; l_3\kappa; q_{||}) = D_{\alpha\beta}^*(l_3\kappa; l_3'\kappa'; q_{||})$,   (4.16)

Translational invariance: $\sum_{l_3'\kappa'} \sqrt{M_{\kappa'}} \, D_{\alpha\beta}(l_3\kappa; l_3'\kappa'; q_{||} = 0) = 0$.   (4.17)

A common physical reason for the appearence of negative eigenvalues is that the *condition of microscopic stability* of the lattice is violated. This condition requires that all the principal minors of the DM be positive, and this requirement can also be used to help identify the cause of the trouble. A violation of the condition of microscopic stability usually implies that the configuration of mean particle positions around which the potential energy of the crystal is expanded (cf. (4.6)) is not stable under the forces acting in the crystal, i.e. is not an equilibrium configuration[2]. This indicates that a proper relaxation or reconstruction calculation should lead to a different equilibrium configuration. This brings up questions of thermodynamic stability of crystal models which go beyond the scope of this article. Nevertheless, the question of stability of surface configurations is of central importance in surface vibrational calculations, because the breaking of symmetry and the free boundary conditions at the surface may lead to instabilities which are absent in the bulk.

4.2.2 Surface Brillouin Zone Sampling

So far we have only given examples of slab dispersion curves along high symmetry directions of the SBZ. Of course, if one wishes to evaluate physical quantities

---

[2] An example is the monatomic bcc lattice which is unstable under the action of a large class of spherically symmetric pair interactions between the particles [4.28]

depending on the entire phonon spectrum of the slab, such as the mean-square amplitudes (MSA) of vibration at or near the surface and the surface vibrational correlation functions (both of which are important in LEED and inelastic surface scattering[3]), one needs to solve the dynamical equations (4.12) over an appropriate sample of $q_\parallel$ points in the SBZ. The same is true if one wishes to evaluate the surface-excess thermodynamic quantities (cf Sect. 4.5.3).

We illustrate the SBZ sampling problems on two quantities: the MSA's and the surface-excess distribution of frequencies. The α component ($\alpha = x, y, z$) of the MSA of particle ($l_3 \kappa$) is given by [4.10]

$$\langle u_\alpha^2(l_3\kappa) \rangle = (\hbar/2\bar{N}M_\kappa) \sideset{}{'}\sum_{q_\parallel p} |\xi_\alpha(l_3\kappa; q_\parallel p)|^2$$
$$\times \coth[\hbar\omega(q_\parallel p)/2k_B T]/\omega(q_\parallel p) , \tag{4.18}$$

where $\bar{N}$ is the number of 2D unit cells contained in the 2D periodic boundary conditions (hence the number of $q_\parallel$ values contained in the summation of (4.18)). The prime on the summation indicates the omission of the three zero-frequency modes corresponding to the uniform rigid-body translation of the slab. The surface-excess distribution of frequencies $f^s(\omega)$ is defined as [4.7, 11]

$$f^s(\omega) = (N/n)[f^{\text{slab}}(\omega) - f^{\text{bulk}}(\omega)] , \tag{4.19}$$

where N and n are, respectively, the total number of ions in the slab and the total number of surface ions, and f^{slab} and f^{bulk} are the spectral densities of the slab and the bulk, each normalized to unity [$f_{(\omega)}^{\text{slab}} = (3sN_3N)^{-1}\sum_{q_\parallel,p} \delta(\omega - \omega(q_\parallel p))$; $\int f(\omega)d\omega = 1$].

For a crystal of macroscopic dimensions, $\bar{N}$ would be of the order 10^{15} and (4.18) would be an integral over the SBZ. In actual calculations the evaluation of (4.18) amounts to a numerical quadrature approximation to this integral, with $\bar{N}$ quadrature coordinates distributed on the regular periodic boundary condition mesh and with $\bar{N}^{-1}$ for the quadrature weights: this is the *root sampling method*. An alternative approach is to replace the sums over the sampling mesh and their weights, by Gaussian-quadrature mesh points and weights (a combination of the two methods was used by *Firey* et al. [4.12] in the evaluation of the MSA's of the (0001) surface of graphite). The question of the sampling method becomes critical for the MSA's because of the appearance of $\omega(q_\parallel p)$ in the denominator of (4.18). For 2D crystals this gives rise to the logarithmic divergence of the MSA's with the lateral dimension ($\propto \sqrt{\bar{N}}$) of the crystal (Landau–Peierls divergence); however, this divergence decreases in inverse proportion to the thickness of the crystal (i.e. proportional to the inverse layer number N_3^{-1}) and disappears for $N_3 \rightarrow \infty$. On the other hand, for fixed $\bar{N}$, the MSA's diverge linearly with N_3,

---

[3] In scattering applications in which individual phonon frequencies are measured, such as in He scattering and EELS, one also needs to know the so-called "surface projected density of states" (cf. Sect. 4.4.2 (4.34)). This is a dynamical quantity not involving SBZ sampling.

because they are then effectively the MSA's of a one-dimensional crystal (see *Allen* and *de Wette* [4.2, Appendix D]). The question then is how one can calculate the MSA's for macroscopically large 3D crystals with surfaces (which means in the limits $\bar{N}, N_3 \to \infty$) while avoiding the divergences inherent in taking these limits for a slab. *Firey* et al. [4.12] dealt with this problem in the following way. Microscopically large 3D crystals (as opposed to microscopic slabs) are systems in the thermodynamic limit, which is the limit of vanishing surface to volume ratio; this ratio is proportional to $\bar{N}/\bar{N}N_3 = N_3^{-1}$. The only way to approach the thermodynamic limit ($N_3 \to \infty$) and avoid the two divergence difficulties is to keep the ratio $\sqrt{\bar{N}}/N_3$ fixed while letting $N_3 \to \infty$. But $\sqrt{\bar{N}}/N_3$ is proportional to the *aspect ratio* of the slab, which characterizes the *shape* of the crystallite. The procedure is thus to choose a certain aspect ratio and evaluate the MSA's for various slabs with this particular aspect ratio, but with increasing layer number N_3 (cf. Fig. 4.17). In view of the nature of the dependencies of the MSA's on $\bar{N}$ and N_3, the MSA's for finite crystals of different aspect ratios are different, but they all approach the same value in the thermodynamic limit ($N_3 \to \infty$). In fact, to an excellent approximation, they approach this limit as a linear function of N_3^{-1}. This circumstance makes the procedure of evaluating the MSA's for a macroscopic crystal quite simple. One chooses an aspect ratio ($\sqrt{\bar{N}}/N_3$) and evaluates the MSA's for two values of N_3; the corresponding $\bar{N}$ determines the $q_\parallel$ mesh in the SBZ (note that this is the real mesh, not a sampling mesh). Since the variation of the MSA's with N_3 comes mainly from the area around the center of the SBZ, in practice it usually suffices to use the real $q_\parallel$ mesh only in the small region around $\bar{\Gamma}$, and use a Gaussian quadrature in the remainder of the zone. Of course, if one is actually interested in crystallites of finite size, then the proper aspect ratio should be chosen, and the limit $N_3 \to \infty$ should *not* be taken. Then, of course, one finds the real size and shape effects exhibited by the MSA's of microcrystals (cf. [4.12]).

We close this section by noting that the SBZ sampling does not, of course, pose any problems of the kind just discussed, for the evaluation of the surface-excess frequency distribution $f^s(\omega)$ (4.19), except for the fact that it is much more difficult to implement the Gilat–Dolling-Raubenheimer interpolation scheme [4.13] for obtaining smooth $f^s(\omega)$ curves (rather than histograms), because of the multitude of degeneracies that are encountered in slab vibrational spectra. Evaluation of the slab and surface-excess thermodynamic quantities, which can be obtained either by integrations over frequency of the appropriate functions weighted with $f^{slab}(\omega)$ and $f^s(\omega)$, or by direct SBZ summation, also does not lead to any difficulties.

4.3 Interaction Models

The choice of interaction models used in slab calculations depends on the kind of solid that is being considered (molecular, ionic, semiconductor, or metal crystal), on the available knowledge about bulk interactions, and how these interactions might be modified by the presence of a surface. For instance, for atoms and ions with closed-shell electronic configurations (noble gas crystals, alkali halides) one expects the electronic configurations to be quite rigid so that the crystalline environment and the presence of a surface will modify the spherical charge distributions in only a minor way; in such a case, parametrized central pair interactions obtained from fitting bulk phonon dispersion curves (e.g. the shell model) form an excellent starting point for surface dynamical calculations. In contrast, in semiconductors and metals the interactions in the bulk are difficult to model because of the complications of the underlying microscopic theory. In addition, near the surface the interactions can be modified significantly because of the presence of dangling bonds or surface electronic states, so that parametrized *bulk* force constant models may be a questionable starting point for surface dynamical calculations. A very important development in the microscopic approach to the lattice dynamics of semiconductors and metals are the *ab initio* self-consistent total energy calculations of frozen phonons and interplanar force constants [4.14]. Although application of these methods is subject to computational restrictions, they are beginning to be applied to surface dynamical problems [4.15], and appear to be the most promising first-principle approaches to surface dynamics. However, this latest development falls outside the scope of this article. Therefore, in this section we restrict ourselves to a brief summary of the main atomic/ionic interaction *models* which have been used in surface dynamics.

4.3.1 Simple Pair Potentials (Molecular Crystals)

For atoms and molecules with closed-shell or saturated electronic configurations, spherically symmetric pair interactions can be quite useful for obtaining general results for monatomic slabs. For example, the original work of *Allen* et al. [4.2, 5, 6] which provided the first detailed description of and established the nomenclature for the surface modes of monatomic fcc crystals, employed the Lennard–Jones pair potential: $\varphi(r) = 4\varepsilon[(\sigma/r)^{12} - (\sigma/r)^6]$. This potential is analytically simple, and the results can be expressed in dimensionless units so that they are not dependent on the mass of the particles or the potential parameters, but only on the dimensionless density parameter σ/a ($\sqrt{2}a$ is the nearest neighbor).

In a more generalized use of pair potentials one can use different potentials for different shells of neighbors, or different particles. For instance, if the interaction between particles κ and κ' in cells l and l' is given by $\varphi_{\kappa\kappa'}(|r(l\kappa) - r(l'\kappa')|)$, then the force constant $\phi_{\alpha\beta}$ is given by (cf. [4.8]):

$$\phi_{\alpha\beta}(l\kappa;l'\kappa') = -\left[\frac{r_\alpha r_\beta}{r^2}\left(\varphi''_{\kappa\kappa'}(r) - \frac{1}{r}\varphi'_{\kappa\kappa'}(r)\right) + \frac{\delta_{\alpha\beta}}{r}\varphi'_{\kappa\kappa'}(r)\right],$$

$$r = r_0(l\kappa,l'\kappa'), \tag{4.20}$$

where r is the distance between the atoms $(l\kappa)$ and $(l'\kappa')$, and r_0 the distance between their mean positions. It is often useful to introduce radial and tangential force constants,

$$\varphi_r^n = (\varphi''(r))_{r=r_n} = \left(\frac{d^2\varphi(r)}{dr^2}\right)_{r=r_n}, \tag{4.21}$$

$$\varphi_t^n = \left(\frac{\varphi'(r)}{r}\right)_{r=r_n} = \left(\frac{1}{r}\frac{d\varphi(r)}{dr}\right)_{r=r_n}, \tag{4.22}$$

where r_n is the distance between atoms which are nth neighbors of each other. The interaction model then specifies the φ_r^n and φ_t^n up to a certain n (force constant models).

The DM elements which are obtained for a monatomic crystal in which all neighbors interact through the same pair potential $\varphi(r)$ are given in [4.2, Appendix B]. The lattice sums appearing in these DM elements are simple and can be evaluated by direct machine summation. However, the general appearance of the slab dispersion curves (e.g. such as given in Fig. 4.6) is not very dependent on the range of interaction.

4.3.2 Shell Models (Ionic Crystals)

The predominant interactions in ionic crystals are the long-range Coulomb interactions and the short-range repulsive overlap interactions. At equilibrium the net attractive and repulsive forces on the particles balance each other. To a good approximation the short-range interactions can be represented by central pair potentials of the Born–Mayer form,

$$V_{\kappa\kappa'}(r) = a_{\kappa\kappa'}\exp(-b_{\kappa\kappa'}r). \tag{4.23}$$

The shell model (SM) was introduced to describe these interactions and account for electron screening in a consistent way [4.16–18]. In the simple SM an ion consists of a point core, carrying the total mass of the ion, and a massless shell (with fixed radius) representing the valence electrons. The total potential energy ϕ of the crystal then depends in a bilinear form on the core (ionic) coordinates $u(l\kappa)$ and polarization coordinates $w(l\kappa)$ (displacements of the shells with respect to their cores). The equations of motion are

$$M_\kappa u_\alpha(l\kappa) = -\frac{\partial\phi}{\partial u_\alpha(l\kappa)}, \tag{4.24}$$

$$m^{el}\omega_\alpha(l\kappa) = -\frac{\partial\phi}{\partial w_\alpha(l\kappa)} = 0 , \tag{4.25}$$

where (4.25) expresses the adiabatic condition for the SM. Introducing the Coulomb interactions, and the short-range interactions between neighboring shells as well as between shells and cores, one obtains after elimination of the polarization coordinates the equations of motion for the core coordinates,

$$\omega^2\xi^c = D\xi^c , \tag{4.26}$$

where the dynamical matrix has the form

$$D = (R + ZCZ) - (T + ZCY)(S + YCY)^{-1}(\tilde{T}^* + YCZ) . \tag{4.27}$$

Here ξ^c are the column matrices for the core eigenvectors, C is the matrix describing the Coulomb interactions, and R, T, and S are the matrices describing the short-range interactions between neighboring cores, between cores and neighboring shells, and between neighboring shells, respectively. Finally, Z and Y are the diagonal matrices for the ion and core charges.

Equations (4.26) and (4.27) are the customary expressions for the equations of motion of the simple SM. A variety of modifications and elaborations of this basic SM have been introduced for a variety of purposes; for a summary of these models we refer to the phonon atlas by *Bilz* and *Kress* [4.19].

The calculations of *Chen* et al. [4.11] for unrelaxed slabs of alkali halides with (001) surfaces were based on bulk SM's obtained from fitting measured bulk phonon dispersion curves. We note the following:

(1) It is usually assumed that the short-range forces act entirely through the shells ($R = T = S$) and extend up to second neighbors. Thus, most of the existing SM's provide six short-range parameters, A^{+-}, B^{+-}, A^{--}, B^{--}, A^{++} and B^{++}, for the cation–anion, anion–anion, and cation–cation interactions, respectively. These parameters A^{ij} and B^{ij} (radial and transverse force constants, respectively) are defined in the usual fasion:

$$A^{ij} = \left(\frac{2V_a}{e^2}\right)\frac{d^2V}{dr_{ij}^2}\Bigg|_{eq} , \tag{4.28}$$

$$B^{ij} = \left(\frac{2V_a}{e^2}\right)\frac{-1}{r_{ij}}\frac{dV}{dr_{ij}}\Bigg|_{eq} , \tag{4.29}$$

where $V_a = 2r_0^3$, r_0 is the nearest neighbor distance, and V the potential function for the short-range interactions (Born–Mayer potential). From (4.28) and (4.29) one can derive an expression for $R_{\alpha\beta}(l_3\kappa, l_3'\kappa'; q_\parallel)$ in terms of A^{ij} and B^{ij}. (Expressions for $R_{\alpha\beta}$ for slabs are given in [4.11, Appendix C].)

(2) If the crystal does not satisfy the Cauchy relation, then the fitted values of B^{ij} usually contain noncentral parts. In this connection it should be kept in mind that the equilibrium condition of the crystal is expressed in terms of the central parts:

$$B_{\text{cent}}^{+-} + 2(B_{\text{cent}}^{++} + B_{\text{cent}}^{--}) = -\frac{2}{3}\alpha_M Z^2 , \qquad (4.30)$$

where α_M is the Madelung constant and Z the ionic charge.

(3) The elements of the Coulomb interaction matrix C contain lattice sums which converge extremely slowly in direct space so that special methods have to be used to improve the convergence. For the special case of the Coulomb coefficients coupling ions in the same plane, *Chen* et al. [4.11] used the method of plane-wise summation [4.20], which gives very rapid convergence of the Coulomb sums (cf. [4.11, Appendix C]).

(4) In addition to the six short-range parameters, the bulk SM usually specifies the electronic and mechanical polarizabilities α_κ and d_κ rather than the shell charge Y_κ and the spring constant k_κ. The latter can be obtained from the relations

$$\alpha_\kappa = \frac{Y_\kappa^2}{k_\kappa + R_0} , \qquad (4.31)$$

$$d_\kappa = -\frac{R_0 Y_\kappa}{k_\kappa + R_0} , \qquad (4.32)$$

where $R_0 = A^{+-} + 2B^{+-}$.

4.3.3 Force Constant Models (Semiconductors, Metals)

In the discussion of simple pair potentials we mentioned that it is often customary to specify force constants (cf. (4.21, 22) and (4.28, 29)) without reference to pair potentials of which these force constants are the derivatives. This practice has been particularly common in Born–von Kármán treatments of dynamical properties of semiconductor [4.21] and metal [4.22–27] surfaces; for both classes of materials the notion of central pair potential is a questionable concept. In addition to central forces it is usually necessary to introduce noncentral forces to account for the nonfulfillment of the Cauchy relations, or also to satisfy stability conditions of the lattice structure [4.28]. Most often this is done through the introduction of three-body angle bending interactions with potential

$$\varphi_{\text{ang}} = \tfrac{1}{2}\gamma(\Theta - \Theta_0)^2 , \qquad (4.33)$$

where γ is the force constant and Θ_0 the equilibrium value of the angle Θ defined by the three particle positions.

A complete force constant model may consist of radial and tangential force constants between a particle and its neighbors in various shells and angular force constants pertaining to various angles involving the particle. Often, in the bulk, some of these force constants vanish or cancel as a result of bulk symmetry; moreover the tangential force constants are always related by the equilibrium condition.

The numerical values of a chosen set of force constants are obtained from optimum fits of predicted quantities (bulk elastic constants, phonon frequencies) to their experimentally measured values. The more input data that are available, the better the model can be made to fit the data through a least-squares fitting procedure, which may allow fitting weights to reflect the importance of the quantity to be fitted.

Surface Force Constant Changes. Because ionic environments in the bulk and at the surface are different, the forces on bulk and surface particles will be different. This may or may not require modifications in the pair interaction models. For instance in noble gas crystals the basic pair interaction is hardly changed at the surface, so that force constant changes at the surface are the result of different coordination and changed neighbor distances; this can be calculated given the basic pair interaction, cf. [4.2, 6, 22]. In metals, on the other hand, the changed electronic states at the surface can modify the effective ion–ion interaction so that, in the context of a force constant model, ad-hoc surface force constant changes may be needed to account for the experimental data, cf. [4.29].

Even force constant models derived from microscopic calculations, such as frozen phonons or force constants obtained from total energy calculations, need ad-hoc extensions to produce full phonon dispersion diagrams, because of the computational limitations of the total energy calculations.

4.4 Results

In this section we review some recent dynamical results for alkali halide (001) slabs, monatomic fcc slabs (mainly metals), and graphite. Progress in these areas has been stimulated by recent experimental results with inelastic He scattering and EELS, and it is in particular the close collaboration of experimentalists and theorists which has helped maintain the momentum in these investigations.

4.4.1 Relaxation and Dynamics of the (001) Surfaces of Alkali Halides

He-scattering measurements of the Rayleigh mode dispersion curves of the light alkali halides [4.30] (NaF, NaCl and KCl) were found to be in excellent agreement with the results of slab calculations [4.11] and Green's function calculations [4.31] for the unrelaxed (001) surfaces. (For a discussion of the latter we refer to Chap. 3.) This agreement indicates that the bulk SM's are valid for surface dynamical calculations and, moreover, that surface relaxation in these compounds is very small or at least that the influence of relaxation on the Rayleigh mode is negligible. To determine how large the surface relaxation actually is, and to evaluate its effects on other surface modes, like the optical modes, *de Wette* et al. [4.32] performed systematic relaxation calculations of (001) oriented slabs of the alkali halides, and dynamical calculations for the relaxed slabs [4.33], using the same bulk SM's as used in the "unrelaxed" calculations of *Chen* et al. [4.11].

Surface relaxation is driven by the imbalance between Coulomb and short range forces if ions (cores and shells) near the surface are held at their unrelaxed bulk positions. This imbalance gives rise to very large net forces, but small shifts in the positions are sufficient to cancel these forces.

The relaxation results for the alkali halides show the following systematic features [4.32]. (1) The relaxation shifts in the surface layer are at most a few percent of the lattice distance, and the relaxation is confined to at most four layers. (2) The relaxed surface is "rumpled": the ion with the larger polarizability moves outward, that with the smaller polarizability inward, such that the overall relaxation of the outer layer is slightly inward.

The effects of these relaxations on the surface vibrations [4.33] are in most cases quite small, usually not more than a few percent in the surface mode frequencies. (A brief description of the surface modes of ionic crystals was given in Sect. 4.1.3.) This is particularly true for the Rayleigh modes, and this explains the excellent agreement between measured and calculated Rayleigh modes for NaF, NaCl, and KCl. The surface mode most affected by relaxation is the transverse optical surface mode (TOSM) S_4. In a number of cases (NaBr, KF, RbF) frequency shifts of over 10% are found. This is illustrated for RbF in Fig. 4.7. The dominant mode in the TO-LO gap, labeled with $Sp_\perp^-$, is the TOSM S_4. The upward shift of S_4 from the unrelaxed case (dashed line) to the relaxed case (solid line) is 10% at $\bar{X}$ and 17% at $\bar{M}$. Unfortunately, so far these higher-energy modes are difficult to measure with He scattering, and S_4 has only been detected in NaF.

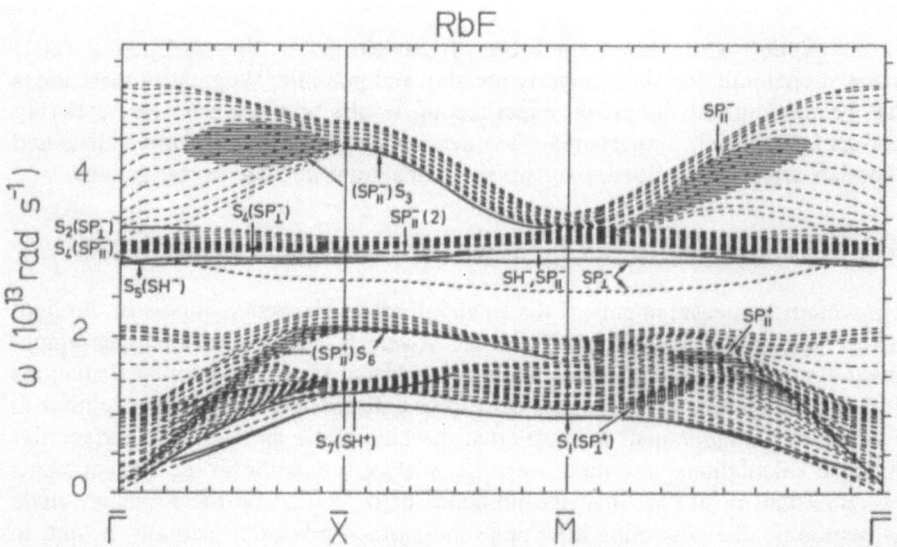

Fig. 4.7. Dispersion curves for a relaxed 15-layer slab of RbF with free (001) surfaces. Because of the large mass ratio there is an absolute gap between the optical bulk bands (top) and acoustical bulk bands (bottom). The transverse optical surface mode $S_4(SP_\perp^-)$ in the relaxed case (solid line) lies significantly higher than in the unrelaxed case (dashed line). For an explanation of the other symbols, see [4.33]

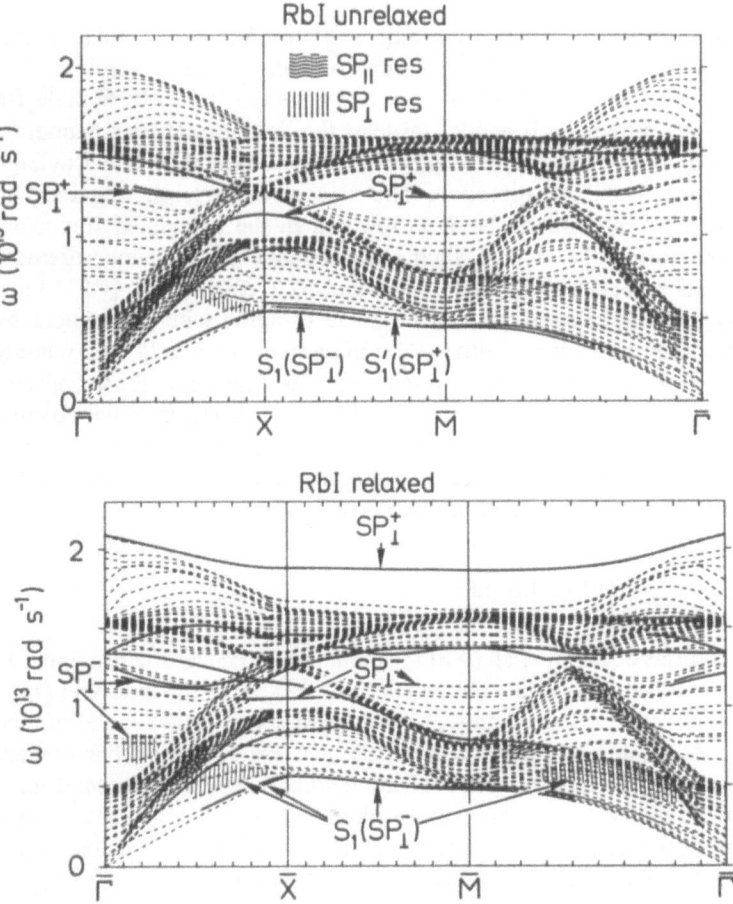

Fig. 4.8. Disperson curves for unrelaxed and relaxed 15-layer slabs of RbI with free surfaces. $SP_\perp$ and $SP_\parallel$ indicate predominantly longitudinal and transverse sagittal plane vibrations; + or − indicate that the positive or negative ion are vibrating in the surface layer. Only surface modes with vibrations perpendicular to the surface have been identified. (After [4.34])

The relaxation results for RbBr and RbI are of particular interest. As a result of the large overall polarizabilities of these compounds, their surface relaxations are the largest among the alkali halides (e.g. an 11% rumpling in RbI) and this causes the TOSM $SP_\perp$ to shift *above* the bulk continuum [4.34]. In Fig. 4.8 we show the dispersion curves for RbI for both the unrelaxed and relaxed surfaces. The high frequency mode in the relaxed case ($SP_\perp^+$) has a vibrational pattern in which the lighter particle vibrates perpendicular to the surface in the top layer, and the heavier particle in the second layer. Since in both RbBr and RbI the mode is so well separated from the bulk continuum over most of the SBZ it should be relatively easy to measure with inelastic He scattering. Experimental detection would confirm that surface relaxation has an important effect on the dynamics of RbBr and RbI, and that the bulk SM's used in the calculation can indeed be relied upon to lead to reliable predictions for surface dynamics.

Although it was argued earlier that bulk SM's should be fairly reliable for surface calculations of the alkali halides, at least the lighter ones, this cannot be taken for granted. An interesting example of this is provided by the Rayleigh mode of LiF along the ⟨100⟩ direction. The measured Rayleigh frequency at the $\bar{M}$ point [4.35] turned out to be 12% lower than the calculated frequency for the unrelaxed surface [4.11, 31]. Such a discrepancy between measurement and calculation does not occur for NaF, NaCl and KCl. *Kress* et al. [4.37] have shown recently that the calculations can be made to agree with experiment by either one or a combination of two surface modifications in the bulk SM, namely a change in the noncentral part of the nearest neighbor transverse force constant B (cf. Sect. 4.3.2), or a small transfer of charge between the shells of the positive and negative ions, both in the surface layer. The two mechanisms make different predictions for the frequencies of the TOSM S_4 at the point $\bar{X}$, so that experiment is needed to make a choice between the two possibilities.

4.4.2 Results for fcc and bcc Metals

The early calculations of Allen et al. [4.6] for monatomic fcc slabs with Lennard–Jones interactions characterized the phonon spectra that occur for the (001) (111) and (011) surfaces of these structures. The rapid increase in available surface phonon spectra of metals measured with EELS and He scattering has created a need for prediction and interpretation of the details of the experimental data. Accordingly, most recent theoretical studies have been based on force constant models fitted to bulk phonon properties, sometimes including ad-hoc force constant adjustments to reproduce special features of the measured surface phonon spectra. We mention here the work of two groups.

Black et al. [4.24, 25] have made a systematic study of the surface modes of the (001) and (111) surfaces of the bcc metals, Cr, Fe, K, Mo, Na, V and W, and the fcc metals Ag, Al, Au, Cu, Ir, Ni, Pd, Pt and Rh, with various force constant models. The models include central force interactions up to third neighbors and angle bending interactions, with constants obtained from fitting bulk phonon data of the various metals. The main conclusions of these studies can be summarized as follows. (1) While overall phonon spectra are similar for all metals of a given structure, each metal has quantitatively different dispersion curves, and in some cases there exist qualitative differences as well. (2) Different bulk models for the same metal can lead to differences in the predicted surface modes, and thus to differences in the surface force constants adjustments that might be needed to obtain agreement with experiment. Of course, for these adjustments to have any physical meaning, the bulk model itself must accurately represent the bulk phonon data [4.38]. (3) *Black* et al. [4.25] also provide wave vector information for selected surface modes, and give examples of the *surface projected density of states*, which is a key quantity in surface scattering cross-section calculations.

Because of its importance we introduce here the layer projected spectral density. It expresses the participation of particle κ in layer l_3 in the α component of the vibrations of wave vector $q_{\parallel}$, and is defined as

$$g_\alpha(l_3\kappa; q_{\parallel}, \omega) = \sum_p |\xi_\alpha(l_3\kappa; q_{\parallel}p)|^2 \delta(\omega - \omega_p(q_{\parallel})) . \tag{4.34}$$

To obtain a smooth representation of g_α rather than the coarse histogram that a calculation for 15–25-layer slab would provide, the authors use the Green's function approach of Black et al. [4.39] (see Chap. 7).

Investigations similar to those of Black et al. have been carried out by *Bortolani* et al. [4.26, 27] in the context of full atom–surface scattering cross-section calculations. These authors carried out calculations for the (001) and (111) surfaces of Ag, Au, Cu, Pt and Ni, based on bulk interaction models containing central forces up to fourth neighbors and angular forces up to second neighbors. In fitting the measured bulk dispersion curves to obtain the model parameters the authors observe that the fitting of certain phonon branches is strongly model dependent and they reach the same conclusion as [4.38] "that it is necessary to have a very good parametrization of the bulk properties in order to study surface dynamics" [4.27a].

Of particular interest is the interpretation given by these authors of a resonance ("anomalous peak") in the inelastic He-scattering results for Ag(111), observed by *Doak* et al. [4.40] with the time of flight technique. Using a bulk force constant model it is possible to account for the position and the decay of intensity of the Rayleigh mode peak as a function of $q_{\parallel}$, along the $\langle 11\bar{2} \rangle$ direction of the SBZ, but not for an additional peak in the bulk continuum, found in the TOF spectra. However, by lowering the nearest neighbor radial force constant in the surface layer β_1^s to 0.48 of its bulk value, β_1^{bulk}, these authors find a well-defined longitudinal resonance embedded in the transverse bulk continuum, in agreement with experiment, while the Rayleigh mode is hardly effected by this charge (cf. Fig. 4.9). The change in the radial force constant at the surface is to be understood as resulting from different screening by the electrons in the surface layer.

A *first-principle approach* to force constant changes at the surface is given by *Ho* and *Bohnen* [4.15] in a calculation of the surface phonons on the Al(110) surface. This is a sequel to their self-consistent total-energy calculation of the structure of Al(110) [4.41]. To calculate the surface force constants these authors start from the zero force geometry [4.41] and give the atoms in the surface layer a variety of displacements from which, through evaluation of the self-consistent electronic configurations associated with these displacements, the corresponding interlayer and surface-layer intralayer force constants can be determined. Using these force constants the dynamical matrix of the slab (51 layers) can be constructed, and the frequencies of the surface phonons corresponding to these displacements can be evaluated. In this fashion it is found that the frequencies of the in-planar shear modes in the top layer are raised substantially above the

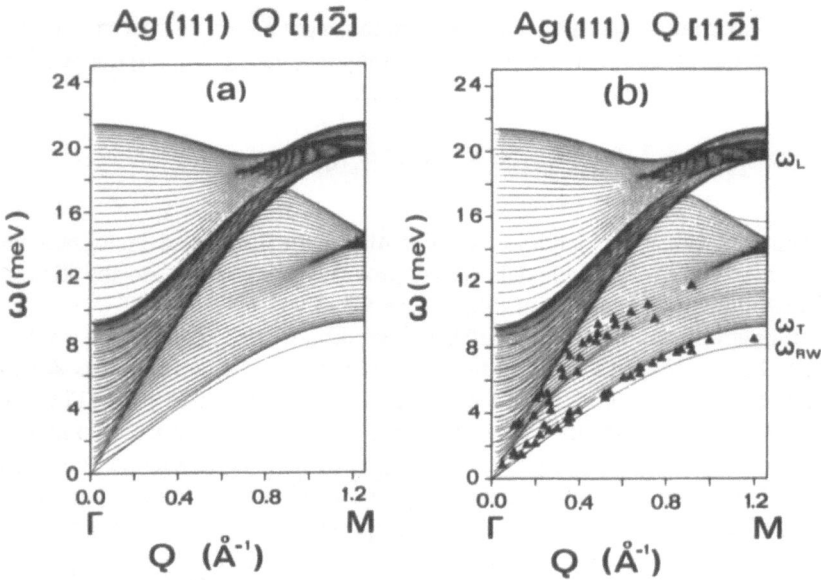

Fig. 4.9. (a) Calculated phonon dispersion curves for a 45-layer Ag(111) slab with $q_\parallel$ in the [11$\bar{2}$] direction. The surface force constants are taken to be equal to their bulk values. (b) Same calculation performed with $\beta_1^s = 0.48\beta_1^{bulk}$. Experimental points (triangles) come from [4.40]. (After [4.27])

frequencies that would be obtained with bulk force constants. The physical origin of the stiffening of these particular surface vibrations is the inward relaxation of the surface layer, which produces charge buildup between surface atoms and their neighbors. This can lead to large increases in the surface force constants (up to 85%).

It should be pointed out that with this method only frequencies of particular surface modes at a few symmetry points of the SBZ can be calculated. To extend these calculations to other parts of the SBZ the authors had to resort to fitting Born–von Kármán central force constants to their first-principles intralayer and interlayer force constants. Figure 4.10a shows the surface phonon dispersion curves (solid lines) and bulk bands (shaded) along symmetry directions in the SBZ. The crosses indicate the frequencies of the surface modes at $\bar{X}$ and $\bar{Y}$ obtained from the first-principles calculation; the remainder of the plot is obtained with the central force Born–von Kármán model. For comparison, Fig. 4.10b shows the results obtained with the bulk force constant model. From a comparison with Fig. 4.10a, it is evident that the force constant changes at the surface have a significant effect on the surface modes.

A different approach to describing the many-body effects in the bulk and surface dynamics of the noble metals has recently been proposed by *Jayanthi* et al. [4.42]. These authors represent the *sp* electron charge density in the lattice in terms of a multipole expansion around an appropriate high symmetry point in the unit cell. The modulation of these multipoles during the vibrations of the lattice corresponds to the deformation of a massless pseudoparticle at this

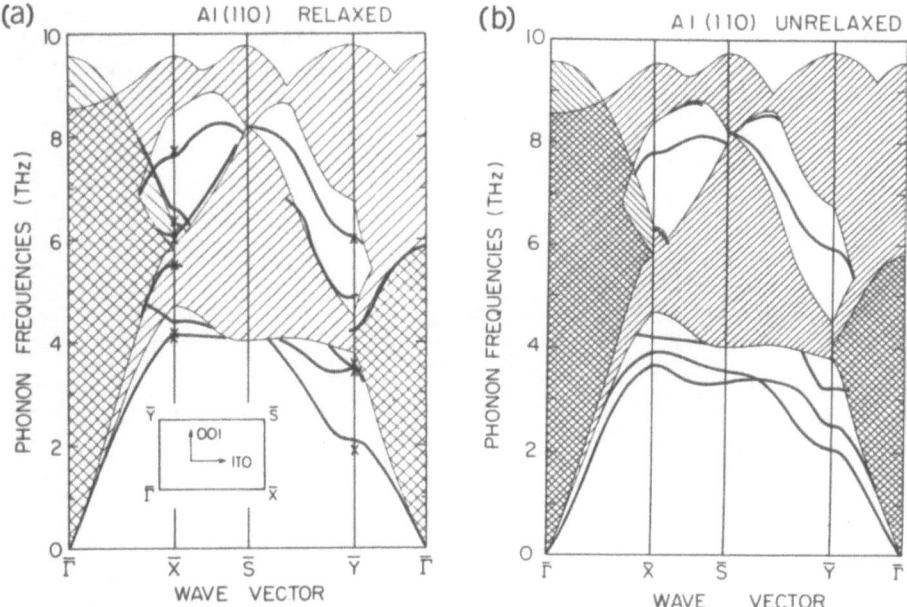

Fig. 4.10. (a) Calculated phonon dispersion curves for a 51-layer Al(110) slab along the symmetry directions of the SBZ. Shaded regions indicate the bulk bands; different shadings indicate different symmetries. The crosses indicate the frequencies of surface modes at $\bar{X}$ and $\bar{Y}$ obtained from first-principles interlayer and intralayer force constants. (b) Slab phonon dispersion curves obtained with bulk force constants. (After [4.41])

position, whose coupling to the ion displacements represents the electron–phonon interaction. This model is an extension of both the shell model and the bond-charge model [4.43]. The method appears to provide an accurate description of both bulk and surface dynamics by introducing only a few parameters with microscopic meaning, in contrast to Born–von Kármán models, which for noble and transition metals require a comparatively large number of ad-hoc long range force constants. The model reproduces the surface phonon dispersion curves of Cu(111), Ag(111) and Au(111) quite well, including the instability on the latter surface.

Finally, a self-consistent calculation of the surface phonon dispersion curves of the Al(110) surface is being carried out by *Eguiluz* et al. [4.44]. The calculation is based on a self-consistent implementation of pseudopotential perturbation theory, in which the screening response of the conduction electrons to the field of the vibrating ions is obtained within the local density approximation of density functional theory.

4.4.3 Layered Structure – Graphite

We conclude this section with a brief discussion of the dispersion curves of a graphite slab, because it is characteristic for what is obtained for layered struc-

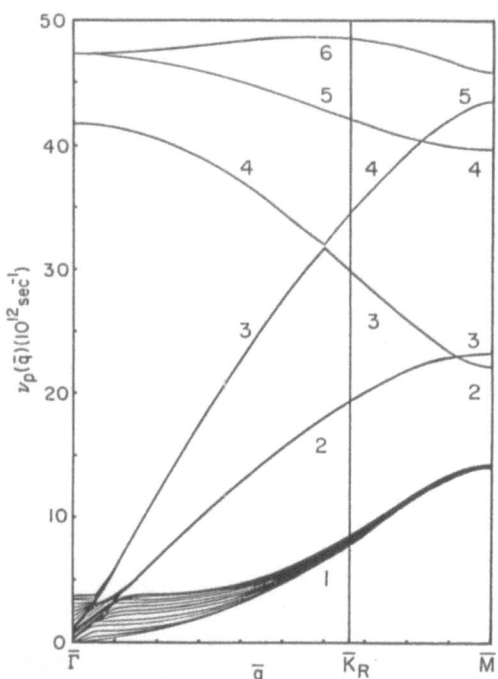

Fig. 4.11. Calculated slab dispersion curves for a 13-layer graphite (0001) slab with free surfaces for $q_{\parallel}$ values along the $\overline{\Gamma M}$ direction in the SBZ. (After [4.45])

tures. The calculation by de *Rouffignac* et al. [4.45] was carried out for a 13-layer slab with (0001) surfaces with intraplanar central interactions up to third neighbors, and interplanar interactions up to second neighbors (cf. (4.21, 22)), fitted to bulk dispersion curves. The SBZ for the graphite (0001) surface is the same as for the fcc(111) surface (cf. Fig. 4.1). In Fig. 4.11 we show the slab dispersion curves for the 13-layer slab for wave vectors along the $\overline{\Gamma M}$ direction of the SBZ. This diagram contains 78 individual dispersion curves (78 degrees of freedom in the slab unit cell), falling into six bands and exhibiting a high degree of near degeneracy (except for band 1). The near degeneracy is due to the almost complete absence of dispersion for bulk modes with wave vectors perpendicular to the (0001) planes.

An examination of the attenuation functions $|\xi(m)|^2$ vs. m [cf (4.15)] for each of the 78 modes shows that each band contains two degenerate surface modes with frequencies slightly below those of the other eleven bulk modes in the band. The absence of the near degeneracy in band 1 is the result of the dispersion of the LO and LA bulk modes in the direction perpendicular to the (0001) planes. The surface mode pair in this band is the Rayleigh mode.

4.5 Derived Physical Quantities

4.5.1 Mean-Square Amplitudes of Vibration and Vibrational Correlation Functions

Mean-square amplitudes of vibration (MSA's) and vibrational correlation functions are of considerable importance for scattering applications. The MSA's enter the Debye–Waller factors which govern the temperature variation of diffraction and scattering intensities [4.46]. For example, many experimental results on the temperature dependence of surface diffraction (LEED, He diffraction) are analyzed in terms of an effective surface Debye temperature $\Theta_\perp$, which can be related to the ratio of surface MSA to bulk MSA at high temperatures by the expression [4.10]

$$\langle u_z^2 \rangle_s / \langle u_z^2 \rangle_{\text{bulk}} = (\Theta_0 / \Theta_\perp)^2 , \tag{4.35}$$

where the z direction is normal to the surface, and Θ_0 the bulk vibrational Debye temperature ($\langle u_z^2 \rangle$ is defined in (4.18)).

Another example concerns the equal-time near-neighbor amplitude correlation functions which are needed for multiple-scattering classical trajectory calculations of, for instance, light ion scattering in cases in which the scattering events are short compared to typical periods of the lattice vibrations. (If the scattering time is comparable to typical periods of lattice vibrations, time dependent correlation functions will be needed.)

We begin with examples of MSA's, MSA ratios and equal-time correlation functions for LiF(001), obtained from the SM calculations of the unrelaxed surfaces of seven alkali halides by *Chen* et al. [4.10, 11, 47]; these calculations use 15-layer slabs and wave vector sampling grids of 66 points in the irreducible (1/8)th part of the SBZ. In Fig. 4.12 we display the temperature variation of the MSA's $\langle u_\alpha^2(\kappa) \rangle$ for the surface layer and the second layer of LiF(001). We note the following. (1) The MSA's perpendicular to the surface (z), are larger than those in the surface (x). (2) The MSA's of the lighter ion (Li$^+$) are larger than those of the heavior ions (F$^-$). (3) The surface MSA's show a larger relative increase with temperature than the second layer MSA's. (4) The classical (i.e., linear in T) behavior of the MSA's sets in at about $T = 300\,\text{K} \simeq 0.4\Theta_0$. (The Debye temperature $\Theta_0 = 727\,\text{K}$ was calculated from the elastic constants deduced from the shell models.) We will see below that the classical behavior of the correlation functions sets in at lower temperatures. In Fig. 4.13 we have plotted the temperature variation of the ratios $\langle u_\alpha^2(\kappa) \rangle_s / \langle u_\alpha^2(\kappa) \rangle_{\text{bulk}}$. Note that this ratio has the strongest T variation for the perpendicular (z) vibration of the negative – i.e. the more polarizable – ion. This is a general feature of the MSA ratios of alkali halides, cf. [4.10].

Comparison of these results with experiment is quite complicated because of numerous uncertainties in unfolding the data to get the MSA. Yet, an analysis by *Goodman* [4.48] of the LiF scattering data of *Hoinkes* et al. [4.49] (based

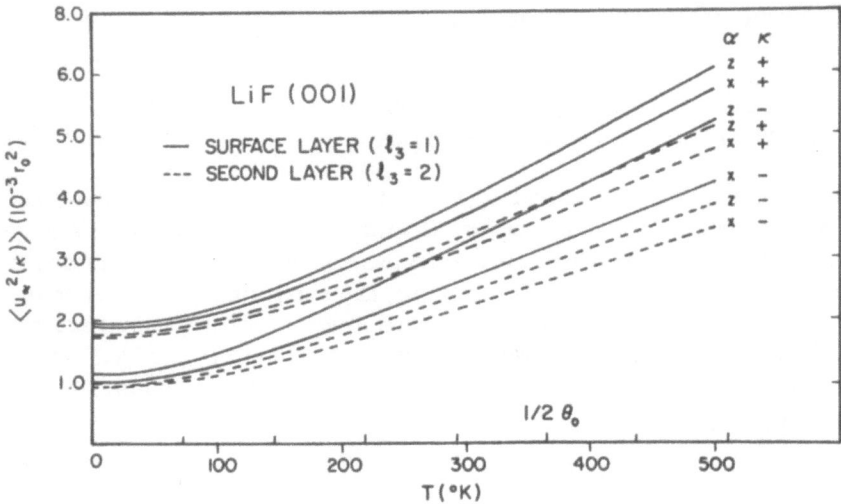

Fig. 4.12. Temperature dependence of the mean-square amplitudes of vibration $\langle u_\alpha^2(\kappa)\rangle$ in units $10^{-3}r_0^2$ (r_0 is the nearst neighbor distance) in the surface layer and second layer of the LiF(001) surface. α and κ indicate the coordinate direction and the charge of the ion, respectively. The z component is normal to the surface; the x component is parallel. $\Theta_0 = 727\,\mathrm{K}$. (After [4.47])

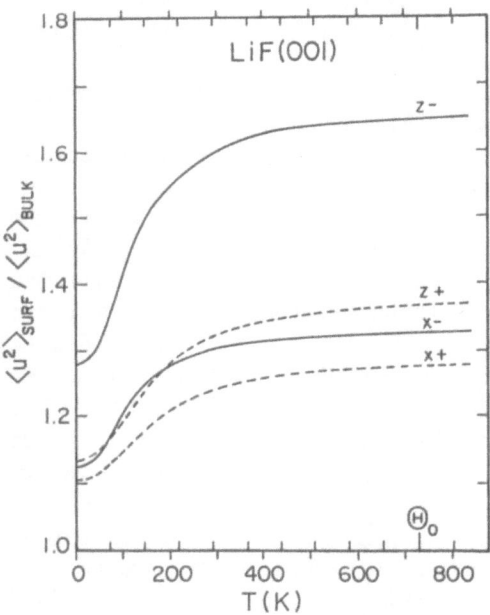

Fig. 4.13. Temperature variation of the ratio of the MSA of a surface ion to that of a bulk ion of the same species for the LiF(001) surface. The solid line stands for the negative ion (F^-) and the dashed line represents the positive ion (Li^+). (After [4.10])

on scattering by the F^- ion) leads to a high temperature MSA ratio of 1.71, which is in good agreement with the calculated value of 1.65. This agreement is another indication that these slab calculations are quite reliable for the alkali halides, and that the calculated results might be useful as a guide if difficulties are encountered in unfolding the experimental data.

Equal-time correlation functions for particle amplitudes of particles $(l_\parallel, l_3, \kappa)$ and $(l'_\parallel, l'_3, \kappa')$ are defined by [4.47],

$$
\langle u_\alpha(l_\parallel, l_3, \kappa) u_\beta(l'_\parallel, l'_3, \kappa') \rangle = \hbar (2\bar{N})^{-1} (M_\kappa M_{\kappa'})^{-1/2}
$$
$$
\times \sum_{q_\parallel p} \xi_\alpha(l_3 \kappa; q_\parallel p) \xi_\beta^*(l'_3 \kappa', q_\parallel p) \coth[\hbar\omega(q_\parallel p)/2k_B T]\omega(q_\parallel p)^{-1}
$$
$$
\times \exp[iq_\parallel \cdot (r(l_\parallel) - r(l'_\parallel)) + iq_\parallel \cdot (r(\kappa) - r(\kappa'))] . \qquad (4.36)
$$

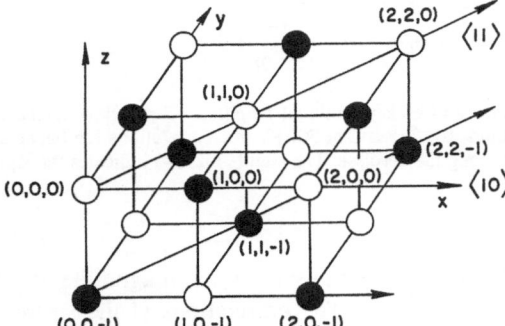

Fig. 4.14. Surface layer and second layer at the (001) surface of a rocksalt-structured crystal. The z direction is perpendicular to the surface, the x and y directions are in the surface. The ion $(x, y, z) = (0, 0, 0)$ is the reference ion for taking correlations with neighboring ions. (After [4.47])

We give some examples of the correlation functions for LiF(001). In Fig. 4.14 we show the top layer ($l_3 = 1$) and the second layer ($l_3 = 2$) of the (001) face of the rocksalt structured slab. Correlations have been evaluated between the ion at the origin (0,0,0) and ions along the $\langle 10 \rangle$ and $\langle 11 \rangle$ directions, both in the surface layer and the subsurface layer. As an example we display in Fig. 4.15 the distance dependence of the function

$$
C_\alpha = \langle u_\alpha(l_\parallel, l_3, \kappa) u_\alpha(l'_\parallel, l'_3, \kappa') \rangle \qquad (4.37)
$$

along the $\langle 10 \rangle$ direction between particles in the surface layer (C_α^s), between a surface particle and particles in the second layer ($C_\alpha^{1,2}$), and in the bulk (C_α^{bulk}), at 300 K. The surface ion $(x, y, z) = (0, 0, 0)$ is the reference ion for taking correlations with neighboring ions; $z = -1$ indicates the first layer below the surface. In the units of C_α, the nearest neighbor distance r_0 is 2.014 Å for LiF. The symbol ($\pm$) indicates that the upper dot refers to the case in which the positive ion is at the origin (0,0,0) and the lower dot for the negative ion at (0,0,0); for ($\mp$) vice versa.

We note the following. (1) The z correlations are larger than the x correlations. (This is reversed along the $\langle 11 \rangle$ direction.) (2) The surface function C_z^s

99

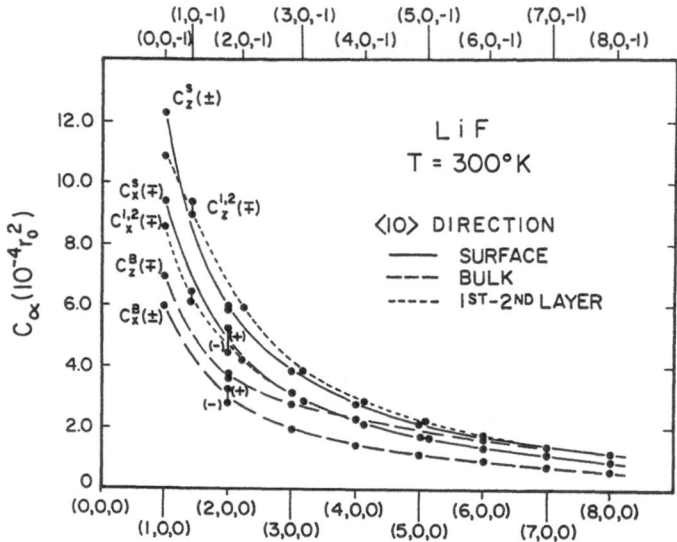

Fig. 4.15. Distance dependence of the surface and bulk vibrational amplitude correlation functions C_α for the $\langle 10 \rangle$ direction of LiF(001) in units $10^{-4} r_0^2$ and at 300 K. The correlations are between the ions at $(0,0,0)$ and (x, y, z) as indicated along the abscissa. The sign $(+)$ or $(-)$ indicates the sign of the reference ion. (After [4.47])

and the 1-st–2-nd layer function $C_z^{1,2}$ have essentially the same magnitude at all distances; the same holds for C_x^s and $C_x^{1,2}$. (3) The correlations at the surface are larger than those in the bulk. In particular, at the first neighbor $(1,0,0)$, C_z^s is 175 percent of C_z^{bulk} and C_x^s is 159 percent of C_x^{bulk}. (4) The feature that C_x^{bulk} and C_z^{bulk} are not equal (as they should be because of the cubic symmetry) and that C_z^{bulk} flattens out at larger distances is a result of the fact that for C_α^{bulk} we have used the values obtained for the center layer of the 15-layer slab instead of values obtained from a real bulk calculation. In conclusion we can say that the overall distance dependence of the correlations is quite similar in all cases.

The temperature dependence of the correlation functions can be conveniently displayed by plotting the *relative* correlation function

$$\varrho_\alpha(T) = C_\alpha [\langle u_\alpha^2(\boldsymbol{l}_\|, l_3, \kappa) \rangle \langle u_\alpha^2(\boldsymbol{l}_\|', l_3', \kappa') \rangle]^{-1/2} , \tag{4.38}$$

which compares the correlations to the auto correlations (MSA's). In Fig. 4.16 we display, again for LiF, the functions $\varrho_z^s(\pm)$ and $\varrho_x^s(\pm)$ for the correlation $(0,0,0) - (1,1,0)$, i.e. for the first neighbor along the $\langle 11 \rangle$ direction; these are correlations between particles of the same kind. In this case the denominator of (4.38) is simply the MSA of the particle in question, so that ϱ gives a direct measure of the magnitude of C relative to the MSA. It is seen that in this example, while at absolute zero the C_α's have values between 2 and 12 percent of the corresponding MSA's, at and above room temperature these values have increased to between 15 and 30 percent of the corresponding MSA's.

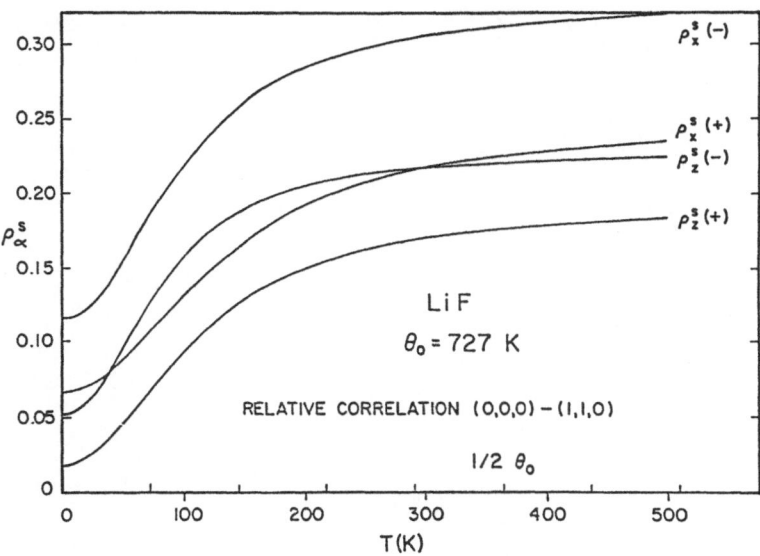

Fig. 4.16. Temperature dependence of the relative surface vibrational amplitude correlation functions ϱ_α for the LiF(001) surface. (After [4.47])

Size Dependence of MSA's. Before leaving the subject of MSA's we present an example of the effect of size and shape dependence of the MSA's. As was discussed in Sect. 4.2.2, for crystals of finite size, the MSA's depend on the aspect ratio (lateral dimension divided by thickness) and, for a given aspect ratio, on the thickness of the crystal. This is of importance for measurements on finite crystals such as graphite crystallites (Grafoil). In addition, this parametrization enables us to carry out, in a unique way, an extrapolation to the thermodynamic limit (infinite half crystal: vanishing surface to volume ratio) in a single parameter, namely the inverse thickness. While for finite crystals the MSA's are dependent on the aspect ratio of the crystal, this dependence disappears in the limit of an infinite crystal, so that the MSA's are unique quantities in this limit.

Firey et al. [4.12] evaluated the MSA's for graphite slabs for aspect ratios 1, 8 and 80 (called cube, slab and sheet, respectively), and for each aspect ratio for four thicknesses $N_3 = 7, 13, 21$, and 27 as well as their extrapolation to infinite thickness. The reason for this particular procedure – fixed aspect ratio and increasing thickness – is that it is the proper way to approach the thermodynamic limit, i.e. the limit of vanishing surface to volume ratio.

From a practical point of view it is most useful to present the MSA data for the sheets. The thickest of these, the 27-layer sheet, has a thickness of 90.6 Å; this is in the range of the actual dimensions of graphite crystallites in Grafoil.

In Fig. 4.17 we show the MSA's of the surface atoms of the 7-, 13-, 21-, and 27-layer and infinitely thick sheets for vibrations perpendicular and parallel to the surface ($\langle u_\perp^2 \rangle_s$ and $\langle u_\parallel^2 \rangle_s$, respectively). It will be noted that $\langle u_\perp^2 \rangle_s$ is strongly dependent on thickness, decreasing as the thickness increases and converging

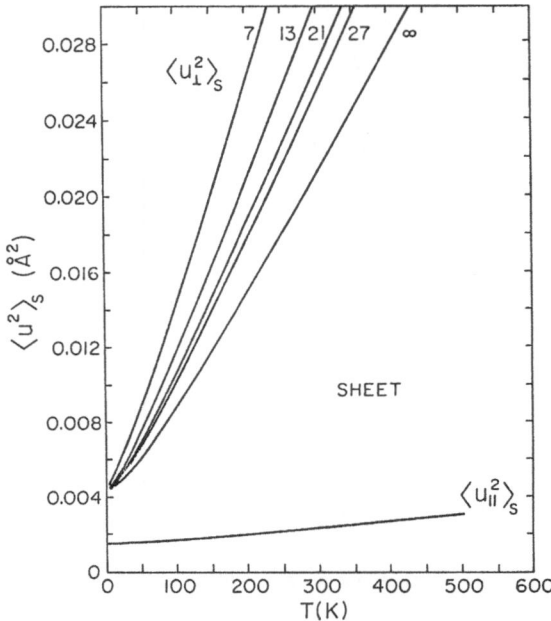

Fig. 4.17. Surface mean-square amplitudes (MSA) for the 7-, 13-, 21-, and 27-layer sheets and the infinitely thick sheet as functions of temperature. $\langle u_\perp^2 \rangle_s$ indicates the MSA's for the vibrations perpendicular to the surface $\langle u_\parallel^2 \rangle_s$ for the vibrations parallel to the surface. The latter coincides within the thickness of the curve for all sheet thicknesses. (After [4.12])

to the infinite-thickness value as N_3^{-1}, i.e., proportionally with the vanishing surface-to-volume ratio as pointed out above. This is a finite-size effect; for the smaller thickness, the surface-to-volume ratio is larger, and the entire system is more strongly perturbed by the surface. The effect is not connected with the Landau–Peierls divergence, since it is of opposite sign: at fixed aspect ratio the lateral extent of the system is smaller for the smaller thickness.

It seems almost redundant to point out here that the thickness dependence of the MSA's is an effect that can only be obtained by calculations on crystals of finite thickness. However, this is important to keep in mind if one considers alternative methods for calculating surface dynamical properties, such as the matching method, the Green's function method or the continued-fraction method. These methods, as usually applied, all give results that apply to half-infinite crystals and the question of crystal shape (aspect ratio) is not addressed. Fortunately, as we find here, the shape dependence of the MSA's disappears for infinite crystals; but, of course, size effects are not treated by these methods when applied to semi-infinite samples.

4.5.2 Surface Debye Temperature

It is often customary (especially in low-energy electron diffraction (LEED) experiments) to express measured MSA's in terms of Debye temperatures Θ. The probable reason for this practice is that one can express in one number the MSA's as a function of temperature, provided the Debye approximation is valid over the entire temperature range. Since this is never the case, one has to specify the function $\Theta(T)$, and so there is no gain over specifying the MSA's themselves as functions of temperature, except that in a plot of $\Theta(T)$ the deviation from Debye-type behavior becomes very visible; for a Debye system $\Theta(T)$ is a constant. At each temperature the value of Θ is obtained from $\langle u_\perp^2 \rangle$ by determining the value of Θ which satisfies the relation

$$\langle u_\perp^2 \rangle = \frac{3\hbar^2}{k_B M \Theta} \left(\frac{1}{4} + \frac{\varphi(x)}{x} \right), \tag{4.39}$$

where

$$x = \frac{\Theta}{T}$$

and

$$\varphi(x) = \frac{1}{x} \int_0^x \frac{t dt}{e^t - 1} . \tag{4.40}$$

This exercise is made convenient by using a rational approximation for the Debye function $\varphi(x)$ [4.50].

In Fig. 4.18 we have plotted the temperature-dependent Debye temperature $\Theta_\perp(T)$ (as obtained from (4.39)) for the vibrations perpendicular to the (0001) planes, for the bulk, and for the 27-layer and infinitely thick sheets. We see a strong temperature variation in $\Theta_\perp(T)$ below 200 K, which indicates a breakdown of the Debye approximation at those temperatures.

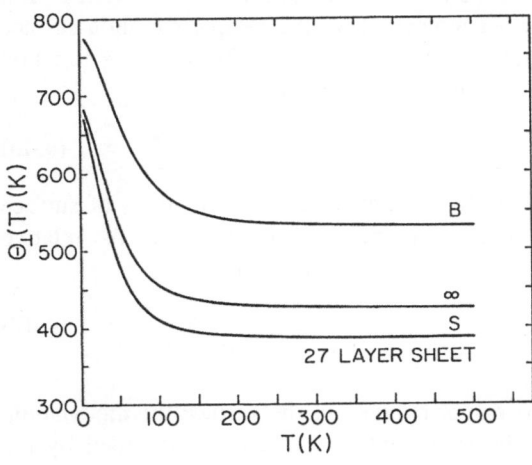

Fig. 4.18. Debye temperatures $\Theta_\perp(T)$ associated with the perpendicular vibrations for the bulk (B) and for the surface (S), the latter for the infinitely thick crystal and the 27-layer sheet. (After [4.12])

4.5.3 Surface Thermodynamic Quantities

Thermodynamic properties of surfaces [4.51] are of importance because they govern the formation and stability of surfaces, play a role in the macroscopic description of adsorption and desorption processes, as well as in a number of other surface processes.

The vibrational contributions to the energy E, the entropy S, the Helmholtz free energy F, and the specific heat at constant volume C_v can be expressed in terms of the vibrational frequencies of the system. If we define

$$x = \hbar\omega(q_{\parallel}p)/k_B T ,\tag{4.41}$$

then

$$E = k_B T \sum_{q_{\parallel}p}[x/2 + x/(e^x - 1)] ,\tag{4.42}$$

$$S = k_B T \sum_{q_{\parallel}p}[-\ln(1 - e^{-x}) + x/(e^x - 1)] ,\tag{4.43}$$

$$F = k_B T \sum_{q_{\parallel}p}[x/2 + \ln(1 - e^{-x})] ,\tag{4.44}$$

$$C_v = k_B \sum_{q_{\parallel}p}[x^2 e^x /(e^x - 1)^2] .\tag{4.45}$$

The thermodynamic functions for an infinite, three-dimensional crystal can be calculated by means of equations similar to (4.42–45), with the only difference that $q_{\parallel}$ is replaced by the three-dimensional wave vector q, and that p takes on $3s$ values, where s is the number of particles in the unit cell. Once the thermodynamic functions for the crystal with and without surface have been calculated, the surface contributions to the thermodynamic function can be obtained in the following way. Let u be any one of the thermodynamic functions reduced to unit quantity of material (per mole or per molecule) for the sample having a surface area per quantity (n moles or molecules) A/n, and u^{bulk} be the same function for a reference bulk sample. We may term the difference

$$\Delta u = u - u^{\text{bulk}}\tag{4.46}$$

the *surface-excess* u, just as is done for experimental determinations of surface thermodynamics. If the surface-excess u is indeed proportional to the relative surface area A/n of the sample, then the *surface-specific* u,

$$U^s = \Delta u/(A/n) ,\tag{4.47}$$

is a meaningful quantity.

An alternative but equivalent way of expressing the surface thermodynamic functions obtains by integrating the summands in (4.42–45) multiplied by the surface frequency distribution $f^s(\omega)$ (cf. (4.19)), e.g.

$$C_v^s = (3sk_B/A_0) \int_0^{\omega_{max}} d\omega\, f^s(\omega)\, \left[x^2 e^x/(e^x - 1)^2\right] , \qquad (4.48)$$

where $A_0 = A/n$ is the area per surface particle. In actual calculations $f^s(\omega)$ will be available in terms of a table (histogram) obtained from the dynamical calculations. Since the formation of a table for $f^s(\omega)$ involves a loss of information about the exact values of the frequencies, it is more accurate to carry out the summations of (4.42–45) for the crystals with and without surfaces, and then use (4.47).

Surface-Excess Specific Heat. Measurements of the heat capacity of samples of small particles are the most direct approach to the surface contributions to vibrational thermodynamic functions, because, in principle, the other thermodynamic functions can be determined from the heat capacity. In Fig. 4.19 we display the surface-excess specific heat $C_v^s(T)$ for the NaCl(001) surface as calculated in the Kellerman rigid-ion model (KRIM) calculations [4.52], without and with relaxation of the surface, and with the SM [4.53]. The shape of these curves is characteristic: $C_v^s(0) = 0$; there is a peak at roughly $0.16\Theta_0$ (Θ_0 is the bulk Debye temperature) and for increasing T, C_v^s decreases to zero because the specific heat tends to its classical value of $3k_B$ per particle, whether the crystal has surface or not. The peak is mainly due to contributions of low lying acoustical surface modes. On the basis of the Debye theory one might expect the low-T slope of the peak to exhibit a T^2 behavior. This is indeed found to a good approximation in the rigid-ion calculation but not in the SM calculation; in the latter case we find for the low temperature behavior, $C_v^s \propto T^{2.4}$. This deviation from Debye-like

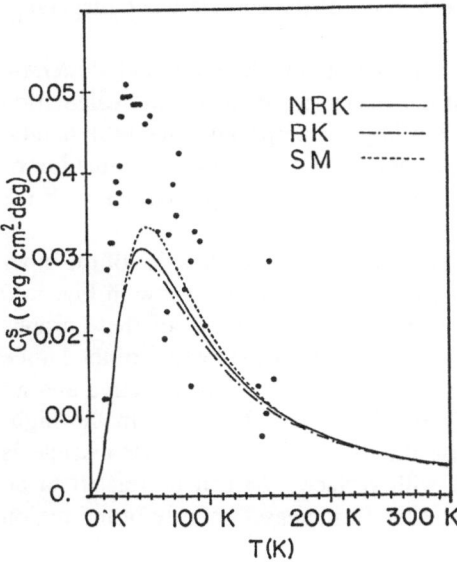

Fig. 4.19. Surface-excess specific heat $C_v^s(T)$ for the NaCl(001) surface. NKR, RK and SM mean: nonrelaxed KRIM-, relaxed KRIM- and SM calculation, respectively. The experimental points displayed are from [4.54]. (After [4.52, 53])

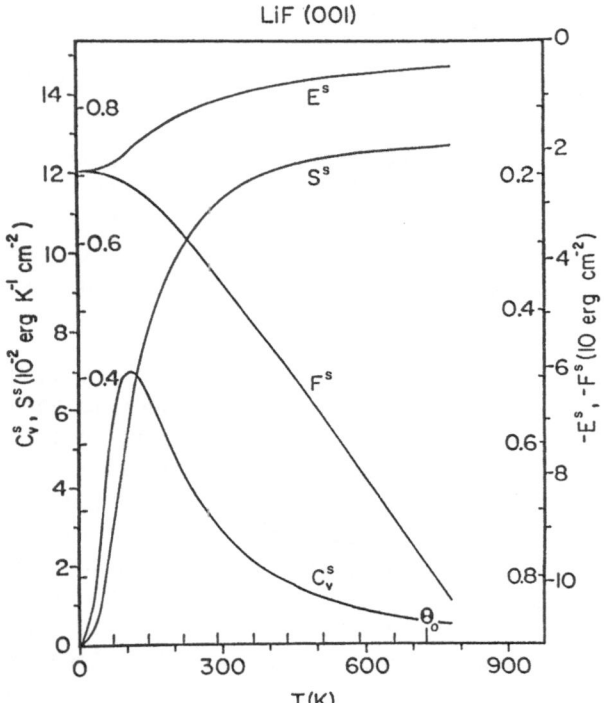

Fig. 4.20. Surface-excess thermodynamic functions E^s, F^s, S^s and C_v^s for the LiF(001) surface. The scales for C_v^s and S^s are given on the left and those for E^s and F^s on the right. Absolute cgs scales are given on the outside of the frame. (After [4.51])

behavior in the SM calculation is a general characteristic of C_v^s and occurs for all seven crystals studied; it appears to be the result of the surface wave dispersion in the lattice.

The points in Fig. 4.19 are experimental points of *Morrison* and *Patterson* [4.54]. There is a reasonable agreement between measurements and calculation in the peak location near 40 K, but there is a large discrepancy in the peak height. The most probable reason for this discrepancy is an erroneous determination of the surface area of the experimental sample, resulting from difficulties with the BET method, cf. [4.51].

Finally, to give an impression of the typical temperature behavior of the other three surface-excess thermodynamic functions E^s, F^s, S^s, we show in Fig. 4.20 a combined plot of these functions for LiF. The gross features of these graphs can be understood using simple arguments [4.52, 55]. The softening of the lattice vibrations in the presence of a surface causes E^s and F^s to be negative and S^s and C_v^s to be positive. At $T = 0$ K, $S^s = C_v^s = 0$ and $F^s = E^s$. In the high-temperature limit, $E^s = C_v^s = 0$ because the vibrational energy per particle is $3k_B T$ in both the bulk and the crystal with surfaces. S^s can be related to an integral over the surface specific heat, so that S^s increases sharply in the region

106

of the peak in C_v^s and then asymptotically approaches a constant value (in the strict harmonic approximation) as C_v^s tends to zero. F^s in turn decreases linearly with temperature at high temperature since E^s tends to zero and S^s tends to a positive constant in the relation $F^s = E^s - TS^s$.

4.6 Concluding Remarks

In this article we have given a short introduction to the slab method of surface dynamics, as well as a brief overview of the surface vibrational properties of ideal low-index surfaces of noble gas crystals, ionic crystals, fcc and bcc metals, and graphite. The article can serve as an introduction to the surface phonon atlas in this volume, compiled by W. Kress (Chap. 8). We have also discussed a number of surface related properties which can be obtained once the vibrational spectrum of the system is known: surface mean-square amplitudes of vibration and vibrational correlation functions, and surface-excess thermodynamic properties. While the former are of importance for surface scattering calculations and the unfolding of experimental scattering data, the latter relate to surface stability, structural transitions and the thermodynamics of adsorption and desorption.

As was noted in the introduction to this volume, up to the early 1980s there existed only a few direct experimental manifestations of surface vibrational properties. Then, as a result of rapid improvements in inelastic He scattering and electron energy loss spectroscopy, it became possible to measure surface phonon dispersion curves over the entire Brillouin zone. This meant that for the first time our theoretical understanding of surface dynamics could be compared in substantial detail with experiment, and this provided a new stimulus and a challenge to the theory, to interpret existing measurements and to indicate areas of interest for future experimental study. Since the interpretation of experimental data on surface dynamics has to rely much more on theoretical support than is the case for bulk dynamics, a close collaboration between experimentalists and theorists has been developing. The role of the theory is to evaluate the relevant surface dynamical quantities and to perform full scattering calculations so that the measured energy loss spectra can be directly compared with calculated spectra.

At the present time the main challenge for the theory is to carry out surface vibrational calculations as much as possible from first principles, i.e. to develop microscopic models which properly account for the effects of the core and valence electrons on the dynamics. This development is of particular importance for covalent crystals and metals.

Acknowledgements. The author wishes to thank Dr. A.D. Kulkarni for a critical reading of the manuscript.

This work was supported in part by the National Science Foundation and the Robert A. Welch Foundation. A NATO travel grant for international collaboration is gratefully acklowledged.

References

4.1   B.C. Clark, R. Herman, R.F. Wallis: Phys. Rev. **139**, A 860 (1965)
4.2   R.E. Allen, F.W. de Wette: Phys. Rev. **179**, 873 (1969)
4.3   R.E. Allen, F.W. de Wette, A. Rahman: Phys. Rev. **179**, 887 (1969)
4.4   S.Y. Tong, A.A. Maradudin: Phys. Rev. **181**, 1318 (1969)
4.5   R.E. Allen, G.P. Alldredge, F.W. de Wette: Phys. Rev. B **4**, 1648 (1971)
4.6   R.E. Allen, G.P. Alldredge, F.W. de Wette: Phys. Rev. B **4**, 1661 (1971)
4.7   F.W. de Wette, G.P. Alldredge: In *Methods of Computational Physics*, ed. by G. Gilat, B. Alder, S. Fernbach, M. Rotenberg (Academic, New York 1976) Vol. 15, Chap. 5, p.163
4.8   A.A. Maradudin, E.W. Montroll, G.H. Weiss, I.D. Ipatova: *Theory of Lattice Dynamics in the Harmonic Approximation*, 2nd ed. (Academic, New York 1971)
4.9   G.P. Alldredge: Phys. Lett. **41A**, 281 (1972);
       V.L. Zoth, G.P. Alldredge, F.W. de Wette: Phys. Lett. **47A**, 247 (1974)
4.10   T.S. Chen, G.P. Alldredge, F.W. de Wette: Surf. Sci. **57**, 25 (1976)
4.11   T.S. Chen, F.W. de Wette, G.P. Alldredge: Phys. Rev. B **15,** 1167 (1977)
4.12   B. Firey, F.W. de Wette, E. de Rouffignac, G.P. Alldredge: Phys. Rev. B **28**, 7210 (1983)
4.13   G. Gilat, G. Dolling: Phys. Lett. **8**, 304 (1964);
       G. Gilat, L.J. Raubenheimer: Phys. Rev. **144**, 320 (1966);
       L.J. Raubenheimer, G. Gilat: Phys. Rev. **157**, 586 (1967)
4.14   See for instance articles by R.M. Martin and K. Kunc: In *Electronic Structure, Dynamics and Quantum Properties of Condensed Matter*, ed. by J.T. Devreese, P. Van Camp, NATO ASI Series B, Vol. 121 (Plenum, New York 1985)
4.15   K.M. Ho, K.P. Bohnen: Phys. Rev. Lett. **56**, 934 (1986)
4.16   A.D.B. Woods, W. Cochran, B.N. Brockhouse: Phys. Rev. **119**, 980 (1960)
4.17   W. Cochran: Adv. Phys. **10**, 401 (1961)
4.18   R.A. Cowley: Proc. Roy. Soc. A **268**, 109 (1962); ibid., 121 (1962)
4.19   H. Bilz, W. Kress: *Phonon Dispersion Relations in Insulators* (Springer, Berlin, Heidelberg 1979)
4.20   F.W. de Wette, G.E. Schacher: Phys. Rev. **137**, A 78 (1965); **138** erratum AB **4** (1965)
4.21   W. Goldammer, W. Ludwig, W. Zierau, C. Falter: Surf. Sci. **141**, 139 (1984)
4.22   D.J. Cheng, R.F. Wallis, L. Dobrzynski: Surf. Sci. **43**, 400 (1974)
4.23   D. Castiel, L. Dobrzynski, D. Spanjaard: Surf. Sci. **59**, 252 (1976)
4.24   J.E. Black, D.A. Campbell, R.F. Wallis: Surf. Sci. **105**, 629 (1981); Surf. Sci. **115**, 161 (1982)
4.25   J.E. Black, F.C. Shanes, R.F. Wallis: Surf. Sci. **133**, 199 (1983)
4.26   V. Bortolani, A. Franchini, F. Nizzoli, G. Santoro: In *Dynamics of the Gas-Surface Interaction*, ed. by G. Benedek, N. Valbusa (Springer, Berlin, Heidelberg 1982) p.196
4.27   (a) V. Bortolani, A. Franchini, G. Santoro: In *Electronic Structure, Dynamics and Quantum Structural Properties of Condensed Matter*, ed. by J.T. Devreese, P. Van Camp, NATO ASI Series B, Vol. 121 (Plenum, New York 1985) p.401
       (b) V. Bortolani, A. Franchini, G. Santoro: In *Dynamical Phenomena at Surfaces, Interfaces and Superlattices*, ed. by F. Nizzoli, K.-H. Rieder, R.F. Willis (Springer, Berlin, Heidelberg 1985) p.66
       (c) V. Bortolani, G. Santoro, U. Harten, J.P. Toennies: Surf. Sci. **148**, 82 (1984)
4.28   For instance, the bcc lattice is unstable for a large class of central pair potentials; see M. Born, K. Huang: *Dynamical Theory of Crystal Lattices* (Oxford University Press, London 1962) p.152
4.29   V. Bortolani, A. Franchini, F. Nizzoli, G. Santoro: Phys. Rev. Lett. **52**, 429 (1984)
4.30   G. Brusdeylins, R.B. Doak, J.P. Toennies: Phys. Rev. B **27**, 3662 (1983);
       G. Benedek, G. Brusdeylins, R.B. Doak, J.G. Skofronick, J.P. Toennies: ibid. B **28**, 2104 (1983)
4.31   G. Benedek, G.P. Brivio, L. Miglio, V.R. Velasco: Phys. Rev. B **26**, 497 (1982)
4.32   F.W. de Wette, W. Kress, U. Schröder: Phys. Rev. B **32**, 4143 (1985)
4.33   W. Kress, F.W. de Wette, A.D. Kulkarni, U. Schröder: Phys. Rev. B **35**, 5783 (1987)
4.34   F.W. de Wette, A.D. Kulkarni, U. Schröder, W. Kress: Phys. Rev. B **35**, 2476 (1987)
4.35   G. Brusdeylins, R.B. Doak, J.P. Toennies: Phys. Rev. Lett. **46**, 437 (1981)
4.36   G. Benedek: Surf. Sci. **61**, 603 (1976)
4.37   W. Kress, F.W. de Wette, A.D. Kulkarni, U. Schröder: Phys. Rev. B **35**, 2467 (1987)

4.38   J.E. Black, R.F. Wallis: Phys. Rev. B **29**, 6972 (1984)
4.39   J.E. Black, T.S. Rahman, D.L. Mills: Phys. Rev. B **27**, 4072 (1983)
4.40   R.B. Doak, N. Harten, J.P. Toennies: Phys. Rev. Lett. **51**, 578 (1983)
4.41   K.M. Ho, K.P. Bohnen: Phys. Rev. B**32**, 3446 (1985)
4.42   C.J. Jayanthi, H. Bilz, W. Kress, G. Benedek: Phys. Rev. Lett. **59**, 795 (1987)
4.43   J.C. Philips: Phys. Rev. **166**, 832 (1968); R.M. Martin: Phys. Rev. **186**, 871 (1969); W. Weber: Phys. Rev. B **15**, 4789 (1977)
4.44   A.G. Eguiluz, A.A. Maradudin, R.F. Wallis: *Proceedings of the 19th Solvay Conference*, Austin, Texas, Dec. 14–18, 1987 (Springer, Berlin, Heidelberg 1988)
4.45   E. de Rouffignac, G.P. Alldredge, F.W. de Wette: Phys. Rev. B **23**, 4208 (1981)
4.46   Cf. T. Engel, K.M. Rieder: In Springer Tracts in Modern Physics, Vol.91, ed. by G. Höhler (Springer, Berlin, Heidelberg 1982) p.55
4.47   T.S. Chen, F.W. de Wette: Phys. Rev. B **17**, 835 (1978)
4.48   F.O. Goodman: Surf. Sci. **46**, 118 (1974)
4.49   H. Hoinkes, H. Nahr, H. Wilsch: Surf. Sci. **33**, 516 (1972); erratum Surf. Sci. **40**, 457 (1973)
4.50   H.C. Thacher: J. Chem. Phys. **32**, 638 (1960)
4.51   T.S. Chen, G.P. Alldredge, F.W. de Wette: Surf. Sci. **62**, 675 (1977)
4.52   T.S. Chen, G.P. Alldredge, F.W. de Wette, R.E. Allen: J. Chem. Phys. **55**, 3121 (1971)
4.53   T.S. Chen: Ph. D. Dissertation, University of Texas at Austin, 1971. [Available from University Microfilms, Inc., Ann Arbor, Michigan 48106 (Order No. 72-15726, T.S. Chen)]
4.54   J.A. Morrison, D. Patterson: Trans. Faraday Soc. **52**, 764 (1956)
4.55   R.E. Allen, F.W. de Wette: J. Chem. Phys. **51**, 4820 (1969)

Additional References: Reviews of Surface Dynamics

R.F. Wallis: Lattice Dynamics of Crystal Surfaces, in *Progress in Surface Science*, ed. by S.G. Davison (Pergamon, New York 1974) Vol. 4, p.234

K.L. Kliewer, R. Fuchs: Theory of Dynamical Properties of Dielectric Surfaces, in *Advances in Chemical Physics*, ed. by I. Prigogine, S.A. Rice (Wiley, New York 1974) Vol. 27, p.355

R.F. Wallis: Effects of Surfaces in Lattice Dynamics, in *Dynamical Properties of Solids*, ed. by G.K. Horton, A.A. Maradudin (North-Holland, Amsterdam 1975) Vol. 2, p.443

M.G. Lagally: Surface Vibrations, in *Surface Physics of Materials*, ed. by J.M. Blakeley (Academic, New York 1975) Vol. II, p.419

F.W. de Wette, G.P. Alldredge: Lattice Dynamics of Surfaces of Solids, in *Methods in Computational Physics*, ed. by G. Gilat, B. Alder, S. Fernbach, M. Rotenberg (Academic, New York 1976) Vol. 15, p.163

G.I. Stegeman, F. Nizzoli: Surface Vibrations, in *Surface Excitations*, ed. by V.M. Agranovich, R. Loudon (North-Holland, Amsterdam 1984) Chap. 2, p.195

J.E. Black: Dynamical Surface Properties in the Harmonic Approximation, in *Structure and Dynamics of Surfaces I*, ed. by W. Schommers, P. von Blanckenhagen (Springer, Berlin, Heidelberg 1986) p.153

F.W. de Wette: Surface Vibrations of Insulators, in *Dynamical Properties of Solids*, ed. by G.K. Horton, A.A. Maradudin (Elsevier Science, Amsterdam 1990) Chap. 5

5. Experimental Determination of Surface Phonons by Helium Atom and Electron Energy Loss Spectroscopy

J.P. Toennies

With 30 Figures

The experimental study of phonon dispersion curves is motivated by the quest for a detailed understanding of the forces between the atoms of condensed matter in the bulk and at the surface. Bulk phonons have been studied by neutron scattering since about 1955, when *Brockhouse* and *Stewart* [5.1] reported on the first measurements on aluminum. The study of surface phonons by particle scattering is a rather recent development. Neutrons are not well suited for surface studies because of their very small cross sections but nevertheless have been used to study adsorbates on layered compounds (graphite) [5.2] or vibrations on powder particles with a large ratio of surface to volume areas [5.3].

The first attempt to study surface vibrations by low energy electron diffraction was discussed in terms of inelastic scattering from Rayleigh modes [5.4]. Stimulated by the successes of inelastic electron scattering in detecting adsorbate vibrations [5.5] the theory of electron scattering from surface phonons was fully developed by *Li* et al. as recently as 1980 [5.6]. These authors pointed out the advantages of carrying out the experiments in the impact regime of high energies ($\approx$ 100–200 eV). In 1983 *Ibach* and coworkers [5.7] first succeeded in measuring a surface phonon dispersion curve in Ni(100) using electron energy loss spectroscopy (EELS) at high impact energies.

The first reported suggestion that surface phonon dispersion curves could be measured by inelastic atom scattering was made in two theoretical studies published by *Cabrera* et al. (1969) [5.8] and *Manson* and *Celli* (1971) [5.9]. These authors adapted the theory for scattering of neutrons from bulk phonons to the surface problem. This work immediately stimulated several experimental groups to carry out the experiments. S.S. *Fisher* and coworkers [5.10] at Virginia, who were at the same institutions as the theoreticians, were the first to clearly resolve energy losses in time of flight measurements for He scattering from LiF(001). The groups of *Miller* [5.11] at La Jolla and *Williams* [5.12] at Ottawa continued this work in the 1970's and succeeded in resolving structures which they were able to attribute to single phonons near the zone origin. These early experiments were, however, hampered by inadequate velocity resolution ($\Delta v/v \cong 5\%$) and detector sensitivity. The first successful measurements of surface phonon dispersion curves out to the zone boundary were carried out by Toennies and collaborators on LiF(001) in 1981 [5.13] and on Ag(111) in 1983 [5.14]. These experiments were made possible by the introduction of a new mode of operation of nozzle beam sources that provides a relative velocity resolution of 1%, which is a factor of 5 to 10 better than that of the earlier experiments, and the use of a carefully designed system of differential pumping to increase the sensitivity of the detector.

Since these initial experiments, many groups have entered the field to study the dispersion curves of surface phonons. The He atom measurements cover the region of energy transfers up to about 30–50 meV with a resolution of about 0.5–1 meV, whereas the EELS measurements extend up to about 500 meV albeit with a much lower resolution of about 7 meV. In this survey we will attempt only to describe the experimental techniques used in these studies and how the data is interpreted. What can be learned from the data will only be touched on briefly in connection with the experimental techniques and in a brief summary. For recent surveys the reader is referred to reviews by *Toennies* [5.15] on He scattering and by *Ibach* [5.16] on EELS experiments.

The article is organized as follows. In the next section we will discuss the differences in the coupling matrix elements which determine the theoretical sensitivity of the two techniques. Then we present the kinematical relations from which the dispersion curves are obtained in both cases (Sect. 5.2). The experimental set up used in helium and electron scattering measurements is described in Sect. 5.3. More details of the important system components used, especially in He atom scattering, are presented. In Sect. 5.5 the factors which affect the intensities in both techniques are discussed at some length and the intensities obtained in the two methods are compared. In the final section we present a compilation of systems studied up to date and survey some important new physical phenomena emerging from experiments using both techniques.

5.1 Theoretical Background

The vibrational modes at the surface of a single crystal can be attributed to a contribution from discrete bulk modes, which when projected onto the surface form a continuum (bulk bands), and another contribution from surface localized modes [5.15–18]. The latter are referred to as pure surface modes when they appear in regions of the $(\hbar\omega, Q)$ dispersion space not filled by bulk bands, or as resonances when they appear in the region occupied by bulk bands with which they may or may not couple (hybridize), depending on the mode polarizations. The polarizations are customarily referred to the sagittal plane, which is defined by the incident wave vector k_i and the surface normal z as shown in Fig. 5.1a. In most experiments up to now the final scattered wave vector k_f has been in the sagittal plane. Surface modes then are classified into three types: transverse, longitudinal and shear horizontal, as indicated in Fig. 5.1b. Note that only the shear horizontal modes have a well-defined polarization normal to the sagittal

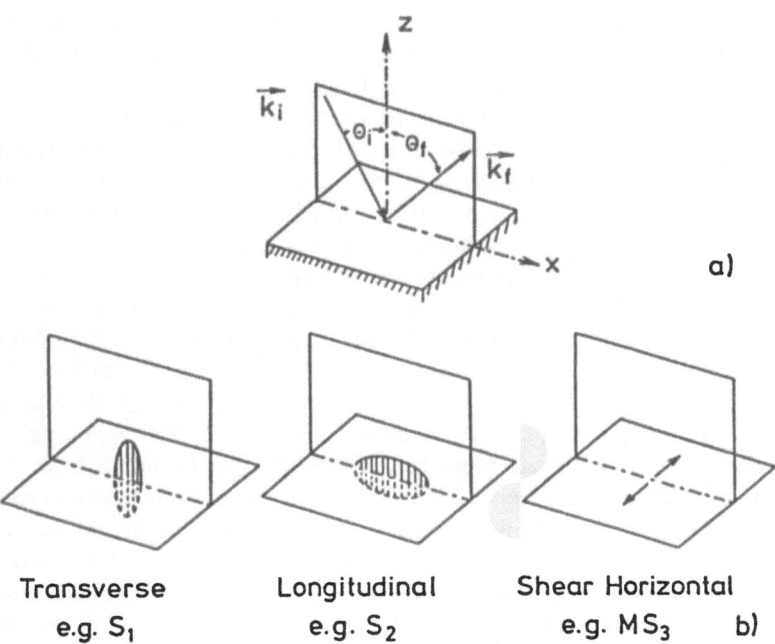

Fig. 5.1. Schematic diagram showing in (a) the typical scattering geometry used in a surface scattering experiment and in (b) the three main types of polarizations of surface phonons. The sagittal plane shown in (a) is the plane containing the incident wave vector k_i and the surface normal z

plane, whereas the so-called transverse and longitudinal modes are actually coherent mixtures of both polarizations, the name indicating which has the greatest relative amplitude. In sagittal plane scattering, symmetry considerations rule out a coupling of atoms or electrons to the shear horizontal modes.

The scattering of He atoms and electrons from the vibrational modes of a crystal surface is fully described by the differential reflection coefficient $d^2R/dE_f d\Omega_f$, which is the probability that a particle striking the surface is scattered with a final energy E_f into a solid angle Ω_f. For He atom scattering the differential reflection coefficient can be calculated from the golden rule [5.8, 19–21]:

$$\frac{d^2R}{dE_f d\theta_f} = \frac{L^4 m^2 k_f}{4\pi^2 \hbar^4 |k_{iz}|} \sum_{\{n_i\}} p(\{n_i\}) \sum_{\{n_f\}} |\langle f|v_{\mathrm{dyn}}|i\rangle|^2 \delta(E_f^{\mathrm{tot}} - E_i^{\mathrm{tot}}) , \qquad (5.1)$$

where the prefactor guarantees the correct normalization, L is a quantization length, which cancels factors in the matrix element, m is the mass of the projectile, k_f is the final wave vector on the surface normal. v_{dyn} is the dynamical coupling potential, which can be treated as a small perturbation. The final and initial states $\langle f|$ and $|i\rangle$ are each products of the vibrational eigenstates $\langle\{n_f\}|$ and $|\{n_i\}\rangle$ of the unperturbed crystal Hamiltonian in the occupation number representation and the final and incident scattering states $|\psi_f^{(-)}\rangle \dots |\psi_i^{(+)}\rangle$. Since the

113

initial and final vibrational states of the crystal are not determined, a sum over the initial states weighted by the temperature dependent Boltzmann factor $p\{n_i\}$ and a sum over all final states both appear in (5.1).

An analogous expression holds for electron scattering as well. The only fundamental difference between the two dynamical theories enters into (5.1) through different expressions for the matrix elements coupling the final and initial scattering states. These differences can be understood in terms of the different masses, collision energies and gas phase scattering cross sections of electrons and He atoms, summarized in Table 5.1. To compensate for their low mass, much greater energies are needed in the case of electrons than for He atoms in order to achieve a parallel momentum comparable with that of the reciprocal lattice. At these high energies the interaction of electrons with the individual atoms of the surface lattice is of a much shorter range than in the case of the low energy He atoms. This can be understood by comparing typical cross sections for scattering of electrons from free Cu atoms and those for the scattering of He atoms from Ar atoms at the nominal energies (see Table 5.1). As a result of their smaller cross section, electrons penetrate into the crystals whereas the He atoms are unable to do so and are reflected at relatively large distances, the order of 3–4Å, from the surface plane. Thus in general, except for highly corrugated surfaces and grazing collisions, multiple interactions of the He atoms with the surface can be excluded. This is not the case in the electron scattering. Here the penetration of the electrons into the crystal makes it mandatory to take into account multiple scattering of the electrons from many atoms of the lattice before they finally leave the crystal. These differences are illustrated in a schematic way in Fig. 5.2.

Table 5.1. Comparison of some kinematic properties of electron and He atom beams used in measuring surface phonon dispersion curves

	Electrons	He atoms
Energies [eV]	200	0.02
ΔK [Å^{-1}]	~ 1	~ 1
Total gas phase scattering cross section [Å^2]	5[a]	297[b]
Energy resolution [meV]	~ 7	0.2
Penetration	$\sim$2–3 layers	0
$\hbar\omega_{max}$ [meV]	~ 200	~ 30

(a) This is the calculated total integral cross section for the scattering of a 100 eV electron from a free Cu atom taken from [5.22].
(b) This is the measured toal integral cross section for scattering of a 60 meV K atom from a free He atom taken from [5.23].

The matrix elements in (5.1) must account for these differences in the nature of the interactions. As we have seen in Fig. 5.2, the He atoms are deflected through large angles by the repulsive interaction. For formal reasons it is assumed in the theory that the interaction is only dominated by the two body potentials $v^{(2)}$, which, because of the comparatively short range of the repulsive forces, has been found to be a good approximation [5.20]. Thus the total potential of

Electron scattering

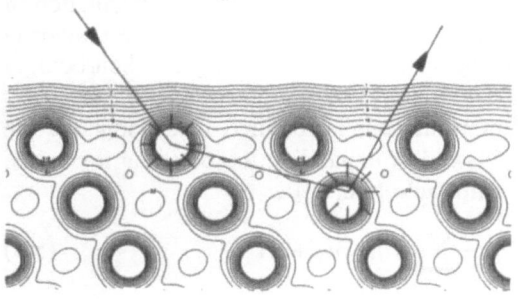

He atom scattering

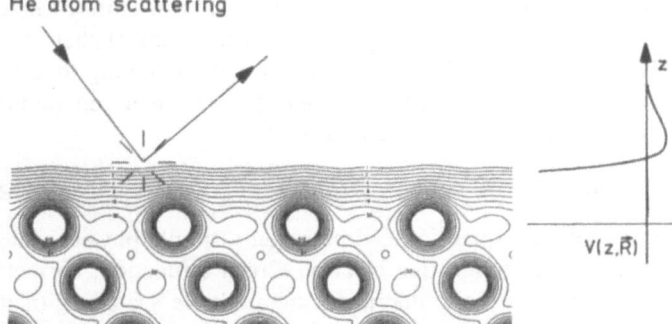

Fig. 5.2. Schematic diagram showing the different nature of the interaction of electrons and He atoms with a crystal surface. Whereas electrons may undergo a complicated multiple scattering involving several layers of the solid, He atoms are deflected far from the top layer and undergo only a single scattering. The contour lines of equal electron density were calculated for fcc Cu(111) cut along the $\langle 112 \rangle$ direction [5.24]. The potential governing the He–solid interaction is indicated schematically at the bottom right

the atom at position r can be expanded in terms of the displacements u_j of the individual atoms at sites j with equilibrium position r_{j0} [5.9]:

$$V_{(r,\{r_j\})} = \sum_j v^{(2)}(r - r_{j0}) + \sum_j \nabla v^{(2)}(r - r_{j0}) \cdot u_j + \dots \qquad (5.2a)$$

$$= V_{st} + v_{dyn} \; . \qquad (5.2b)$$

The first term, which is called the static potential, does not depend on the small displacements (typically 0.1 to 0.2Å) and therefore is much larger than the second term. V_{st} is used to describe the elastic diffraction from the surface. The second smaller term v_{dyn} provides for a linear coupling to the phonons and accounts for single phonon processes. As we shall discuss in greater length in Sect. 5.5.1 this separation of the interaction into a large static and a small dynamical term makes it possible to apply the Distorted Wave Born Approximation to the calculation of the matrix elements. Thus in the matrix elements of v_{dyn} in (5.1) the scattering wave states $|\psi_i^{(+)}\rangle$ and $|\psi_f^{(-)}\rangle$ may be calculated by solving the Schrödinger equations for the static potential V_{st}.

In electron scattering a different formalism is required to account for the multiple scattering of the electron. Instead of expanding the potential it is more expedient to expand the scattering amplitude [5.6, 23].

$$f(\boldsymbol{k}_f, \boldsymbol{k}_i, \{\boldsymbol{r}_j\}) = f_0(\boldsymbol{k}_f, \boldsymbol{k}_i, \{\boldsymbol{r}_{j0}\}) + \sum_j \nabla f_0 \cdot \boldsymbol{u}_j + \dots , \tag{5.3}$$

since this quantity does not depend on the exact specification of the interaction with the individual atoms of the solid. As in (5.2), the first term is much larger since it describes the scattering from the rigid lattice with atoms at their equilibrium positions. The smaller second term accounts for the coupling to single phonons. The final expression for the matrix element ($\langle\ \rangle$) for the electron–phonon coupling to be inserted into (5.1) is more complicated:

$$\langle\ \rangle = \langle k_f|(G + GT_0G)|\nabla v^{(2)}(\boldsymbol{r} - \boldsymbol{r}_j)|(1 + GT_0)|k_i\rangle . \tag{5.4}$$

This rather formidable looking expression can be given a simple physical interpretation. First we note that G is the free electron Green's function, corrected for the inner potential, and T_0 is the T-matrix for the multiple scattering of the electron from the nuclei at the equilibrium sites. Moreover the factors containing these operators imply the integration over coordinate space so as to include each possible path of the electron. Thus the combination $(1 + GT_0)|k_i\rangle$ describes the propagation of the electron during its multiple scattering path to the point where it interacts with one particular atom. This term has to be calculated in the same way as in LEED theory. The inelastic interaction of the electron with the atom is governed by an analogous term as for He scattering with, of course, a potential appropriate to electron scattering. The combination $\langle k_f|(G+G\,T_0\,G)$ describes the propagation of the electron from the point of interaction to the detector. The propagation may either be direct, $\langle k_f|G$, or via multiple scattering, $\langle k_f|(GT_0G)$. The calculation of these complicated propagators has been extensively studied since they also appear in the theory of photoemission. In Sect. 5.5.2 we will discuss in more length some consequences of the multiple scattering on the intensities of inelastically scattered electrons.

5.2 Kinematics

The dynamical theory of single phonon inelastic scattering from a periodic lattice requires that both the total energy of the entire system, consisting of the free particle and the whole crystal, as well as the total momentum projected on to the surface plane are conserved:

$$\hbar\omega(Q) = \frac{\hbar^2}{2m}(k_{\mathrm{f}}^2 - k_{\mathrm{i}}^2) \tag{5.5}$$

and

$$\Delta K = G + Q = k_{\mathrm{f}} \sin \theta_{\mathrm{f}} \begin{pmatrix} \cos \phi_{\mathrm{f}} \\ \sin \phi_{\mathrm{f}} \end{pmatrix} - k_{\mathrm{i}} \sin \theta_{\mathrm{i}} \begin{pmatrix} \cos \phi_{\mathrm{i}} \\ \sin \phi_{\mathrm{i}} \end{pmatrix} . \tag{5.6}$$

Here G is a reciprocal lattice vector, $\hbar\omega$ is the energy of the phonon and Q its wave vector projected onto the surface plane. The polar angles θ_{i} and θ_{f} are measured from the surface normal z (see Fig. 5.1) and ϕ_{f} and ϕ_{i} are azimuthal angles in the surface plane, usually chosen to be zero. Equation (5.6) implies that the momentum of the phonon Q is always measured to the nearest reciprocal lattice vector. Throughout we will use the convention that a positive $\hbar\omega(Q)$ corresponds to annihilation and a negative $\hbar\omega(Q)$ to creation of a phonon. Similarly we will refer to positive Q as "forward-directed" and negative Q as "backward-directed" phonons. As an aside we note that the same conservation equations hold for the scattering of any two wave motions from each other. For example the scattering of light from a sound wave in a condensed medium (Rayleigh scattering) also obeys the same conservation equations, only in this special case the change in energy of the light beam can be neglected. Thus these conservation equation imply a *coherent* interaction just as in ordinary elastic diffraction, with the only difference being that in inelastic scattering all wave vectors Q are possible whereas in inelastic diffraction only the fixed reciprocal lattice vectors are involved.

In the usual experiment, values of $|k_{\mathrm{f}}|$ are determined from an energy charge spectrum measured for a fixed $|k_{\mathrm{i}}|$ and θ_{i}, where k_{f} and k_{i} are in the sagittal plane. The observed peaks in the spectrum provide the phonon energy via (5.5). Then for each value of the observed magnitude of k_{f}, ΔK is determined from (5.6) and lies on the intersection of the surface plane and the sagittal plane. In order to compare the experimental results with the theoretical dispersion curves the crystal is usually lined up so that ΔK is parallel to one of the symmetry directions as in Fig. 5.3.

For the special case of sagittal plane scattering the conservation equations can be conveniently visualized with the aid of an Ewald diagram in momentum space, which can easily be adapted to surface scattering. Figure 5.3 shows a side view of the Ewald diagram for sagittal plane scattering of He atoms from the LiF(001) surface along the $\langle 100 \rangle$ direction. The dashed vertical rods are projections of the reciprocal lattice vectors in the z direction out of the surface plane. The dashed circle shows the final momentum states accessible in elastic scattering in the sagittal plane and its intersection with the vertical rods shows the possible directions for elastic diffraction. In the example shown the angle between the incident and scattered beam, with directions indicated by the solid straight lines, is held fixed at $90°$. The points (1) to (6) along the k_{f} vector indicate a number of different possible inelastic surface phonon events. By drawing a horizontal line to the nearest reciprocal lattice vector the value of Q for each of the events can be determined. These different possible events can be separated if

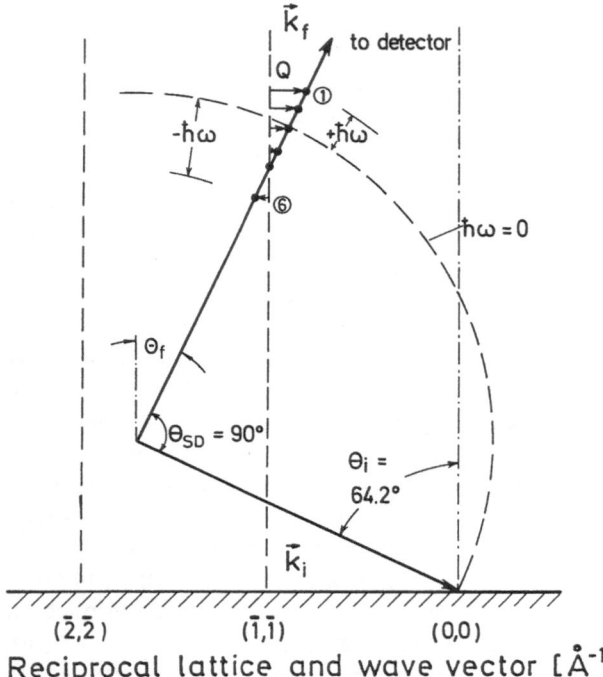

Reciprocal lattice and wave vector [Å^{-1}]

Fig. 5.3. Diagram showing the side view of an in-plane Ewald diagram along the $\langle 100 \rangle$ direction of a LiF(001) crystal [5.26] for an Umklapp phonon process centered at the $(\bar{1}, \bar{1})$ reciprocal lattice vector. The angle between incident and final beams is $\theta_{SD} = 90°$ and the incident beam angle is $\theta_i = 64.2°$, the incident wave vector is $k_i = 6.0 \, \text{Å}^{-1}$. The points (*1*) to (*4*) and (*6*) indicate different inelastic events observable at this angular setting. Only 6 events are shown since depending on the geometry only certain discrete phonons are kinematically accessible. Their resolution requires a measurement of the final velocity distribution

the final momentum k_f is measured. Note that, in the example shown, annihilation events (1) and (2) have momentum transfer Q in the forward direction, while the creation events (3) to (6) have momentum transfer both in the forward and backward directions. Event (3) is elastic with a forward Q. A similar diagram can be drawn for electrons, in which, however, $\hbar\omega$ is much smaller in relation to the length of the vector depicting k_f.

As noted earlier, if the scattering angles are restricted to the sagittal plane, only sagittally polarized modes can be excited. Shear horizontal phonons near the zone boundary may however become accessible with the same geometry when the momentum transfer extends beyond the Brillouin zone boundary, as illustrated in Fig. 5.4 for the (111) surface of a fcc crystal. In the example shown Q is perpendicular to ΔK and the polarization vector of a shear horizontal mode is then parallel to ΔK and a coupling is possible. Such phonons have as yet not been detected in this way. Another way to detect shear horizontal phonons is to rotate the detector out of the sagittal plane. Such experiments have only been attempted in a few cases [5.27] and the kinematic constraints appear to be rather complicated.

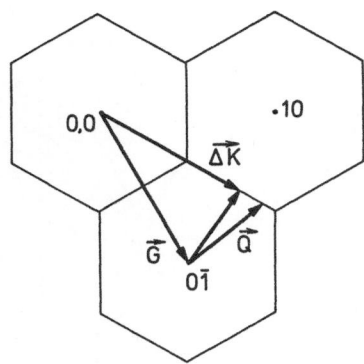

Fig. 5.4. Schematic diagram showing momentum vectors in the surface plane involved in a proposed experiment for detecting shear horizontal phonons in a sagittal plane arrangement. With ΔK in the sagittal plane Umklapp phonons centered at G with Q nearly perpendicular to ΔK should become observable

A more useful representation of the sagittal plane Ewald diagram for helium atom scattering is to project all the possible $(\omega, \Delta k)$ values which can be observed for a given k_i, θ_i and θ_f onto an extended zone diagram in $(\hbar\omega, \Delta k)$ space. The resulting curve is called a 'scan curve' and depends on the geometrical constraints of the apparatus. From the conservation equations we derive the following expression for the scan curve:

$$\frac{\hbar\omega(\Delta k)}{E_i} = \frac{\sin^2\theta_i}{\sin^2\theta_f}\left[1 + \frac{\Delta K}{K_i}\right]^2 - 1 . \tag{5.7}$$

Thus in $(\hbar\omega, \Delta K)$ space the scan curve is a parabola with origin at $(-E_i, -K_i)$ and with a curvature which depends on the angles θ_i and θ_f and the energy E_i. Figure 5.5 shows scan curves for two different arrangements, one in which $\theta_i = $ constant and θ_f is varied and one in which $\theta_i + \theta_f = 90°$. In addition, a typical dispersion curve is shown for a Rayleigh mode. Note that in both arrangements the scan curve $\theta_f = \theta_i$, which corresponds to specular scattering, passes through the origin of the dispersion curves. The intersections of the scan curve with the Rayleigh mode dispersion curve determine the energies at which maxima would be expected to appear in an energy loss spectrum measured for the given angles. The corresponding ΔK values can also be read from the diagram.

A number of interesting special situations can be seen from Fig. 5.5 for the case of a fixed θ_i. For $\theta_f = 0°$ the parabolas become vertical lines, extending upwards from k_i. Thus the energy losses provide all values of $\hbar\omega$ for constant $Q = K_i$ and this is called a 'constant Q scan'. By changing θ_i the value of Q can be varied. One problem with the constant Q-scan mode of operation is that in order to probe the dispersion curve close to the origin, θ_i and θ_f go to zero, corresponding to an angle between incident and final direction denoted by θ_{SD} of $0°$. For practical reasons it is hard to build an apparatus for which measurements with $\theta_{SD} < 50°$ are possible. Thus for He the constant Q scan is only feasible in the vicinity of a diffraction peak. The more common mode of operation, and that used in all the experiments to be discussed here, is a constant θ_{SD} scan in which, by rotating the crystal, θ_i is varied, and $\theta_f = \theta_{SD} - \theta_i$. The scan curves for this case are shown in Fig. 5.5b. They are somewhat different than for fixed θ_i

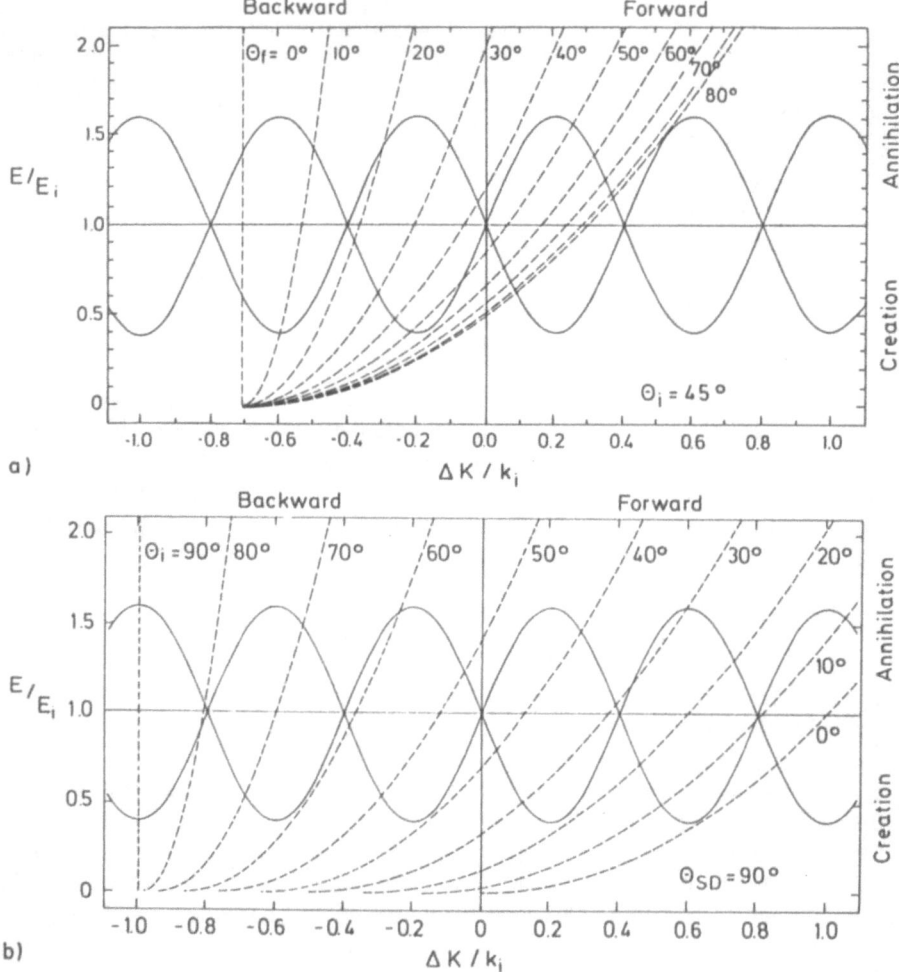

Fig. 5.5. Extended zone diagram showing some typical scan curves (- - -) for He scattering for (a) an arrangement in which the incident angle θ_i is fixed at 45° and θ_f is varied and for (b) an arrangement in which the angle between incident and final beams is $\theta_{SD} = \theta_i + \theta_f = 90°$. The dark solid sine curves (——) represent a possible Rayleigh mode dispersion curve. The intersections with the scan curves indicate possible inelastic scattering events

since K_i goes from $K_i = k_i$ at $\theta_i = 90°$ to $K_i = 0$ at $\theta_i = 0°$. A value $\theta_{SD} = 90°$ is used in most measurements.

From Fig. 5.5 it is apparent that it is kinematically possible to observe both backward and forward phonons in the same time of flight spectrum. To see both types of features in the vicinity of the specular peak, one would preferably choose θ_i smaller than 45°; and in the other mode of operation θ_{SD} should be chosen smaller than 90°, to get a scan curve which is steeper. The intensities of the two events will depend on a number of kinematic and dynamic factors to be

120

discussed later. Several examples of the time-of-flight spectrum showing both forward and backward phonons have been reported [5.28, 29].

The scan curves for electrons are much simpler than those of He atoms since $\hbar\omega(Q) \ll E_i$ [see (5.7)]. To account for this the ordinate of the scan curve in Fig. 5.5 would have to be expanded by a factor proportional to the ratio of the masses of the He atom to that of an electron (factor 10^4). Thus the scan curves will pass through the dispersion curves nearly vertically at positions of ΔK depending only on the scattering angles. Consequently for electrons, the constant Q condition is always satisfied.

For certain special angles a kinematic enhancement of the inelastic intensitiy has been predicted [5.30] and this effect is therefore called 'kinematic focussing'. In Fig. 5.5a we note that at the angles $\theta_i = 42°$ and $\theta_i = 80°$ the scan curves are tangential to different segments of the Rayleigh dispersion curve. In Fig. 5.5b the same occurs at $\theta_i = 15°$, $32°$ and $46°$ for annihilation and at $\theta_i = 21°$, $40°$ and $60°$ for creation. Thus at these angles an extended section of the dispersion curve is seen by the detector and the following additional kinematic condition is fulfilled:

$$\frac{\partial \omega_d(\Delta K)}{\partial \Delta K} = \frac{\partial \omega_s(\Delta K)}{\partial \Delta K} , \tag{5.8}$$

at special points $\bar{\omega}$, $\Delta\bar{K}$ in the $(\omega, \Delta K)$ plane. Here $\omega_d(\Delta K)$ is the dispersion curve in an extented zone scheme and $\omega_s(\Delta K)$ the scan curve (5.7). The intensity enhancement at the kinematic focussing angles can be calculated by integrating the differential reflection coefficient [see (5.1)] using ω in place of E_f along the scan curve [5.32]:

$$\frac{dR}{d\Omega} = \int_{\omega = \omega_s(\Delta K)} \frac{d^2 R}{d\omega d\Omega} d\omega . \tag{5.9}$$

For further calculation it is convenient to convert to ΔK space:

$$\frac{dR}{d\Omega} = \int_{\omega_s(\Delta k)} A(\Delta K) \delta\left[\omega_s(\Delta K) - \omega_d(\Delta k)\right] \frac{\partial \omega_s}{\partial(\Delta K)} d\Delta K , \tag{5.10}$$

where $A(\Delta K)$ stands for all the factors outside the δ function in (5.1) and is assumed to be a smooth function of ΔK. Expanding ω_s and ω_d in ΔK about the kinematic focussing points $(\bar{\omega}, \Delta\bar{K}, \bar{\theta})$ one obtains for the usual arrangement with $\theta_i + \theta_f = 90°$

$$\frac{dR}{d\Omega} = \frac{\bar{\omega}_d A(\Delta\bar{K}) \tan^{1/2} \theta_f}{[\omega_f(1 + k_i/k_f)(\bar{\omega}_d - \bar{\omega})(\bar{\theta}_f - \theta_f)]^{1/2}} . \tag{5.11}$$

This expression exhibits an inverse square root singularity at the special angles $\bar{\theta}_f$. Of course the finite resolution of the apparatus will tend to smear out the singularity. These singularities are routinely seen in the scattering from alkali halides [5.28, 31, 32] and have also been observed in metals [e.g. Al(111)] [5.33]. They show up particularly strongly at azimuthal angles away from one of the symmetry directions since the large intensity from diffraction peaks is greatly

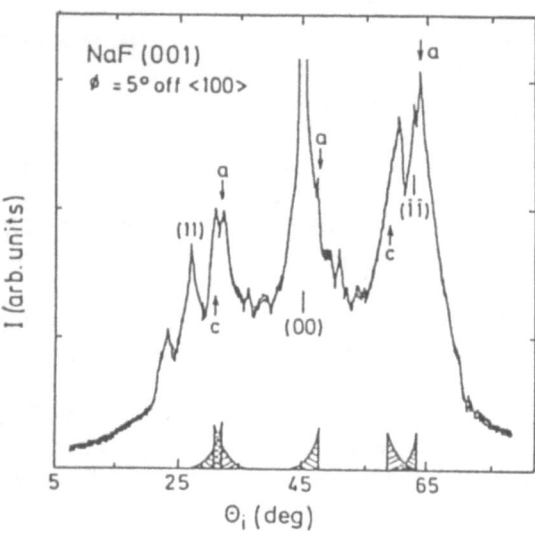

Fig. 5.6. Angular distribution of He (E_i = 18.8 meV, k_i = 6.0Å^{-1}) scattered from NaF(001) at an azimuth ϕ = 5° rotated away from the $\langle 100 \rangle$ direction [5.32]. Incident angles corresponding to kinematic focussing by annihilation (a) and creation (c) of phonons are indicated. Such off-symmetry distributions may allow the kinematic focussing features along symmetry directions to be unambiguously identified without time of flight analysis

reduced. An angular scan for NaF(001) measured at $\phi_i = \phi_f = 5°$ from the $\langle 100 \rangle$ direction is shown in Fig. 5.6.

The kinematic focussing condition provides an alternative method for determining dispersion curves of discrete modes. In this method all the scan curves for the angles at which enhancement occurs are plotted in a reduced zone diagram. For a corrugated crystal surface in which many zones are probed the total number of such curves can be quite large and contain many tangential segments of the Rayleigh dispersion curve. In this way a fairly complete dispersion curve could be determined for NaF(001) [5.31]. Recently *Nichols* and *Weare* [5.34] have suggested that some of the structures seen in the angular distributions measured with He-LiF(001) could be attributed to kinematic focussing involving pseudosurface branches within the acoustic bulk bands. Their interpretation must be questioned because of the low density of bulk modes used in their calculations. *Avila* and *Lagos* [5.35] have proposed that out-of-plane kinematic focussing could also be used to measure dispersion curves. Out-of-plane kinematic focussing has also been observed experimentally [5.28]. Unfortunately, there are a number of other effects (see Sect. 5.3) that can also lead to structures in the angular distribution. These have to be carefully filtered out in order to identify the true kinematic focussing angles. These kinematic effects are not expected to occur in electron scattering because of the different behavior of the scan curves.

5.3 Helium Scattering

5.3.1 General Considerations

In discussing atom scattering, the first question which arises concerns the choice of the probe particles. In order to suppress multiphonon events (see Sect. 5.5.1) light particles are to be preferred. Of the possibilities H, H_2, ^{3}He and ^{4}He, ^{4}He has definite theoretical and experimental advantages. The closed shell structure of helium, the resulting weak binding energy with the surface, as well as the small atomic radius are big advantages for the interpretation of the results. The experimental advantages are that (1) nozzle beams of ^{4}He provide the sharpest velocity distributions, which for example cannot be achieved for ^{3}He [5.36] and the other light particles, and (2) the residual gas background pressure in the mass spectrometer can be reduced much further than for other gases, especially H_2. Essentially two different approaches have been adopted in the measurement of momentum and energy transfer to surface phonons. In the one approach the incident beam is pulsed and the time-of-flight of the scattered atoms is measured. In the other method first introduced in 1932 [5.37], and now widely used in neutron spectroscopy, the momentum of the scattered atoms is selected and analyzed using an LiF crystal which serves as a grating monochromator.

The latter technique has been further developed by *Mason* and *Williams* [5.38] and successfully used to study surface phonons on clean Cu(001) as well as the surface vibrations of an ordered adsorbed layer of Xe atoms at 16 K [5.39]. In many respects a crystal grating spectrometer is equivalent to the time-of-flight technique. In both cases the experimental energy resolution decreases with increasing final beam energy. This is not a serious problem since the energy half width of the nozzle beam increases roughly proportionally to the beam energy anyway. The biggest advantage of the crystal grating spectrometer is that the apparatus can be made very compact. Probably the main disadvantage is the greater mechanical complexity since in addition to the target manipulator another precise rotational manipulation of the spectrometer crystal is required. Since most results, so far, have been obtained with the time-of-flight technique, this approach will be discussed here in more detail.

Figure 5.7 shows a schematic diagram of a time-of-flight apparatus with the essential components and some important coordinates. The ^{4}He atom is produced in a nozzle source, then chopped into short pulses and scattered off of the target surface and detected by a detector set at some fixed angle θ_{SD} with respect to the incident beam direction. The total flight time of the atoms from chopper to detector is measured and from a set of such time-of-flight spectra for different angles θ_i and θ_f a dispersion curve is obtained. To achieve sufficient time-of-flight resolution of the scattered atoms the flight path L_{TD} should be chosen to be sufficiently long relative to the length of the ionization region, which is of the order of 1 cm. For similar reasons it is desirable to keep the distance between chopper and target L_{CT} as small as possible compared to L_{TD}. In the apparatus

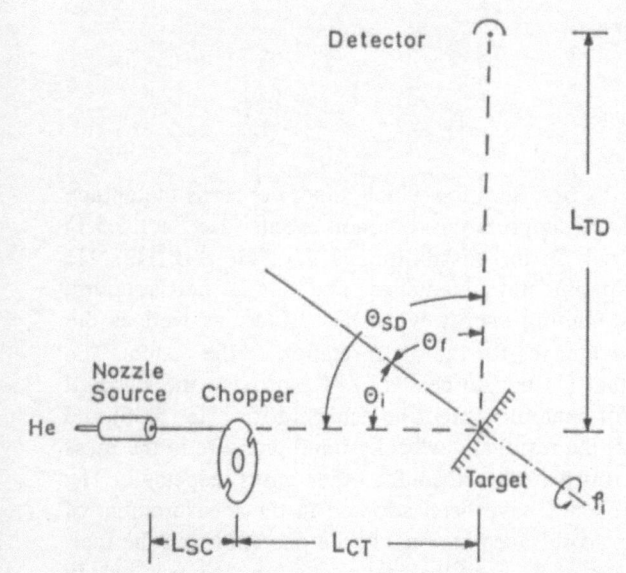

Fig. 5.7. Schematic diagram showing the essential components and important angles and distances of a time-of-flight apparatus

used in Göttingen [5.40, 41] and shown in Fig. 5.8 it has been convenient to have both the source and detector situated external to the target chamber.

An increase in the sensitivity of the time-of-flight apparatus, possibly with some loss in resolution, can be achieved by using pseudorandom chopping [5.42–44]. The duty cycle with this technique is about 50% compared to about 1% for single pulse time-of-flight. A gain in signal to noise ratio is however only expected when the detector background is comparable to or larger than the signal [5.42]. This is not the case in the helium apparatus of Fig. 5.8, but is expected for an apparatus with fewer pumping stages. Pseudorandom chopping is also of considerable advantage in experiments with other beam particles where the detector background can be much larger. The disadvantage of this technique is that the desired time-of-flight spectrum is only obtained after deconvolution of the measurements and depends sensitively on the accuracy of the slits in the chopper and the stability of the rotor frequency.

Table 5.2 provides an extensive compilation of the important operating parameters, dimensions and chamber pressures of the apparatus shown in Fig. 5.8. In the following, the beam source, target chamber and detector are discussed in detail.

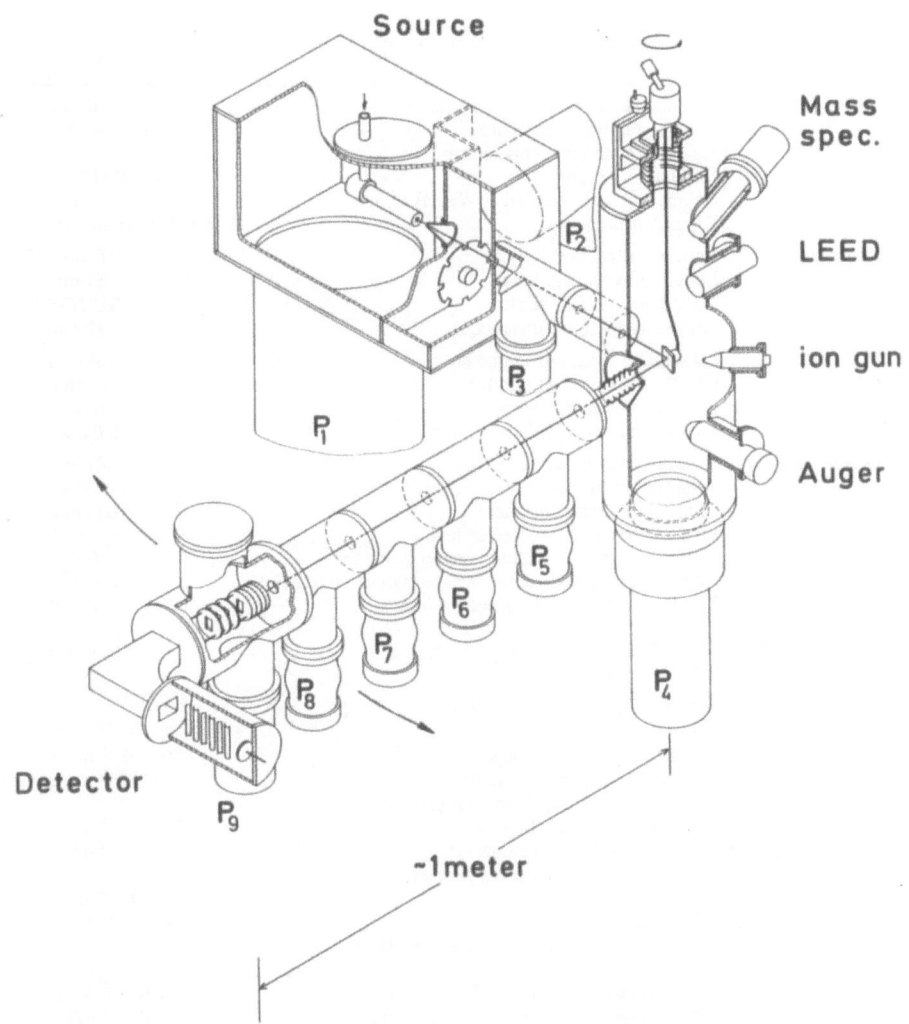

Source

Mass
spec.

LEED

ion gun

Auger

P_2

P_3

P_1

P_5

P_6

P_7

P_8

P_9

P_4

Detector

~1meter

Fig. 5.8. Perspective view of a He atom time of flight apparatus used in Göttingen for measurements of phonon dispersion curves [5.40, 41]. The differentially pumped vacuum stages are denoted by P_1 to P_9. The target chamber is equipped with a residual gas mass spectrometer analyzer, LEED, Auger and an ion gun

Table 5.2. Important apparatus parameters shown in Fig. 5.8

Source	Nozzle-orifice diameter d, nominal	$10\,\mu$m
	Nozzle stagnation pressure	20 bar–200 bar
	Nozzle stagnation temperature	40 K–400 K
	He-beam velocity	644–2037 m/s
	Relative velocity spread (FWHM) $\Delta v/v$	0.01
	Estimated intensity	3×10^{19} atoms/sr sec
Skimmer	Entrance diameter	0.62 mm
	Length: entrance to base	37 mm
	Outside/inside full angles	32°/25°
	Orifice-to-skimmer distance	10 mm
Chopper	Double-slit disk, slit diameter	150 mm
	Double-slit disk, slit width	1.5 mm
	Disk rotational frequency	≈ 300 Hz
	Shutter function (FWHM), effective	$10.6\,\mu$sec
Dimensions	Source–target distance	501 mm
	Target–detector distance	1202 mm
	Chopper–detector distance	1635 mm
	Fixed-source–target-detector angle	90°
	Incident-beam full angular spread, nominal	0.81°
	Angle subtended by detector from target, nominal	0.42°
	Angular resolution, (00) peak, nominal (scanning θ_i)	0.42°
	Angular resolution, (00) peak, measured (scanning θ_i)	$(= 0.2°)$
Pressures	Target chamber P_5, total pressure	
	Base	1×10^{-10} mbar
	Operation, beam chopped	2×10^{-9} mbar
	Operation, beam unchopped	1×10^{-7} mbar
	Target chamber P_5, He partial pressure	
	Operation, beam chopped	2×10^{-9} mbar
	Operation, beam unchopped	1×10^{-7} mbar
	Detector chamber P_9, total pressure	
	Base	2×10^{-10} mbar
	Operation	2×10^{-10} mbar
	Detector chamber P_9, He partial pressure	
	Base	$\approx 10^{-14}$ mbar
	Operation, beam chopped, specular	$\approx 10^{-12}$ mbar
	Operation, beam unchopped, specular	$\approx 10^{-10}$ mbar

5.3.2 Helium Nozzle Beam Source

The atomic beam of helium is formed by an adiabatic expansion from a high pressure (1–300 bar) source chamber through an orifice with a diameter of typi-cally $d = 5$–$30\,\mu$m, as shown schematically in Fig. 5.9 [5.45–48]. In the course of the extremely rapid expansion, with a typical cooling rate of 10^9 K/sec, collisions between the atoms transform the randomly distributed thermal velocities in the source into a motion of uniform velocity directed normal to the orifice. As a

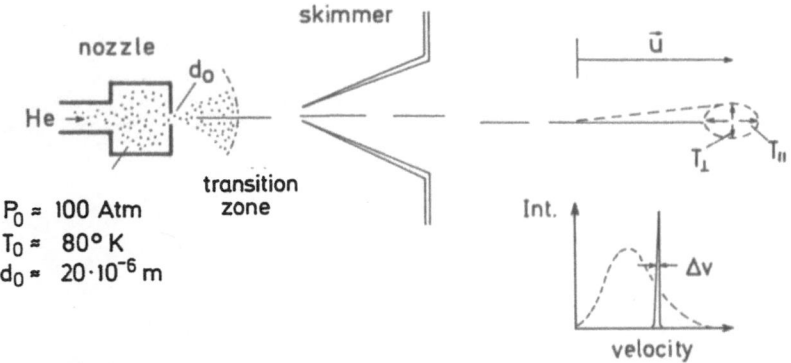

Fig. 5.9. Schematic diagram showing the arrangement used to produce nozzle beams and the velocity collimation of the fully expanded beam. The meaning of the different temperatures $T_\parallel$ and $T_\perp$ of the assumed elliptical velocity distribution is also indicated

result of conservation of enthalpy the energy in the directed motion is built up at the expense of the thermal motion in the moving coordinate system. In the extreme case the enthalpy of the thermal motion is completely converted and the temperature approaches $T_\infty = 0$. In this limit the average velocity of the atoms is given by

$$u_\infty = [5kT_0/m]^{1/2} \ . \tag{5.12}$$

For helium, typical values of directed velocity are $v_\infty = 1765\,\text{m/sec}$ ($k = 11.2\,\text{Å}^{-1}$) at $T = 300\,\text{K}$ and $v_\infty = 911\,\text{m/sec}$ ($k = 5.78\,\text{Å}^{-1}$) at $T_0 = 80\,\text{K}$. Equation (5.12) is an excellent approximation for helium under most conditions except at very low source temperatures $T_0 \lesssim 30\,\text{K}$, where cluster formation sets in [5.49]. For other beams, especially of heavier atoms, the cooling may not be so complete because of the energy released in condensation which will tend to heat up the beam and increase the average velocity. According to (5.12) the velocity of the beam is easily varied by simply changing the source temperature. In the apparatus of Fig. 5.8 the source is attached to a closed cycle helium refrigerator so that its temperature can be varied continuously from below 40 K to 300 K and above. In this way the optimal beam energy can be chosen to match the energy of the phonons of interest and thereby achieve optimal resolution.

As noted above, the temperature of the thermal motion in the beam decreases during the expansion until collisions finally cease to occur. The final temperature describing the velocity distribution parallel to the beam direction, with respect to u_∞, is denoted by $T_{\parallel\infty}$. This velocity distribution is characterized by a most probable velocity α,

$$\alpha = [2kT_{\parallel\infty/m}]^{1/2} \ , \tag{5.13}$$

and the "strength" of the expansion is given by the speed ratio S,

$$S = \frac{u_\infty}{\alpha} = [(5/2)T_0/T_{\parallel\infty}]^{1/2} \ . \tag{5.14}$$

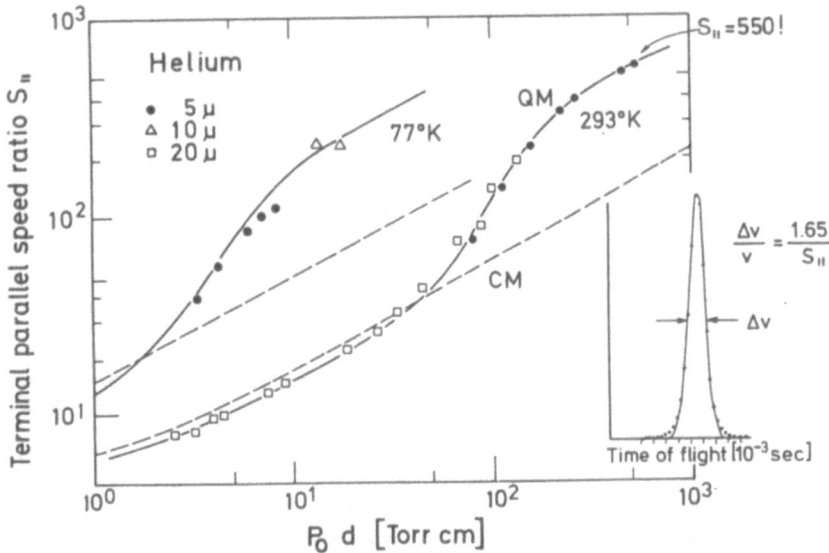

Fig. 5.10. Measured terminal speed ratios [5.46] are compared with Boltzmann equation calculations [5.36] based on classical cross sections (CM) and quantum mechanical cross sections (QM). Because of a zero energy resonance the calculation for the quantum mechanical cross sections predicts much greater speed ratios. The insert shows a typical time of flight spectrum and the definition of the speed ratio. The highest speed ratio achieved so far is $S = 550$ [5.53]

The velocity half width (FWHM) in the beam can easily be shown to be

$$\frac{\Delta v}{v} = \frac{1.65}{S} . \tag{5.15}$$

The conversion of enthalpy in a He nozzle beam can be computed by solving numerically the Boltzmann equation in an approximate way for an assumed interaction potential [5.36]. Figure 5.10 shows some measured and theoretical curves for $S_{\parallel}$ versus $p_0 d$, where p_0 is the pressure in the source and d the nozzle diameter. For a given source temperature, $p_0 d$ is directly proportional to the number of collisions in the expansion. At small $p_0 d$ the helium expansion follows a curve calculated using a classical viscosity cross section, calculated from classical mechanics, which at $T = 80\,\mathrm{K}$ is typically $30\,\text{Å}^2$. However, with increasing $p_0 d$ the cross section shows a sudden increase which has been attributed to a quantum effect known as zero energy resonance [5.36, 50]. This zero energy resonance arises because the ground vibrational state of the He dimer has a bound state energy which is very close to zero. At present the exact bound state energy is not precisely known but is estimated to be about $< 10^{-7}\,\mathrm{eV}$ [5.51, 52]. As a result of this resonance the scattering cross section rises to several $1000\,\text{Å}^2$ as the temperature approaches $0\,\mathrm{K}$. This quantum mechanical enhancement of the cross section has a profound effect on the speed ratios. For example at $T_0 = 300\,\mathrm{K}$ and $p_0 d = 150\,\mathrm{Torr\cdot cm}$ without the quantum effect the speed ratio would only be $S = 80$, but because of the quantum effect they are in fact of the order of

$S = 270$, corresponding to a temperature of $T_{\parallel\infty} = 10^{-2}\,\mathrm{K}$ in the beam. Such a beam has a relative velocity half width of $\Delta v/v \cong 0.006$ corresponding to an energy spread ΔE of about 0.70 meV. An even higher resolution is achieved at $T_0 = 80\,\mathrm{K}$, where for a $p_0 d$ of only 20 Torr·cm, $S = 270$ can also be achieved. This corresponds to a half width ΔE of only about 0.23 meV. The highest speed ratio ever reached is $S = 550$ at $T_0 = 300\,\mathrm{K}$ [5.53]. Note that because of its significantly smaller mass ^{3}He will not have a bound state and consequently does not exhibit a zero energy resonance. Its terminal speed ratios are therefore not expected [5.36] to have the same high values as in Fig. 5.10.

The intensity of a nozzle beam is directly proportional to the pressure in the source and the size of the orifice. A simple rule of thumb predicts that the density in the beam at a distance z from the orifice will be

$$n_z = 0.15 n_0 (d/z)^2 . \tag{5.16}$$

For a typical beam with $d = 20\,\mu$, $T_0 = 80\,\mathrm{K}$ and $p = 200$ bar, n_z at a distance $z = 50\,\mathrm{cm}$ is calculated to be 3×10^7 He atoms/cm^3.

One problem with nozzle beam sources, first encountered in He phonon studies, comes from the presence of a very weak additional beam component with a broad Maxwellian velocity distribution. This undesired component is difficult to detect with ordinary techniques and was only discovered by accident in the first inelastic scattering studies of He from LiF(100) [5.28, 54]. The elastic diffraction of this component from the crystal shows up as what appears to be an inelastic signal comparable to that produced by the phonons. Since its wave vector k_i^* is different from the nominal wave vector k_i it appears at angles removed from the diffraction peak as given by the Bragg condition

$$|G| = k_i^* (\sin \theta_f^* - \sin \theta_i^*) . \tag{5.17}$$

Since the narrow slits permit only a narrow range dk_i^* to arrive at the detector it appears in the time of flight spectra as additional sharp maxima (see Fig. 5.13). These peaks are therefore easily confused with true inelastic structures and for this reason are called "deceptons", "spurions" or "phonions".

Their effect can be calculated by noting that incident atoms with a wave vector k_i^* not equal to the nominal wave vector k_i, will appear as scattered atoms with a wave vector k_f given by [5.28]

$$k_f = k_i^* \left\{ \frac{L_{CD}}{L_{TD}} \left(\frac{k_i^*}{k_i} - 1 \right) + 1 \right\}^{-1} , \tag{5.18}$$

where the apparatus lengths L_{CD} and L_{TD} are defined as in Fig. 5.7. This expression simply accounts for the fact that the observed apparent time delay is attributed to a final wave vector of particles which are assumed to have changed their velocities at the target. From the conservation equations we would then assign the atoms with k_f to the following false frequencies (ω^*) and wave vectors ΔK^*:

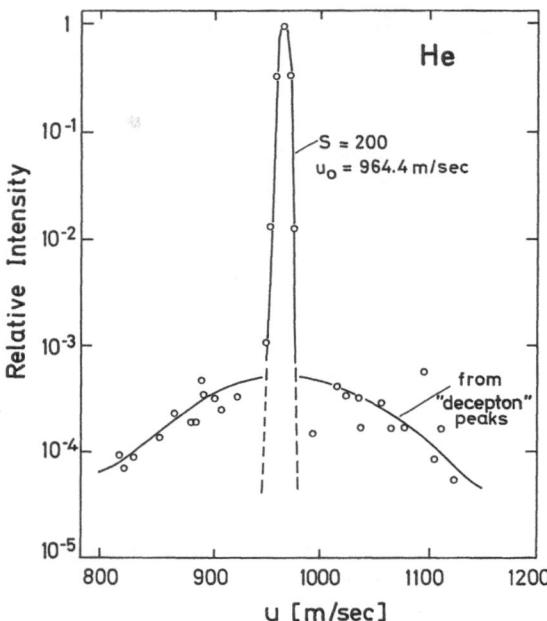

Fig. 5.11. The velocity distribution of the incident beam as determined from an analysis of the scattered beam and decepton peaks [5.54]. The broad decepton velocity distribution is attributed to a diffuse beam originating upstream of the beam chopper. The source temperature was $T_0 \cong 80\,\mathrm{K}$

$$\hbar\omega^* = \frac{\hbar^2}{2m}(k_f^2 - k_i^2) \ , \tag{5.19a}$$

$$\frac{|\Delta K^*|}{k_i} = \frac{k_f}{k_i^*}\sin\theta_f - \sin\theta_i \ . \tag{5.19b}$$

To obtain an effective dispersion curve, θ_f and θ_i are eliminated from (5.19b) by utilizing the Bragg condition (5.17) and the value for θ_{SD}. Note that since the deceptons are produced solely by diffraction they only occur in the vicinity of diffraction peaks. Figure 5.11 shows the result of an evaluation of the intensities of decepton peaks observed in scattering from LiF(001) [5.54]. The broad velocity distribution appears to be Maxwellian. At velocities near that of the nozzle beam the relative intensity of this broad distribution amounts to only about 5×10^{-4} of the major component.

The origin of the beam component producing the deceptons is not fully understood. It does not appear to come from the nozzle chamber since the pressure of helium at the skimmer is considerably greater than in the nozzle beam. Thus any background beam from this chamber should have a larger intensity. This suggests that any residual gas molecules emanating from the nozzle chamber are strongly attenuated by scattering from the intense cone of the nozzle beam. It is probably for this same reason that it has been possible to use oil diffusion pumps in the source chamber without problems from any oil contamination in the beam.

The undesired component must therefore originate in vacuum chambers used for differential pumping further downstream from the source. Assuming this is correct, then the use of other nozzle sources, such as the Campargue source [5.48], which does not use diffusion pumps in the nozzle-skimmer region and which operates at higher residual gas pressures, may have the disadvantage of having larger decepton intensities.

Pulsed nozzles [5.55, 56] have also been used for phonon scattering experiments [5.57, 58]. Since the source is only "on" for short periods (< 1 msec) much larger orifices and consequently much higher peak intensities can be achieved. A gain in average beam flux is not expected, however, since the usual repetition rates of less than 10^2 Hz are much smaller than the 10^3 Hz typically used with mechanical chopping. Moreover the desired pulse durations of about 10 μsec are hard to achieve with pulsed nozzles and mechanical chopping is still required. Furthermore even if sufficiently short pulses would be achieved, they may not be long enough to produce a well developed beam [5.59]. Also, in order to produce a completely expanded beam much larger distances L are needed between source and skimmer [5.60]. This is explained by the fact that the total number of collisions occuring during the expansion scales with L/d and for pulsed nozzles the nozzle diameter d is usually one to two orders of magnitude larger than for continuous nozzles. One big advantage of pulsed beams is that the pulsed background gas will diffuse into the detector chamber after the beam pulse has arrived and will not appear in the time-of-flight spectrum. Another advantage is that, since much larger values of $p_0 d$ can be achieved momentarily, nozzle beams with even larger values of S may become accessible (see Fig. 5.10) [5.61].

5.3.3 Target Chamber

The next important component is the target chamber. For studies of metals and semiconductors, base pressures of $\leq 10^{-10}$ mbar are mandatory in order to provide measuring times of at least 20 minutes before a 10% monolayer is formed from the residual gas. The additional pressure build up from He of about 10^{-8}–10^{-9} mbar is of no concern since He is too weakly bound to be physisorbed at target temperatures of $T_t > 40$ K. In the apparatus of Fig. 5.8 the target chamber is equipped with an electron diffraction device (LEED) to determine the surface structure and a cylindrical mirror analyzer using Auger electron spectroscopy to study the cleanliness of the surface. In addition, a differentially pumped ion sputter is provided for removing surface contamination and a mass spectrometer for residual gas analysis. Because of the need for high angular resolution $\Delta\theta \leq 0.1°$, a very precise manipulator having a θ rotational accuracy of $\delta\theta \leq 0.01°$ is required. This manipulator also provides x, y, z translations as well as rotations about two axes orthogonal to z (azimuth and tilt). The target holder is usually provided with a heater for annealing the crystal and a cooling arrangement to facilitate scattering experiments at low target temperatures of 25–100 K. The θ rotation of the detector together with the time-of-flight tube is achieved in two different ways as shown in Fig. 5.12.

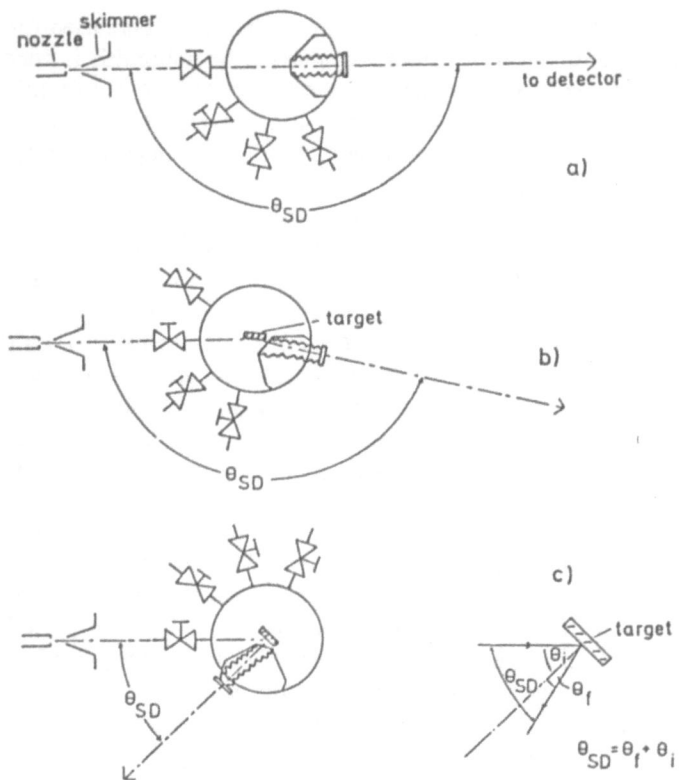

Fig. 5.12a–c. Schematic diagram of the bellows and valve arrangement used to change scattering angles [5.40]. Three different positions of the target chamber and detector time of flight tube assembly are shown. By a combination of continuous rotations and switching entrance ports to the target chamber it is possible to probe continuously a wide range of angles θ_{SD} from 50° to 180° and greater

Small continuous angle adjustments of up to $\pm22°$ are made possible by a bellows. Additional large angle θ_{SD} changes are achieved by rotating the entire target chamber in steps of 36° each in order to permit the primary beam to enter the target chamber via different entrance ports. During the changes of entry ports the target and chopper chambers are isolated by valves and the entire procedure takes less than one hour without a significant disturbance of the target chamber vacuum.

5.3.4 Detector

The detector consists of an efficient electron bombardment ionizer followed by a mass spectrometer to separate off ions from the residual gas background. For the light He ions a simple permanent magnet with radius 5–10 cm can be used for mass selection and since the mass resolution need not be large it is possible to accept a beam with a large area of typically $3 \times 8\,mm^2$ [5.62]. A major design consideration of such an apparatus is the suppression of the He partial pressure in the detector chamber. This is achieved by using a large number of pumping stages as shown in Fig. 5.8. Experience has shown that 9 stages between source and detector are enough to reduce the residual helium partial pressure at the detector to about 10^{-15}–10^{-16} Torr (10–1 atoms/cm^3). Several differential pumping stages are placed between the source and target chambers (1) to prevent hydrocarbons emanating from the diffusion pumps used to evacuate the source chamber from diffusing into the target chamber but also (2) to prevent too large a He pressure build up in the target chamber which would otherwise produce a weak molecular beam of He atoms effusing into the detector. It is interesting to note that so far there is no evidence that any hydrocarbons are picked up by the beam in one of these pumping stages and deposited onto the crystal surface.

5.3.5 Typical Measurements and Resolution

To facilitate their interpretation, measured time-of-flight spectra are first transformed to an energy change abscissa. This has two effects on the spectra [5.28]. Firstly, the change in scale leads to a distortion of the shape of the distribution. From conservation of energy we calculate

$$\frac{\Delta\hbar\omega}{E_i} = -2\left\{1 + \frac{\hbar\omega}{E_i}\right\}^{3/2}\frac{\Delta t}{t_{el}} , \tag{5.20}$$

where t_{el} is the time of flight for elastic scattering and Δt is a measure of the time-of-flight uncertainty. If $\hbar\omega = 0$, then (5.20) yields $\Delta\hbar\omega/E = -2\Delta t/t_{el}$. For inelastic scattering with finite $\hbar\omega$ (5.20) predicts, however, that the resulting relative energy spread $\Delta\hbar\omega/E_i$ approaches zero for extreme creation events ($\hbar\omega/E_i \approx -1$) and has a maximum of about $\Delta t/t_{el}$ for an annihilation event with ($\hbar\omega/E_i \approx +1$). Secondly we note that the intensities will also be affected by the Jacobian in the transformation of the distributions

$$f_\omega(\omega) = f_t[t(\omega)]\frac{dt(\omega)}{d\omega} . \tag{5.21}$$

The Jacobian is given by

$$|\frac{dt}{d\omega}| = \frac{\hbar t_{TD}^3}{m L_{TD}^3} , \tag{5.22}$$

where t_{TD} is the inelastic time of flight and L_{TD} is the distance from target to detector (see Fig. 5.6). Both of these effects can be seen in parts (a) and (b) of Fig. 5.13. There, the energy resolution is indicated by intervals shown by vertical lines at the bottom. The Jacobian leads to a relative enhancement of creation event amplitudes and a relative suppression of annihilation event amplitudes in the energy loss distribution. Unfortunately the Jacobian also tends to amplify strongly the noise of the time of flight spectra in the region of extreme creation energy losses. The two effects make the resolution better for creation

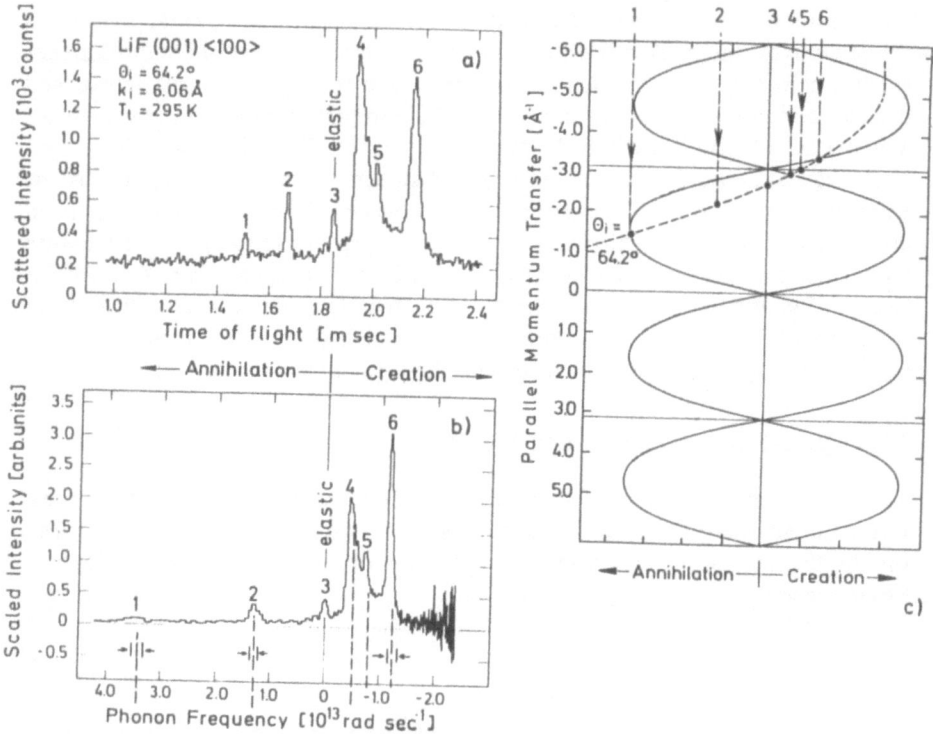

Fig. 5.13. (a) TOF trace taken under typical conditions. Maxima due to three creation events – peaks 4–6 – and two annihilation events – peaks 1 and 2 – appear to be present under this incident angle. In fact, peak 2 may be attributed to a "decepton". (b) The same data transformed to a plot of signal intensity as a function of phonon frequency (energy transfer). As a result of the transformation the energy resolution varies with phonon frequency and estimated values are shown below the different peaks. Note also that the Jacobian factor greatly affects the relative peak heights. (c) The locations of the maxima are interpreted with the aid of a scan curve (- - -) for $\theta_i = 64.2°$ in an extended zone diagram. From the intersection of the energy loss peak locations with the scan curve the corresponding momentum transfers are determined

than for annihilation. Figure 5.13c illustrates the assignments of peaks in the energy loss distribution to the dispersion curve in the case of LiF(001). The peaks 1–6 correspond to those in Fig. 5.3. Note that peak 2 is due to a decepton and peak 3 is due to a incoherent diffuse elastic scattering from surface defects.

In discussing the energy and momentum resolution of a time-of-flight apparatus it is useful to distinguish between the following two effects [5.63]: (1) the smearing resulting from the time of flight measurement and (2) the kinematical effects, such as kinematical focussing, resulting from the finite velocity distributions and angular resolution. Kinematic focussing can be suppressed by choosing appropriate scattering angles such that the scan curve crosses the dispersion curve as close to right angles as possible. For this reason it is customary to probe the dispersion curves on either the annihilation-backward or creation-forward branches (see Fig. 5.5) adjacent to the specular peak. In the following analysis of resolution factors we will implicitly assume this to be the case.

In the time-of-flight spectrometer the measured total time of flight of a single atom consists of four contributions:

$$t_{TOF} = t_{chop} + t_{CT}(v_i) + t_{TD}(v_f) + t_{ion}(v_f) \, , \tag{5.23}$$

where the following effects are taken into account, respectively (1) the finite time of the pulse transmitted by the chopper slit, (2) the flight time to pass from chopper to target, (3) the flight time from the target to detector and (4) the finite time of passage through the ionizer before ionization. The spread resulting from the first term is simply the chopper gate time Δt_{chop}. The spread from the second term is given by

$$\Delta t_{CT} = \frac{L_{CT}}{v_i} \frac{\Delta v_i}{v_i} \, , \tag{5.24}$$

where Δv_i is the half width (FWHM) of the primary beam velocity distribution. The contribution from the third term is

$$\Delta t_{TD} = \frac{1}{2} \frac{L_{TD}}{v_f} \frac{\Delta E_f}{E_f} \, , \tag{5.25}$$

where L_{CT} and L_{TD} are the distances indicated in Fig. 5.6 and finally Δt_{ion} is given by the length of the ionizer:

$$\Delta t_{ion} = \frac{L_{ion}}{v_f} \, . \tag{5.26}$$

The total spread in the time-of-flight distribution can then be calculated from the law of propagation of small errors:

$$\Delta t_{TOF} = [\Delta t_{chop}^2 + \Delta t_{CT}^2 + \Delta t_{TD}^2 + \Delta t_{ion}^2]^{1/2} \, . \tag{5.27}$$

Whereas the first, second and last terms are simply related to mechanical features of the apparatus components, the third term Δ_{TD} depends in a more complicated way on the finite velocity and angular resolution.

To calculate Δt_{TD} from (5.25) we find from conservation of energy,

$$dE_f = dE_i + d\hbar\omega(Q) \, . \tag{5.28}$$

The first term takes account of the inherent spread in velocities of the incident beam

$$dE_i = 2E_i \frac{dv_i}{v_i} \, , \tag{5.29}$$

while the second term takes account of the effect of smearing on the scan curve which leads to an uncertainty in $\hbar\omega(Q)$ called kinematic smearing. It can be estimated by taking the total differential of the scan curve relationship [e.g. (5.7)], for which a given dispersion curve $\hbar\omega(Q)$ can be written as

$$\frac{\hbar\omega(Q)}{E_i} = \frac{\sin^2 \theta_i}{\sin^2 \theta_f} \left[1 + \frac{Q(\hbar\omega)}{K_i} \right]^2 - 1 \tag{5.30a}$$

$$= S[k_i, \theta_i, \theta_f, Q(\hbar\omega)]/E_i . \tag{5.30b}$$

Thus the uncertainty in $\hbar\omega(Q)$ is given by

$$d\hbar\omega(Q) = \frac{\partial S}{\partial k_i}dk_i + \frac{\partial S}{\partial \theta_i}d\theta_i + \frac{\partial S}{\partial \theta_f}d\theta f + \frac{\partial S}{\partial Q}\frac{\partial Q}{\partial \hbar\omega_d}d\hbar\omega(Q) , \tag{5.31}$$

where $\partial Q/\partial \hbar\omega_d$ is the inverse slope of the dispersion curve. Solving (5.31) for $d\hbar\omega(Q)$ yields the following equation for the uncertainty in the phonon energy for a special point on the dispersion curve corresponding to given k_i, θ_i and θ_f:

$$d\hbar\omega = \frac{\left(\dfrac{\partial S}{\partial k_i}dk_i + \dfrac{\partial S}{\partial \theta_i}d\theta_i + \dfrac{\partial S}{\partial \theta_f}d\theta_f\right)}{\left(1 - \dfrac{\partial S}{\partial \Delta K}\bigg/\dfrac{\delta\hbar\omega(Q)}{\partial \Delta K}\right)} . \tag{5.32}$$

Under kinematic focussing conditions this first order approximation leads to a singularity in $d\hbar\omega(Q)$.

Figure 5.14a shows an example of an analytical calculation of the energy resolution for the apparatus of Fig. 5.8 in the investigation of Rayleigh phonons on the Ag(111) surface at $E_i = 17.5\,\text{meV}$ in the vicinity of the specular peak [5.63]. The calculations show three main contributions to the squared half width of the energy loss peak ΔE^2 as a function of the parallel momentum transfer Q: the finite length of the chopper pulse Δt_{chop}, the finite length of the ionizer Δl_{det} and all the kinematic effects. Figure 5.14b shows the different contributions to the kinematics term which includes Δt_{CT} (5.24), dE_i (5.28) and $d\hbar\omega(Q)$ (5.30). The major contribution comes from the finite velocity spread and only a relatively small contribution comes from the angular spread. From Fig. 5.14a we see that the relative contributions from the chopper smearing, detector length smearing

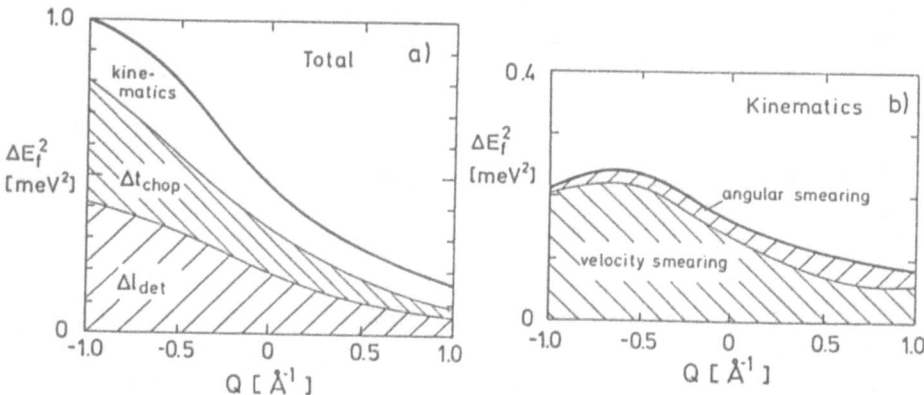

Fig. 5.14. The three major contributions to the half widths of the inelastic peaks are plotted in (a) as a function of the momentum wave vector [5.63]. Only phonons in the annihilation-backward and creation-forward branches are considered. In (b) the two different contributions to the kinematic term in (a) are shown on the expanded scale. The calculations are for $E_i = 17.5\,\text{meV}$, $\theta_{SD} = 90°$ and other operating conditions similar to those of Table 5.2

and kinematic smearing are all nearly the same. This implies that with regard to these three terms the transmission of the apparatus for a given resolution is nearly optimal. However a closer examination of the kinematics suggests that an improvement in the transmission with only a small decrease in the overall resolution could be achieved if in fact the angular resolution could be reduced by increasing the solid angle of acceptance of the detector. The calculations indicate furthermore that the effective total energy resolution is best for extreme creation events where it is less than 0.40 meV. For extreme annihilation events the energy resolution is considerably worse and approaches a value of 1.0 meV.

Finally we note that the uncertainty in the parallel momentum is given by the projection of the energy spread onto the scan curve. The relative resolution of ΔK is usually comparable to that of ΔE.

The quantiative results on resolution discussed here have been checked by a complete Monte Carlo simulation of the apparatus [5.63]. This simulation avoids the assumption implicit in (5.27) that the various contributions may be treated independently. The Monte Carlo simulation yields results for the total energy resolution which closely follow the trend of Fig. 5.14 but for ΔE_f are somewhat smaller (5–10%).

These considerations on resolution have been confirmed in several experiments. Figure 5.15 shows a series of time-of-flight spectra tranformed to an energy loss scale taken along the $\langle 112 \rangle$ direction of the Al(111) surface [5.33, 64]. Aluminum is one of the few surfaces which reveals only a sharp Rayleigh mode and the spectra are not confused by nearby anomalous modes. The scan curves corresponding to the series of spectra plotted in Fig. 5.15 are shown in Fig. 5.16. With the aid of the scan curve a number of features discussed previously can be identified. At $\theta = 43.0°$ the energy loss distributions show additional structures to the left of the sharp peak at $\Delta E = -3.0$ meV which arise from the crossing of the scan curve with the backward-creation branches of the Rayleigh mode in the extended zone diagram. The peak marked 2 is broadened because of the small angle between the scan curve and dispersion curve so that the conditions are near to those for kinematic focussing. In the case of Al(111) it has been possible to test lattice dynamical calculations of the Rayleigh mode very precisely by comparing simulated spectra with time-of-flight spectra taken under kinematic focussing conditions [5.33, 65]. Even small shifts in the frequency were found to have very large effects on the intensities under these kinematic conditions.

With further decreasing incident angles the spectra show broad tails extending to larger energy losses. At these angles a second crossing of the Rayleigh mode dispersion curve is no longer possible and thus these tails are attributed to excitation of modes within the bulk bands which are present in this region.

The change in resolution of the apparatus with increasing Q can also be seen in Fig. 5.15 by noting the trend in peak widths with decreasing scattering angle. At $\theta_i = 43°$ the peak 1 has a half width (FWHM) of 1.3 meV which drops to 0.75 meV at $\theta_i = 29°$. The computer simulation discussed above predicts half widths of 1.5 and 1.0 meV, respectively. The experimental peak half widths are compared with the calculated half widths in Fig. 5.17. One finds reasonably

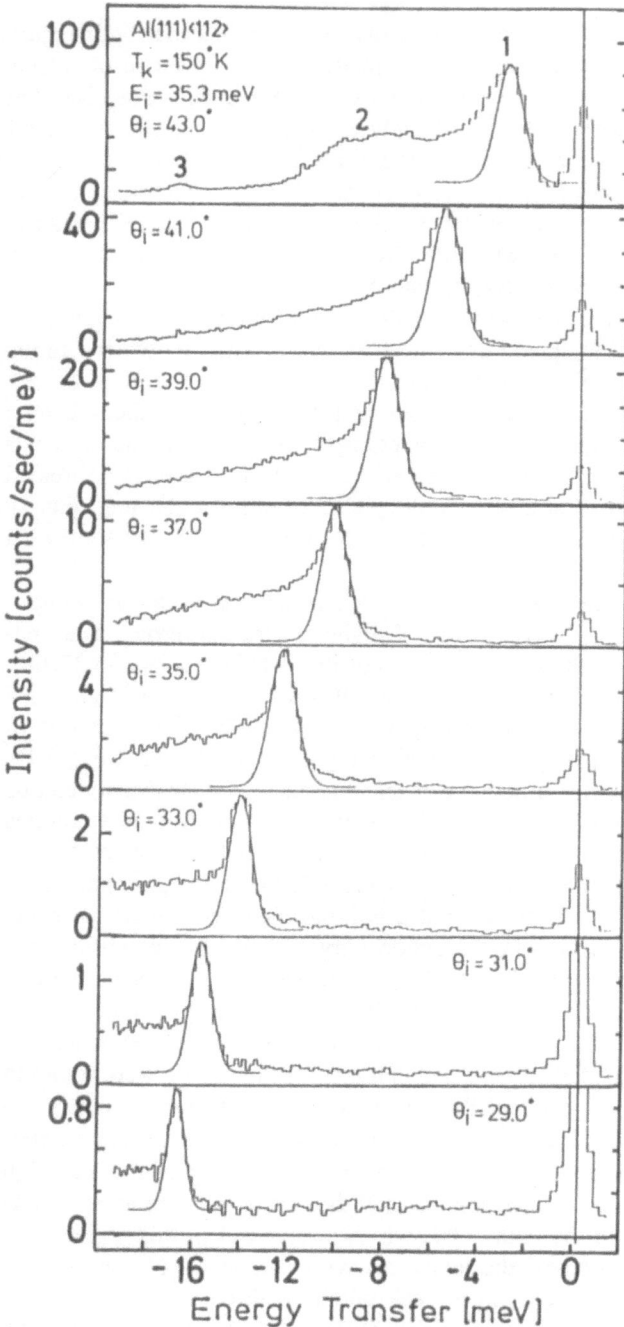

Fig. 5.15. Time of flight spectra taken on the forward-creation branch of the Rayleigh mode dispersion curve of Al(111), $\langle 112 \rangle$, $T_r = 150$ K, $E_i = 35.3$ meV, $\theta_{SD} = 90°$ [5.64]. The numbers in the spectrum taken for $\theta_i = 43.0°$ refer to the corresponding scan curve shown in Fig. 5.16. Note the increase in resolution with decreasing angle corresponding to increasing Q

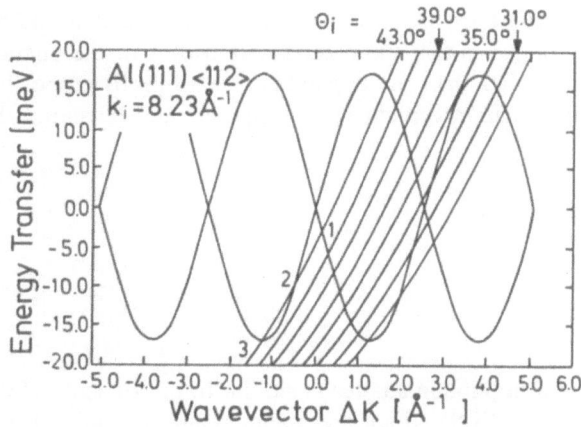

Fig. 5.16. Scan curves and Rayleigh dispersion curves in an extended zone diagram for the measure-ments in Fig. 5.15. The incident energy is $E_i = 35.4\,\mathrm{meV}$ ($k_i = 8.23\,\text{Å}^{-1}$) and $\theta_{SD} = 90°$. Note that the time of flight spectra of Fig. 5.15 do not show features corresponding to dispersion curve scan curve interactions beyond $\Delta K \cong 1.4\,\text{Å}^{-1}$. This is expected since the first order diffraction peak intensities are very small compared to the specular peak ($\sim 10^{-4}$)

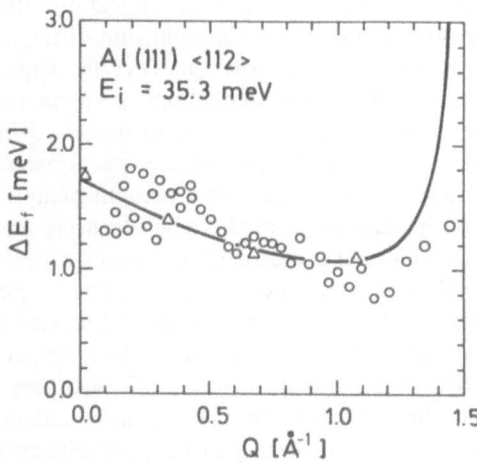

Fig. 5.17. Comparison of experimental Ray-leigh phonon peak half widths taken from the data in Fig. 5.15 with calculations based on Sect. 5.3.5. The singularity at $Q = 1.5\,\text{Å}^{-1}$ is attributed to kinematic focussing (scan curve tangential to dispersion curve; see Fig. 5.16)

good agreement. The large peak at $Q = 1.5\,\text{Å}^{-1}$ in the calculations is attributed to kinematic focussing. The best resolution reported so far is 0.2 meV for $Q > 0$ measured on a Cu(111) surface at a He beam energy of 9.2 meV [5.66].

Finally we note that in all the spectra of Fig. 5.15 a small peak can be seen at $\Delta E = 0$. This was also seen with LiF (Fig. 5.13a) and as discussed there is attributed to elastic incoherent scattering from surface defects. This peak is relatively small compared to other (111) metal surfaces since the inelastic intensities are considerably larger than observed in other crystals. Note also the rapid decrease in the intensity of the Rayleigh peak with increasing phonon wave vector; the reason for this drastic fall-off will be discussed in Sect. 5.5.1.

The good agreement between measured and calculated half widths found in all systems examined so far suggests that the natural half widths of the energy loss peaks are significantly smaller than the apparatus smearing. To probe the natural half widths a significant improvement in the experimental resolution is required.

5.4 Electron Scattering

5.4.1 Apparatus

In principle the apparatus used in phonon studies is similar to the standard electron energy loss spectrometers (EELS) which are commercially available from at least six manufacturers. Since EELS is a standard tool in surface scince its components have been extensively discussed in the literature and for this reason we will not discuss the apparatus components in the same detail as for the He scattering.

Figure 5.18 shows a schematic view of the EELS apparatus and vacuum chamber developed in Jülich for phonon studies [5.67]. In this apparatus the target is mounted on a carousel so that the crystal can be brought into different positions for cleaning and characterization by a simple rotation. Thus the apparatus contains the same LEED and Auger equipment as the helium spectrometer described previously. One reason for the different arrangement to that used in the helium apparatus is the importance of magnetic shielding of the spectrometer components. A second reason is that there is no need for large vacuum pumps.

The electron source is a lanthanum hexaboride cathode. The electrons are first decelerated to energies of less than 1 eV and preselected in a smaller 127° sector field analyzer before the final selection is achieved in the second larger sector field. The electrons are then accelerated to the desired energy and focussed onto the target. The scattered electrons are collected by a similar lens system which slows them down before they pass through two identical 127° analyzers for energy analysis. The detector is a channeltron multiplier. Some machines use only single monochromators for the selector and analyzer or a combination of two monochromators in series as selector and a single monochromator as analyzer.

The major difference in the design of a phonon spectrometer compared to standard EELS is the much higher energy of 50 to 300 eV needed for impact scattering compared to energies of only about 2–5 eV used in the dipole dominated region. Thus the energy resolution ΔE relative to the impact energy is in fact much smaller than in dipole scattering. This places very severe requirements on the electron optics, because of the need for deceleration and acceleration of the electrons by large factors in both the selector and analyzer. Moreover in order to compensate for the loss in intensity in the impact regime it is necessary to increase the acceptance angle of the spectrometer [5.68]. This is achieved by

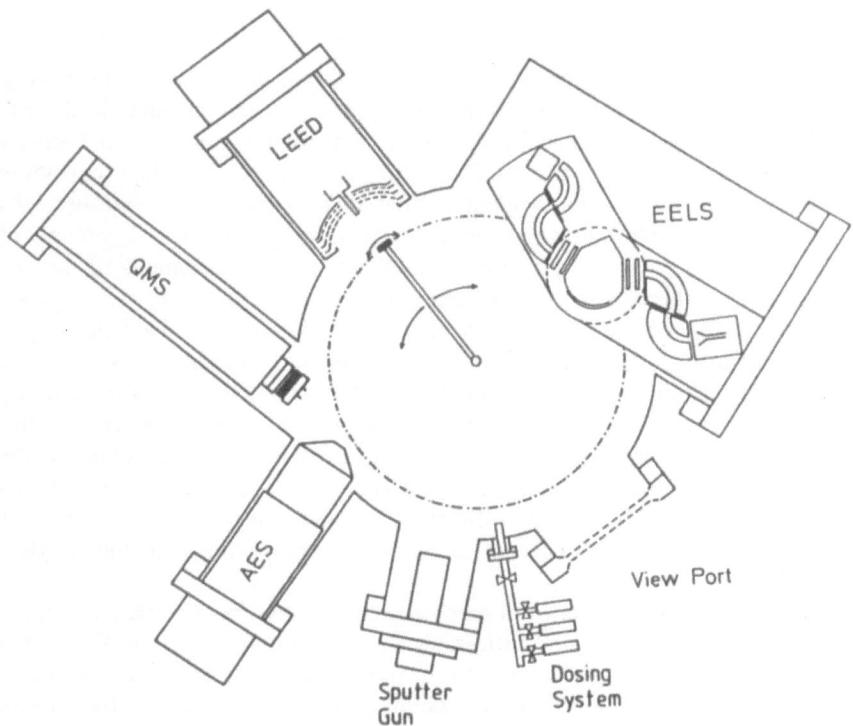

Fig. 5.18. Schematic diagram of a vacuum chamber used for EELS studies of surface phonons [5.67]. The EELS experiments are carried out in one of the positions of the carousel mounted target manipulator. Other positions are used for preparing and characterizing the target surface by LEED, Auger Electron Spectroscopy (AES) etc.

bringing the focal plane of the collector lenses as close to the sample as possible. The electron optics must be carefully designed to prevent electric field penetration through the lenses into the vicinity of the surface. Such field penetration would otherwise affect the acceptance angle in an undefined manner. The proximity of the analyzer and selector monochromators to the target imposes additional space limitations and consequently most spectrometers in use are restricted to angles between the incident and outgoing beams greater than $\theta_{SD} = 90°$. At the resulting large incident angles the incoming beam is more strongly attenuated because of the increased path of the electrons through the surface layers of the sample.

Despite these complications it has been possible to design spectrometers with an energy resolution of the order of 7 meV at $E_i = 100$ eV. The best resolution achieved recently is of the order of 3–4 meV [5.16, 69]. The range of angles which have been probed are typically $\theta_f = 55$–$65°$ and θ_i between $135 - \theta_f$ and $90 - \theta_f$. Compared to He atoms the detection of electrons presents no problems and nearly every scattered electron passing through the analyzer is counted.

5.4.2 Typical Measurements

We recall from the discussion of Sect. 5.2 that the scan curves are especially simple in the case of electron scattering, being straight vertical lines in the extended zone dispersion diagram. In the experiment the incident and final angles are first calibrated by searching for Bragg diffraction peaks and then depending on the arrangement either the incident or final angle is changed according to the desired value of the phonon wave vector Q. Because of small misalignments of the spectrometer axis and that of the manipulator, small adjustments of the target position as well as of the lens voltages are needed to optimize the scattered intensity. Because of the focussing of the electrons onto and from the target, small uncertainties in the trajectories of the order of $1°$ can easily arise in the complicated electron trajectories through the lenses. For this reason the angular resolution is also difficult to determine accurately. These problems are accentuated by the need to change the incident energy frequently in the course of the measurements. The necessary readjustments of the lens voltages can easily lead to small changes in the angular collimation. In addition for kinematic reasons the effective apertures are expected to be linearly proportional to the incident energy.

Figure 5.19 shows two series of measured energy loss spectra taken along the $\overline{\Gamma}-\overline{X}$ direction of Ni(100) [5.67,70]. In each set of measurements the final angle and incident beam energy are held fixed and the incident angle was adjusted to the desired wave vector. As in the helium scattering experiments the spectra all show a well defined peak of zero energy transfer due to elastic scattering from defects. The half width (FWHM) of this peak amounts to about $75\,\mathrm{cm}^{-1}$ or $9.3\,\mathrm{meV}$. The loss peaks are only slightly broader. By fitting the peaks to Gauss distributions a relative accuracy of $\pm2\,\mathrm{cm}^{-1}$ ($\pm0.428\,\mathrm{meV}$) in the determination of the locations of strong peaks is claimed [5.67]. The losses at smaller energy transfers are attributed to the Rayleigh mode S_4 whereas the larger energy transfers are attributed to resonances R_5 and R_6 in the bulk band region. Further discussion of these energy loss spectra will be taken up in Sect. 5.5.2.

5.5 Intensities

The differential reflection coefficient $d^2R/dE_f d\Omega_f$ is defined experimentally in terms of measurable quantities in the following way [5.19]:

$$\dot{N}(\Omega_f, E_f) = n_i \cdot v_i \cdot A_i \cdot \alpha \cdot \frac{d^2R}{dE_f d\Omega_f} [\theta_i, \theta_{\mathrm{SD}}, E_i; T] dE_f d\Omega_f , \qquad (5.33)$$

where $\dot{N}$ is the number of particles scattered per unit time into the direction k_f in the solid angle $d\Omega_f$ and in the energy interval dE_f centered about $E_f (= \hbar^2 k_f^2/2m)$, n_i is the particle density in the incident beam, v_i its velocity ($v_i = \hbar k_i/m$) and A_i

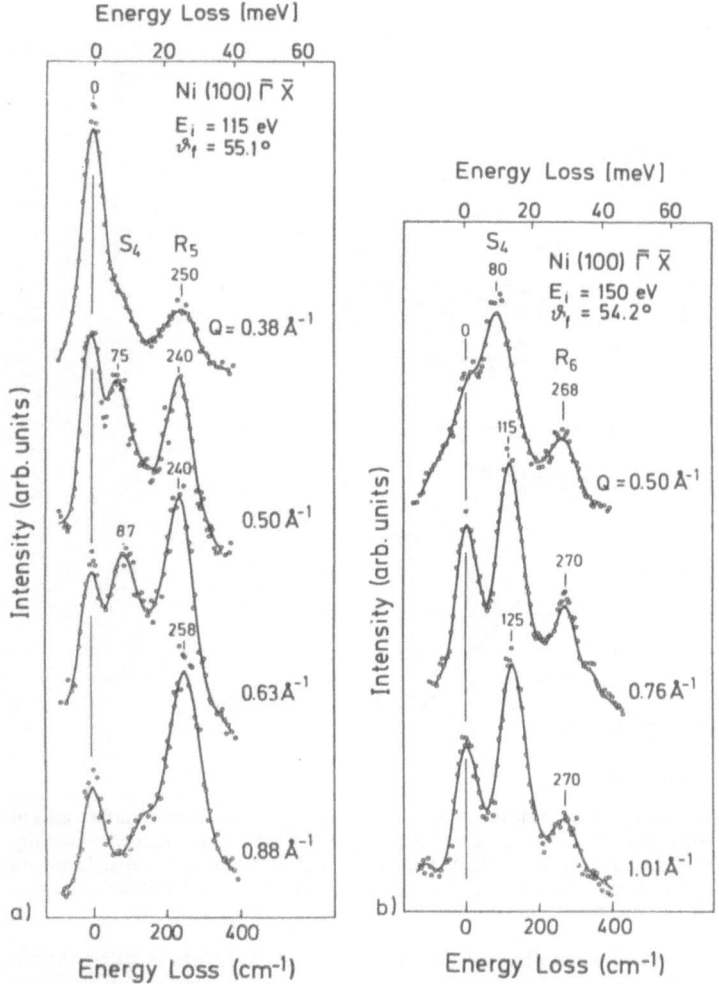

Fig. 5.19. Electron energy loss spectrum along the $\overline{\Gamma}\text{-}\overline{X}$ direction of Ni(100) [5.67, 70]. Note the appearence of incoherent elastic scattering at $\Delta E = 0$. These spectra were recorded under conditions where the S_4 Rayleigh mode is suppressed; however, it still appears as the lowest energy loss peak. The peak with the largest energy loss is the R_5 resonance in (a) and the R_6 resonance in (b)

its cross-sectional area. The area of the surface which is illuminated by the beam is given by $A_i / \cos \theta_i$, where θ_i is measured with respect to the surface normal. The factor α is introduced to take account of a possible geometrical reduction in the effective surface area resulting from additional constraints imposed by either the detector or the incident beam geometries. Figure 5.20a illustrates the situation in which the effective surface area is limited by the area exposed to the incident beam. Thus the factor α is given by

$$\alpha = \frac{\min(A_i / \cos \theta_i, \, A_f / \cos \theta_f)}{A_i / \cos \theta_i} \,, \tag{5.34}$$

143

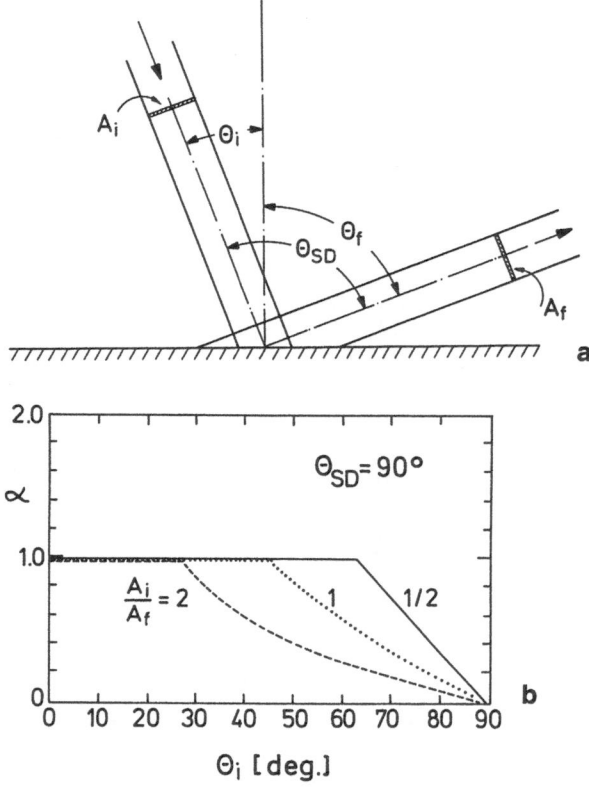

Fig. 5.20. (a) Schematic diagram illustrating the geometrical reduction in the effective surface area in a surface scattering experiment. (b) The factor α defining the relative reduction in scattered intensity is plotted as a function of the incident angle. Depending on the ratio of the areas of the incident and final beams A_i and A_f, the effective area seen by the detector may be less than the illuminated area

where "min" means that the smallest of the two values in brackets is to be taken. The geometrical evaluation is straightforward and the results of illustrative calculations are shown in Fig. 5.20b. It is seen that the distributions are characterized by a critical angle θ_c depending on the ratio of A_f to A_i. For $\theta_i < \theta_c$ α equals one and for $\theta_i > \theta_c$ α is smaller than one.

5.5.1 Helium Atom Scattering

The differential reflection coefficient depends in a rather complicated way on a number of parameters such as incident beam energy E_i, the apparatus geometry via θ_{SD}, and the surface temperature T_s. Since an experimental study to optimize these parameters would be very time consuming it is desirable to simulate these effects theoretically. As already discussed in Sect. 5.1 such a simulation can be carried out using the distorted wave Born approximation [5.20, 21]. We recall that such calculations are only possible if an effective two-body potential describing the interaction between the helium atom and each one of the atoms of the crystal is assumed. For He-LiF(001) a simple additive potential has been devised to reproduce (1) the known bound states of the potential well, (2) the diffraction intensities and (3) the single phonon inelastic scattering intensities [5.71]. In the case of an insulator the success of a two body additive potential is not too unexpected [5.72]. However in a metal the smearing of the surface corrugation by the conduction electrons (Smoluchowski effect [5.73]) is expected to rule out the use of such a simple model. Nevertheless recently an effective two-body additive potential, based largely on ab initio potential parameters, was also found for the metals Cu and Ag [5.20]. This potential is not only able to fit the Q dependence of the measured inelastic reflection coefficient for Cu(111) and Ag(111), but also the bound state energy levels which have been measured for the more corrugated (110) faces of the same metals. By the inclusion of a semi-empirical damping term to account for the smoothing of the surface potential by the conduction electrons it has also been possible to fit the diffraction intensities measured for the (110) surfaces. With the availability of this potential the dependence of the differential reflection coefficient on the above parameters can be realistically simulated in a theoretical calculation.

Figure 5.21 shows a comparison of the calculated differential reflection coefficients with measured relative Rayleigh phonon intensities for He-Cu(111) along the $\langle 110 \rangle$ direction for a beam energy of 9.15 meV [5.20]. The overall agreement is seen to be well within the scatter of the experimental points. A similarly good agreement could also be found in the case of Ag(111).

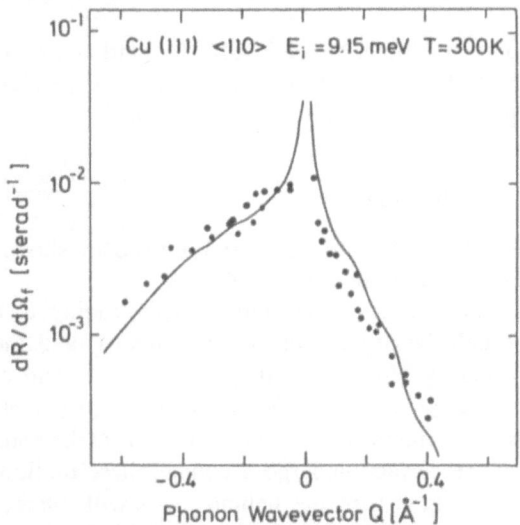

Fig. 5.21. Comparison of experimental (dots) and calculated results (full line) for the Rayleigh peak intensity in He-Cu(111) scattering with $\theta_{SD} = 90°$ along the $\langle 110 \rangle$ direction as a function of the phonon wave vector [5.20]. The experimental data have been taken for the surface temperature $T = 300\,\mathrm{K}$ and the incident energy $E_i = 9.15\,\mathrm{meV}$, whereas in the calculation the slightly different values of $T = 297\,\mathrm{K}$ and $E_i = 9.00\,\mathrm{meV}$ were used. The vertical scale of the theoretical curve has been adjusted to give the best fit to the experimental results

From Fig. 5.21 we note that both the forward and backward experimental reflection coefficients are characterized by a nearly exponential decrease with increasing parallel momentum transfer $|\Delta K|$. This phenomenon can be understood by recalling the close similarity between the dynamics of surface phonon excitation and the more thoroughly studied problem of vibrational excitation of free molecules by atom impact. This correspondence has already been pointed out by *Jackson* and *Mott* (1932) [5.74] in the first application of the distorted wave approximation to both problems. In both cases the scattering probability is determined by what is commonly called the Massey criterion:

$$\frac{t_{coll}}{\tau_{vib}} > 1 \qquad \text{adiabatic collisions: small probability ,} \qquad (5.35a)$$

$$\frac{t_{coll}}{\tau_{vib}} \lesssim 1 \qquad \text{nonadiabatic: large probability ,} \qquad (5.35b)$$

where t_{coll} is a characteristic time of the collision and $\tau_{vib} = \hbar/\Delta E$ is the effective period of vibration. The enhanced probability implied by (5.35b) arises because of a near resonance between the collision and the vibrations when $t_{coll} \approx \tau_{vib}$, while for $t_{coll} < \tau_{vib}$ the interaction becomes impulsive. Although t_{coll} is difficult to estimate absolutely, it is intuitively clear that steep potentials will lead to smaller t_{coll} than soft potentials. In the scattering of atoms from surface phonons the condition (5.35b) is best fulfilled near the zone origin where τ_{vib} is largest. The simple first order time-dependent perturbation theory at the basis (5.35) predicts an exponential fall-off with increasing values of the t_{coll}/τ_{vib} as is shown in Fig. 5.21.

The rather large differences in the reflection coefficient for forward and backward momentum transfer can be largely attributed to the temperature dependent Bose factor. This is shown in Fig. 5.22, where both the Bose factor

$$n(\omega) = \frac{1}{2}[\coth\frac{\hbar}{2kT} \pm 1] \qquad \begin{array}{l} +, \text{ creation} \\ -, \text{ annihilation} \end{array}, \qquad (5.36)$$

and the differential reflection coefficient with the Bose factor removed are shown. The latter shows a more nearly symmetric shape with ΔK.

Although the extent of adiabaticity (5.35) is the major effect leading to a decrease in intensity with Q, the calculations reveal another effect called the "cut-off" effect [5.75] first suggested by *Hoinkes* et al. [5.76] and *Armand* et al. [5.77] in connection with estimates of the Debye–Waller factor. The cut-off effect can be understood best by considering the excitation of phonons at the zone boundary where adjacent atoms at the surface undergo a counterphase motion. If the potential is sufficiently short ranged then the helium atom will interact with only one of the surface atoms. However, with an increase in the range of interaction the helium atom will only be able to interact with several surface atoms at a time and the excitation of the shortest wavelength phonons will be suppressed, hence the designation cut-off effect. Although much discussed in the literature, the additional attenuation coming from this effect amounts to at most

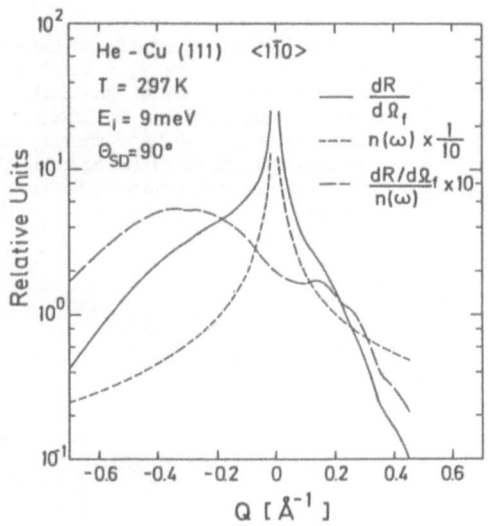

Fig. 5.22. Comparison of the relative contributions of the Bose factor and dynamical factors to the wave-vector dependence of the Rayleigh differential reflection coefficient [5.20]. The dotted line (...) shows the Bose factor alone and the dashed line (- - -) the calculated results after dividing out the effect of the Bose factor. Note that this curve has its maximum at $Q = -0.3 \, \text{Å}^{-1}$

only about a factor of 3 at $Q = 1.0 \, \text{Å}$ in He-Cu(111) at 9 meV. It is expected to be even less for LiF(001) where the repulsive potential is steeper.

Figure 5.23 shows the predicted effect of beam energy on the differential reflection coefficient. An analysis of the reflection coefficient at a fixed value of Q reveals a nearly exponential dependence on the collision energy. Note however that in these and in the next set of calculations the Debye–Waller factor, which will attenuate this increase, has been neglected. Figure 5.24 shows that the reflection coefficient can also be increased by reducing the angle θ_{SD}. For grazing angles $\theta_{SD} = 130°$, unexpected secondary maxima are observed near $\Delta K = -0.15 \, \text{Å}^{-1}$ and $\Delta K = +0.15 \, \text{Å}^{-1}$. They have been attributed to an interference like phenomenon arising from the presence of the attractive well.

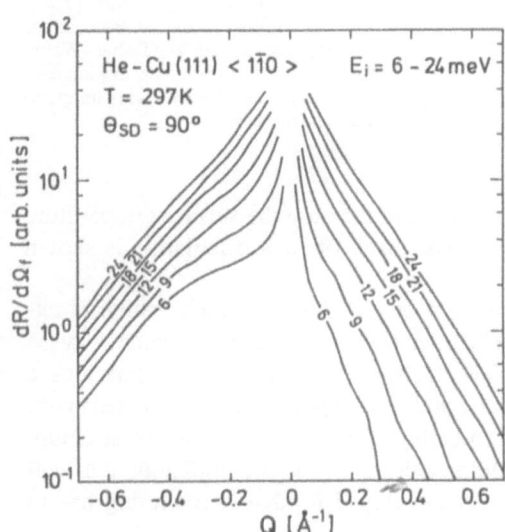

Fig. 5.23. Effect of the incident beam energy on the reflection coefficients for $\theta_{SD} = 90°$ [5.20]. The numbers on the individual curves are beam energy E_i in meV

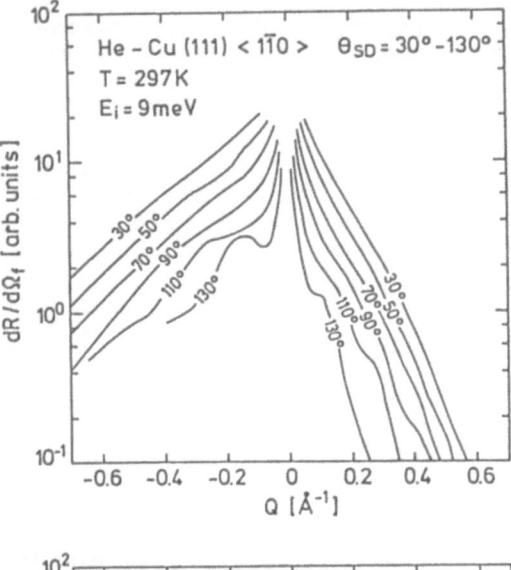

Fig. 5.24. Effect of the angle θ_{SD} on the reflection coefficient for $E_i = 9\,meV$ [5.20]. The numbers on the individual curves are values for θ_{SD}. The strong undulations at $Q = -0.1\,\text{Å}^{-1}$ are attributed to an interference resulting from the attractive and repulsive terms in the potential

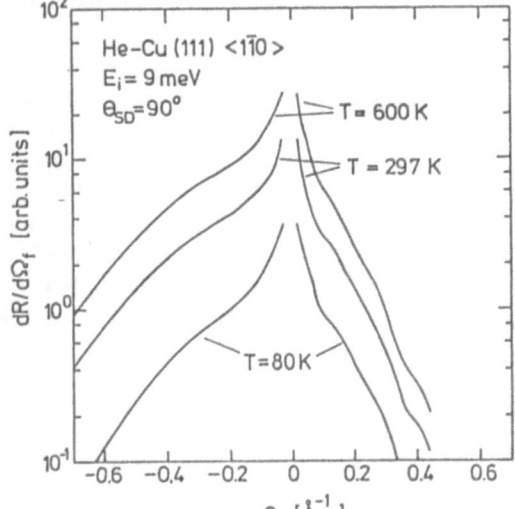

Fig. 5.25. Effect of the surface temperature on the reflection coefficient for $E_i = 9\,meV$ and $\theta_{SD} = 90°$ [5.20]. Only the influence of the Bose factor has been considered. The Debye–Waller factor will tend to reduce the increase in reflection coefficient with temperature shown here

Another way to increase the inelastic signals is to increase the surface temperature which enters in through the Bose factor. The result of such calculation is shown in Fig. 5.25.

Next we discuss the thermal attenuation of phonon intensities, which is expected to be described by the Debye–Waller factor in much the same way as the attenuation of diffraction peaks. The Debye–Waller factor also provides a convenient means for determining the conditions under which an experiment must be carried out in order to avoid smearing of the energy loss/gain structure from multiphonon processes. For dealing with atom–surface collisions the well known Debye–Waller factor as originally developed for X-ray scattering has to be considerably modified [5.75]. Recently considerable progress has been made in resolving this issue [5.78–80]. For the present discussion, however, it will suffice to use a simple impulsive approximation.

First we recall that the Debye–Waller factor W is defined by

$$I/I_0 = \exp[-2W(T)] , \tag{5.37}$$

where I_0 is the scattering intensity expected if the surface atoms were perfectly rigid and I is the reduced intensity resulting from their motion. W can be simply estimated from the Baule formula [5.81] by assuming a fast impulsive collision ($t_{coll} \ll \tau_{vib}$) involving only single atoms of the surface and a surface temperature T large compared to the surface Debye temperature θ_D [5.75]:

$$W = 24 \frac{m}{M_s} \frac{E_{iz}kT}{(k\theta_D)^2} , \tag{5.38}$$

where m is the projectile mass, M_s is the mass of a surface atom and E_{iz} is the normal component of the energy of the incident atom. Note that in calculating E_{iz} the additional attraction due to the atom–surface potential well must be accounted for (Beeby correction [5.82]). For He-atom scattering from most materials the physisorption well depth is typically 5–8 meV and thus can be quite important especially at low collision energies.

In the measurement of phonon dispersion curves it is imperative that only single phonons are excited. The probability of n-phonon scattering events can be estimated from the Debye–Waller factor via an assumed Poisson distribution [5.75]:

$$P(0 \rightarrow n) = \exp(-2W) \frac{(2W)^n}{n!} , \tag{5.39}$$

Accordingly, single phonon processes will dominate if $W < 0.5$. This leads to the following "rule of thumb" criteria for one phonon scattering:

$$\frac{m}{M_s} \frac{E_{iz}kT}{(k\theta_D)^2} \lesssim 0.02 \quad \text{for} \quad \frac{T}{\theta_D} > 1 , \tag{5.40a}$$

$$\frac{m}{M_s} \frac{E_{iz}}{k\theta_D} \lesssim 0.08 \quad \text{for} \quad \frac{T}{\theta_D} < 0.02 . \tag{5.40b}$$

The condition (5.40a) is frequently called the Weare criterion [5.83] and appears to provide a conservative estimate for predicting the onset of strong multiphonon contributions. At much higher surface temperatures the inelastic intensity will begin to become strongly attenuated by the Debye–Waller factor as can be seen in [5.28, Fig. 8].

As mentioned earlier the one-phonon criteria (5.40) also impose bounds on the gain in signal achieved with increasing collision energy or from reducing θ_{SD}. Thus the choice of optimal condition of E_i, θ_{SD} and T_s within which one phonon scattering is assumed will also depend on the mass of the surface atoms M_s and the surface Debye temperature θ_D.

Finally we point out that for highly corrugated surfaces, resonance processes can have a strong and apparently erratic effect on the inelastic intensities [5.40, 84]. These resonances, commonly referred to as selective adsorption, arise at special combinations of incident angles and energies when the atoms are temporarily trapped into bound vibrational states of the long-range van der Waals potential. These resonances can be used to enhance the scattering from otherwise weakly coupled modes [5.85, 86]. On the other hand they can be quite a nuisance since by accentuating bulk phonons at the special resonant conditions they can lead to sharp peaks in the spectra which are hard to distinguish from surface phonons. Fortunately these resonances do not show up strongly on the metal surfaces studied so far.

5.5.2 Electron Scattering

In electron scattering the intensities behave quite differently. Because of the much higher energies the extent of nonadiabaticity is not expected to change significantly for different phonon energies. As a result the inelastic differential reflection probabilities are expected to be nearly independent of $\hbar\omega$ and ΔK. This means that the important region in Q space near the zone boundary is just as easily studied as other regions.

The in-depth probing of the surface layer by electrons opens up the possibility of also obtaining information about lattice motion in deeper layers. Some simple estimates of the relative fraction of electrons which emerge from the surface after interacting with the n-th layer is contained in Table 5.3 [5.67]. Depending on the incident energy and scattering angles, information about phonons in layers up to $n = 2$ and 3 can be extracted.

Table 5.3. Relative fraction of electrons of the incident beam I/I_0 which emerge from the surface after interacting with the n-th layer. The angles are chosen to correspond to a parallel momentum transfer $\Delta K = 1.78\,\text{Å}^{-1}$ [5.67]. λ is the mean free path of the electrons in the crystal

Scattering conditions					I/I_0		
E_i [eV]	θ_i [deg.]	θ_f [deg.]	λ [Å]	penetration [Å]	$n = 2$ [%]	$n = 3$ [%]	$n = 4$ [%]
100	28	55	4	5.0	37	14	5
100	37	72	4	7.8	14	2	0.3
180	44	72	6	8.0	26	7	1.8
220	36	55	8	5.2	52	28	15

As already noted in Sect. 5.1, the in-depth probing by electrons greatly complicates the theoretical description of electron scattering. Whereas He atoms are sensitive to the vibrations of only the topmost layer and therefore provide a direct projection of the surface density of states, electrons interact in a more complex fashion with not only the surface modes and surface resonances but also with the bulk modes as they pass into and out of the bulk region near the surface. Although basically similar to He scattering the theory outlined briefly in Sect. 5.1

is extremely involved because of the necessity to account for multiple scattering of the incident and outgoing electron. As pointed out previously, the theory was formulated in 1980 by Li, Tong and Mills and collaborators [5.6], who took full advantage of the extensive experience in dealing with $I(V)$ curves in LEED.

The most important additional assumption in applying LEED theory to phonon excitation by electrons enters into the calculation of the derivative of the potential with displacement of the atom $\nabla v^{(2)}$, see (5.3) and (5.4). This term is calculated by assuming a type of muffin-tin potential with a fixed radius r_0 surrounding each atom. The nonstructured region between the muffin tin potentials is described by an inner potential with real and imaginary parts as in LEED theory. The atom muffin tin potential is assumed to be rigidly attached to the nucleus and cannot allow for distortions of the electron shell as the nucleus moves. Thus, for example, the theory cannot account for the induced dipole moment generated in the vibrations. These dipoles are responsible for the electron coupling in the dipole regime. Nevertheless we will see that the theory is surprisingly successful in explaining the experimental results.

Figure 5.26 shows two comparisons of measurements of the phonon loss peak intensities as a function of incident energy [5.67, 87] with theoretical calculations. The peaks are for the S_4 and S_6 surface modes at the $\bar{X}$ point of Ni(100). Figure 5.26a and b differ in the value of the final angle which was fixed for each set of measurements. First we note the complicated interference patterns of the intensity for the S_4 mode excitation which show order of magnitude oscillations. They are somewhat reminiscent of $I(V)$ curves but not identical to these. The results for the two values of θ_f differ dramatically, indicating the complicated dependence on the geometry. In addition the intensity of the S_6 mode relative to the S_4 mode fluctuates wildly with energy. As a consequence the intensities from the two modes become comparable in narrow energy windows at 150–180 eV and 205–230 eV for $\theta_f = 65°$, whereas for $\theta_f = 60°$ this only occurs at 160 eV. It is gratifying to see that the overall agreement between theory and experiment curves is satisfactory.

The dramatic sensitivity of the intensities to geometry has the advantage that it is possible to fine tune special features, but also the disadvantage that without a good theory it is easily possible to make erroneous assignments [5.88]. This dilemma is nicely illustrated in Fig. 5.27 where two theoretical spectra are compared for identical $Q = 1/2 Q_B(\bar{X})$ and $E_i = 155$ eV but different θ_f. The solid line spectra show the results for $\theta_f = 65°$. The lowest energy mode which appears as a weak shoulder is attributed to the S_4 Rayleigh mode, the resonance R_3 produces the largest peak and the resonances R_5 and R_6 produce some barely resolvable structure at greater energy losses. The dashed line spectrum is calculated for $\theta_f = 54.2°$ and presents an entirely different picture. The S_4 mode is greatly enhanced and the combination of R_5 and R_6 now lead to a large peak at large energy losses, while R_3 is no longer feasible. This example illustrates how the relative contribution of surface and bulk modes can conspire to suppress or even shift the locations of important features. Thus the assignment of frequencies to surface phonons can even be severely in error.

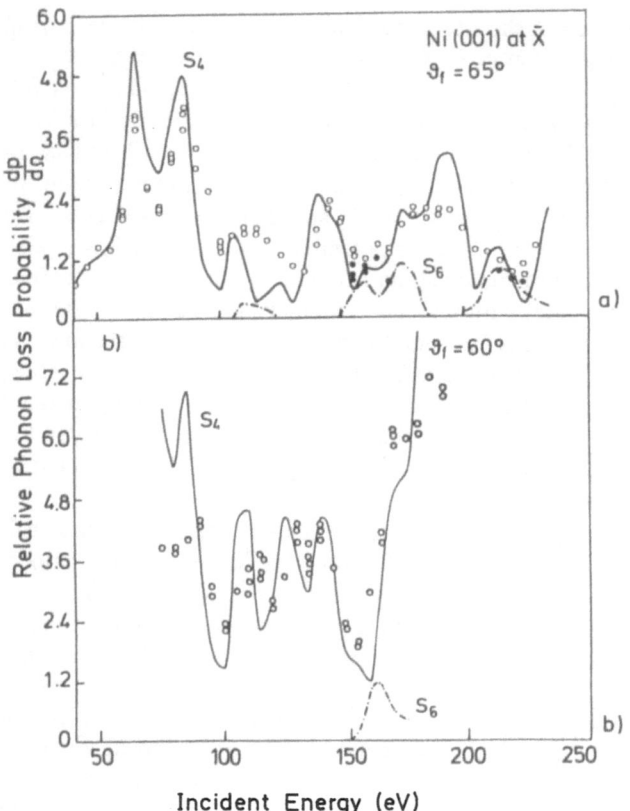

Fig. 5.26. Comparison of measured inelastic intensities with a theoretical calculation for the S_4 and S_6 modes at the $\bar{X}$ point on Ni(001) as a function of incident energy [5.67, 86]. The results in (**a**) are for a final angle of $\theta_f = 65°$ and those in (**b**) for $\theta_f = 60°$. The large oscillations are attributed to complex interferences of the electrons caused by multiple scattering. At special energies and angles the inelastic intensity of the S_6 mode is comparable to that of the S_4 mode

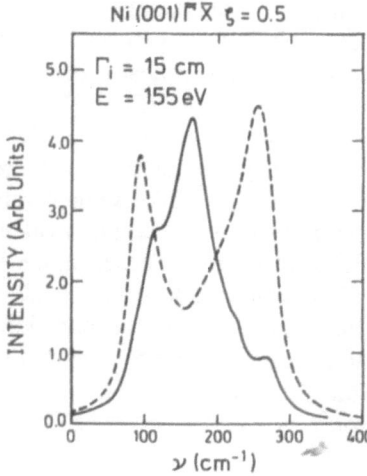

Fig. 5.27. A comparison of two theoretical calculations at $\xi = 0.5$ ($Q = 0.63\,\text{Å}^{-1}$) on the $\overline{\Gamma}\,\overline{X}$ azimuth for $E_i = 155\,\text{eV}$ [5.88]. The solid line is for $\theta_f = 65°$ and the dashed line for $\theta_f = 54.2°$. Note the large differences in the shapes of the two spectra

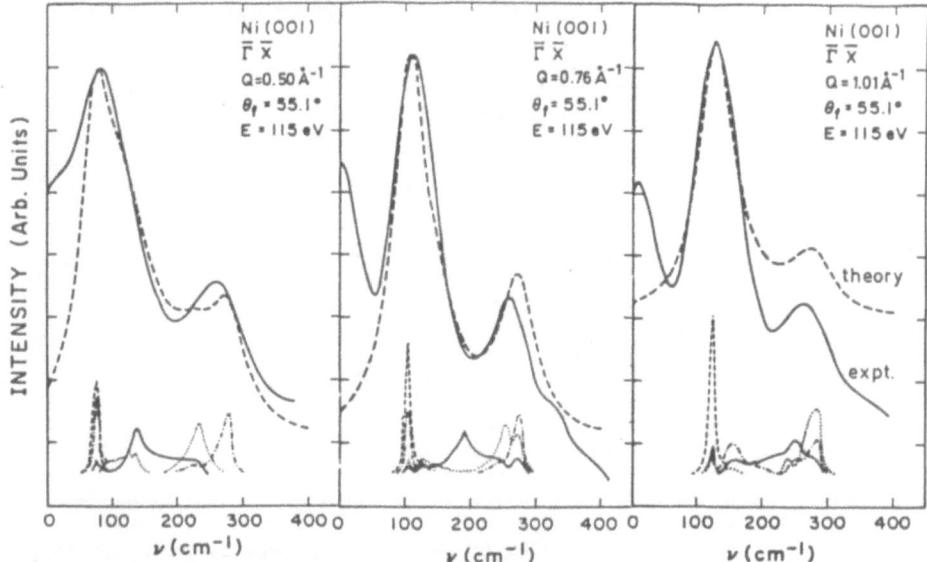

Fig. 5.28. Comparison of simulated energy loss spectra (- - -) with measured curves (——) for Ni(001) for three values of Q along the $\bar{\Gamma}$–$\bar{X}$ direction [5.88]. At the bottom several phonon state spectral densities are shown. The solid line is for longitudinal motion in the outermost layer, the dashed line for vertical motion in the outermost layer. The dotted line and dot-dashed line are for longitudinal and vertical motions in the second layer, respectively

Finally in Fig. 5.28 we present some comparison of recent calculations with some of the spectra presented in Fig. 5.19. The overall agreement, in the examples shown, is remarkably good. Also shown at the bottom of each panel of Fig. 5.28 are the lattice dynamical phonon spectral density functions. In each case it is seen that there is little connection between the phonon density of states and the observed spectra. This illustrates again the complicated nature of the electron–surface interaction.

5.5.3 Comparison of Experimental Intensities

A major consideration in assessing the relative merits of the two methods is the experimental intensities. Table 5.4 presents a comparison of typical intensities for the two apparatuses discussed in Sect. 5.3. It is surprising to find that the detector intensities are similar both with respect to total signal and background count rate. In the case of He atoms the fluxes are much larger since they are not limited by space charge. This advantage is entirely compensated for by the low detection efficiency for He atoms. Finally we note that since the matrix elements governing the coupling with phonons are different for the two projectiles, large derivations from the simple estimates in Table 5.4 can occur for specific systems.

Table 5.4. Comparison of intensities in He atom and electron scattering from surface phonons

	He atoms	Electrons[b]
Incident onto target	3×10^{14} atoms/sec[a]	10^9 el/sec
In specular peak (I_{00})		
smooth surface	1×10^{14} atoms/sec	10^5 el/sec
corrugated surface	1×10^{12} atoms/sec	10^5 el/sec
Inelastic peak		
near zone origin		
(10^{-3} of I_{00})	1×10^9 atoms/sec	10 el/sec
at zone boundary		
(10^{-5} of I_{00})	1×10^7 atoms/sec	10 el/sec
Detection efficiency	10^{-6}	1
Count rate		
near zone origin	10^3 cts/sec	10 cts/sec
at zone boundary	10 cts/sec	10 cts/sec
Background count rate		
at zone boundary	0.3 cts/sec[c]	0.3 cts/sec

(a) Assuming a beam flux of 3×10^{19} atoms/sec sr and a target area of $1\,\text{mm}^2$ for the apparatus described in Table 5.2
(b) Private communication M. Rocca (1986) (c) The total background count rate is typically 10 counts/sec. It is distributed uniformly over the entire time-of-flight spectrum and typically only a small fraction amounting to about 3×10^{-2} is at the inelastic peak.

5.6 Discussion of Experimental Results and Summary

In the following we attempt to summarize some of the salient physical phenomena emerging from experimental studies of surface phonons using these two techniques, emphasizing their relative advantages. The discussion will not attempt to be comprehensive but will only try to show a number of typical cases. A complete discussion of the large amount of data would go far beyond the bounds and aims of this review.

The high energy resolution available with He scattering has lead to the discovery of a new anomalous mode at low frequencies on metal surfaces which was first seen on Ag(111) [5.14]. Since then the same or similar anomalies have been observed on the other noble metals Cu and Au and on the transition metals Pt and Pd. As an example of a crystal exhibiting the anomaly we refer to the time-of-flight spectra for Pt(111) [5.89, 90] shown in Fig. 5.29 taken under similar geometrical conditions as those for Al(111) (see Fig. 5.15). Instead of revealing single sharp inelastic peaks these spectra exhibit much broader loss peaks with a width significantly greater than expected from the resolution of the apparatus. The correct interpretation of these spectra and similar spectra obtained for the other transition and noble metals, which also show two adjacent well separated peaks, was only possible by carrying out a complete theoretical simulation of the measured time-of-flight spectra and varying parameters until a best fit was achieved [5.92]. The simulation starts by accurately fitting the neutron bulk dispersion curves to within a few percent by including radial, tangential and angular force constants and interactions with up to 6 nearest neighbors (typically more than 16 force constants in all) [5.93]. With these force constants, energy loss spectra are calculated with a potential model for the He–surface atom interaction similar to that discussed in Sect. 5.5.1. For all of the above-mentioned metals it was possible to fit the Rayleigh mode using only bulk force constants. Additional modifications of the surface force constants were necessary to fit the other modes seen in the spectra of Fig. 5.29a. As first found for Ag(111) [5.91] a satisfactory fit could only be achieved by substantially lowering one of the radial force constants, namely that which describes the lateral forces within the surface layer, by as much as 50 to 75%. In the fit procedure all other force constants, also those describing the interaction of the first with the second layer, were kept the same. This lateral softening leads to the appearance of a new longitudianl polarized resonance which peels off from the longitudinal bulk band in much the same way as the Rayleigh mode peels off from the transverse bulk bands.

An example of the good quality of a "best fit" achieved in this way between the measured and simulated time-of-flight spectra, which were smeared in accordance with the experimental conditions, is shown in Fig. 5.29a [5.90]. Figure 5.29b shows finally the calculated best fit dispersion curves for the Pt(111) surface. It is seen that whereas along the $\langle 112 \rangle$ direction two modes are present, the Rayleigh mode and the anomalous longitudinal resonace modes, in the $\langle 110 \rangle$ direction an additional third mode is predicted [5.90]. This mode is called the pseudosurface mode since it has a significant bulk character [5.94]. The appeerence of this mode is related to the breakdown in mirror symmetry with respect to a sagittal plane through the crystal in this direction. The softening of the lateral force constant, which was required to produce this best fit in Pt(111), amounts to 65% [5.90]. The longitudinal mode in the other metals could be explained by a softening of 70, 52 and 78% in Cu(111) [5.95], Ag(111) [5.92] and (23×1) reconstructed Au(111) [5.96], respectively.

A similar anomaly has been found in the more corrugated Pd(110) surface, where it amounts to 30% along the close packed rows of the crystal [5.97]. Surprisingly in Pd(110) the vertical force constant between the top layer and the second layer is unaffected by the significant inward surface relaxation of $\Delta d_{12}/d_{12} \cong -6\%$, where d_{12} is the distance between the first two layers.

The origin of this anomalous longitudinal mode is not fully understood at the present time. It is presently attributed to a surface localized reduction in the sp-d hybridization which has, in the past, been introduced in pseudopotential calculations to account for a stiffening of the forces in the bulk in the noble and transition metals [5.92]. The reduction in sp-d hybridization is explained by the spill-out of electrons and the reduction in the number of nearest neighbors. This interpretation is supported by the fact that the anomalous longitudinal mode is not seen in Al(111) [5.33] and disappears in creating the Pt(111) + H(1 × 1) surface [5.90]. Other explanations have been presented and the true mechanism behind the longitudinal softening is by no means completely established [5.98, 99].

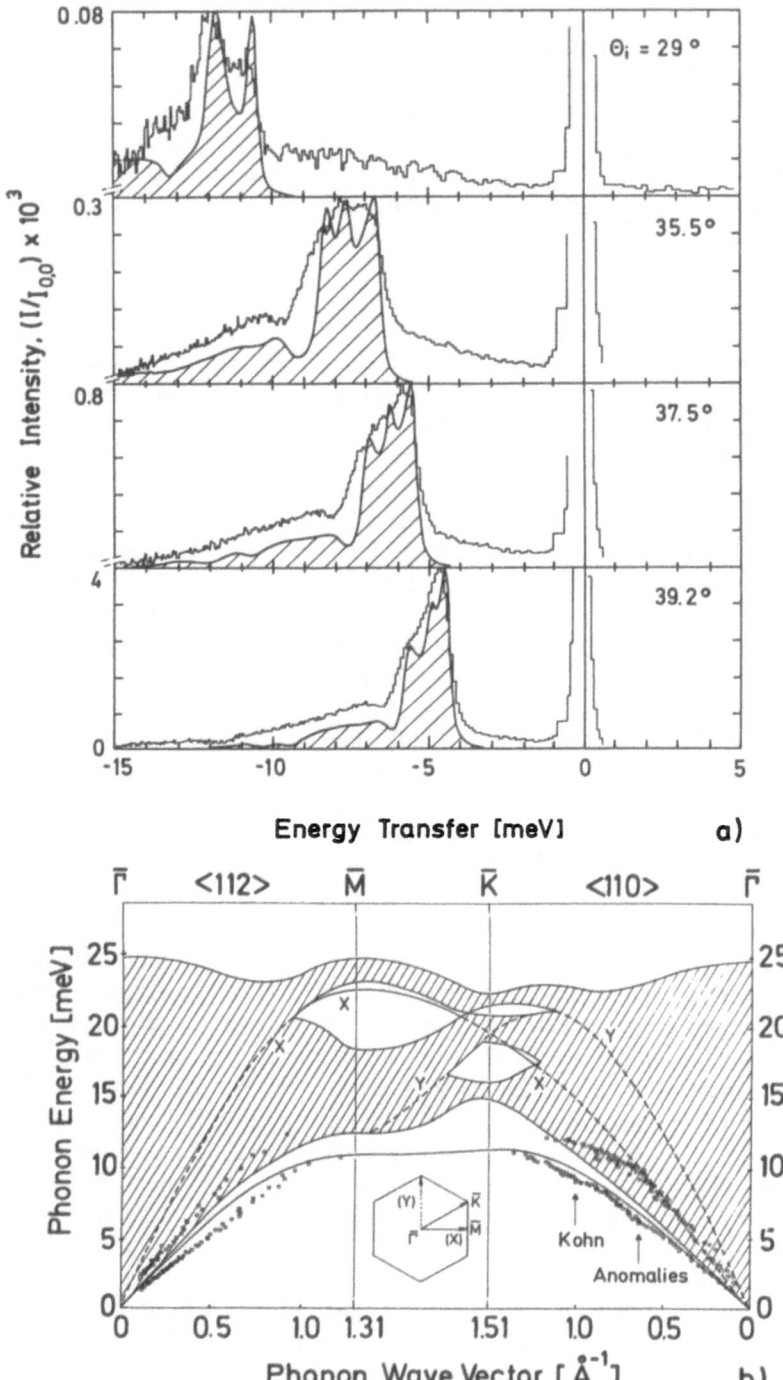

a)

b)

The higher resolution of He atom scattering has also made it possible to observe another type of anomaly which is observed at special values of Q in the Rayleigh mode dispersion curves. In Pt(111) over 500 time-of-flight spectra were measured along the $\langle 110 \rangle$ azimuth [5.89]. These were then fitted to a 9 term Fourier expansion with a mean deviation better than 0.1 meV. The group velocity obtained from the derivative of the dispersion curve shows two sharp minima at $Q \cong 0.5\,\text{Å}^{-1}$ and $\cong 1.0\,\text{Å}^{-1}$. The first of these could be attributed by Bortolani and coworkers to a projection of the bulk anomaly onto the surface [5.90].

The excellent resolution and sensitivity to low frequency vibrations of He scattering have also made it possible to detect a much greater anomaly in the high temperature phase of W(001) [5.100]. This bcc crystal undergoes a reconstruction on lowering the temperature to below 300 K into a $(\sqrt{2} \times \sqrt{2})$–$R45°$ structure [5.101]. The exact mechanism for the reconstruction has lead to considerable theoretical speculation [5.102]. The recent He atom experiments reveal an appreciable softening of what appears to be a sagittal plane mode in the high temperature phase similar to, but much greater than, the anomalies observed in the noble and transition metals. On reducing the temperature to the transition region the frequency of the mode is slightly reduced further although at a position slightly shifted to a smaller Q than the zone boundary. This experiment is the first observation of a soft mode in a two-dimensional phase transition. These experiments are particularly difficult to perform because of the effect of even small amounts of hydrogen on the reconstruction. Thus the time-of-flight measurements were all made within 10–15 minutes after flashing the crystal. Fortunately, the cross section for phonon excitation appears to be fairly large in this crystal making such measurements possible over a wide range of Q values.

Another type of anomaly has been found in Si(111) 2×1 [5.103] and in GaAs(001) [5.104]. In both cases one or two unexpected frequencies of 10 meV in the case of Si and 10 and 13 meV for GaAs are found in addition to the Rayleigh mode. These modes are almost independent of the wave vector. The origin of these modes has not yet been explained but the experimental conditions rule out an effect from surface defects or impurities.

Finally, we mention applications of He atom scattering to the study of adsorbates where the high resolution and great sensitivity to low frequency modes

---

Fig. 5.29. Energy loss histograms obtained from time-of-flight spectra measured at $E_i = 25$ meV on a clean Pt(111) surface are shown in (a) at four incident angles ($\theta_{\text{SD}} = 90°$). The shaded smooth curves are obtained from theoretical simulations [5.90], which have been averaged with a resolution better than that of the experiment to bring out the contributions from the three surface phonons (see text). The high energy tails in the theory are due to bulk phonons. In (b) experimental phonon dispersion curves (squares) [5.89] are compared with the predictions of a one force constant lattice dynamical calculation [5.91]. The letters X and Y on the theory curves indicate the polarization of the edge of the bulk bands. Along $\langle 110 \rangle$ X is a noncoupling shear horizontal mode. Thus the upper experimental mode in both directions cannot be explained by the simple theory. Arrows indicate the locations of Kohn anomalies

has been of great advantage. *Sibener* and coworkers at Chicago were the first to study adlayers of the heavy rare gases on a metal surface [5.105–107]. They were able to measure low frequency modes of about 3 meV, which did not change with wave vector, for monolayers of Xe on Ag(111). With increasing coverage corresponding to bilayers, trilayers and thicker films, they could monitor the transition to the bulk behavior as signaled by the shift in the frequency of the 3 meV mode down to zero. These results have recently been repeated for Kr and Xe on Pt(111), with improved resolution [5.108, 109].

Lahee et al. at Göttingen have studied chemisorbed CO on Pt(111) [5.110]. They observed a strong 6 meV mode which shifted to somewhat higher frequencies (7.4 meV) with increasing CO dosage. This increase corresponds to the transition from pure on-top sites in the $\sqrt{3} \times \sqrt{3}$ structure to a mixture of on-top and bridge sites in the $c(4 \times 2)$ structure. This mode was attributed to a frustrated translation and is the first direct observation of this mode. The spectra also contain evidence for another mode which could be a frustrated rotation mode at 26.5 meV.

In concluding this discussion we point out that many of the phenomena discovered by helium atom scattering derive from the excellent resolution and the unique ability to detect vibrations with frequencies down to 1 meV. Probably the biggest disadvantage of He scattering is the large apparatus. The size, however, is dictated entirely by vacuum considerations; many large pumps are needed to handle the large helium gas flux admitted to the source chamber. These pumping stages require, however, hardly any maintenance and once the apparatus is running only minor adjustments are needed. This contrasts to an EELS machine where a considerable amount of operator expertise and experience is needed to tune up the electron optics and where surface charging frequently presents problems.

The pumping requirements for He scattering may be reduced in the future as new detection schemes are developed. One promising development is the metastable detection scheme developed by El-Batanouny and coworkers [5.111]. Here the He beam is crossed at right angles by a well collimated monoenergetic electron beam at $E_{el} = 20$ eV. This beam excites the molecules into the first electronic $2^3 S_1$ and $2^1 S_0$ metastable states exactly at the threshold where the cross section is at a maximum. Thus the long lived helium metastable atoms are deflected through an angle which depends only on their initial momentum. Since they contain 20 eV of internal energy they are easily detected by an open multiplier or channel plate. Since the solid angle seen by this arrangement, which fully utilizes the well defined direction of the scattered beams, is very small, the background can be reduced greatly in comparison to conventional mass spectrometer detection schemes.

Next we turn to electron scattering and review some of the results obtained. We limit the discussion to the first two metals extensively studied, Ni(001) [5.7, 70, 112] and Cu(001) [5.113, 114]. In clean Ni(001) no anomalous modes similar to the longitudinal mode could be determined by electron scattering. A

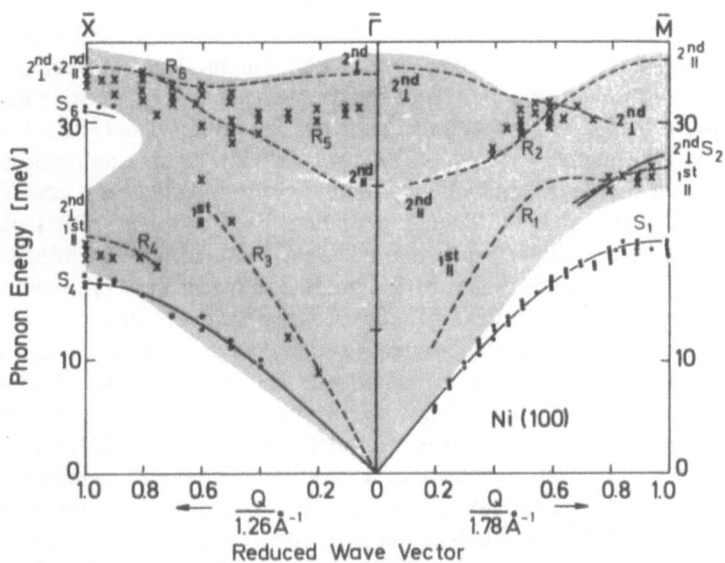

Fig. 5.30. Comparison between measured EELS loss peaks and calculations on the clean Ni(001) surface [5.70]. Loss peaks attributed to surface phonons (S) are indicated by open circles (o) and structures attributed to resonances (R) within the bulk bands are indicated by crosses (×). The best fit calculations reveal bulk bands indicated by the shaded area and the best fit surface phonon and volume resonances are shown as solid (—) and dashed (- - -) curves, respectively. The designations 1st and 2nd refer to the layer mostly involved in the motion and the indices $\perp$, $\parallel$ refer to the polarizations

direct comparison between EELS and helium scattering results could recently be achieved [5.115] and the results are found to be in essential agreement within about 0.5 meV. Figure 5.30 shows a complete dispersion curve measured by EELS for the Ni(001) surface and a comparison with theoretical results [5.70]. Note that the results contain two types of modes, pure surface modes indicated by an S and resonances within the volume bands indicated by an R. Energy loss spectra showing some of these resonances were shown in Fig. 5.19. The solid curves show calculated dispersion curves for the surface modes and the dashed lines in the calculated displacement of the resonances. The richness in structure in Fig. 5.19 makes the analysis more complicated than in the case of the He measurements since a complete set of measurements for a wide range of energies and different geometries in combination with a dynamical simulation is usually required before the modes can be definitely assigned.

The electron scattering experiments on Ni(001) reveal an increase in the frequencies of the zone boundary S_1 Rayleigh mode at $\bar{M}$ as well as the S_4 Rayleigh modes at $\bar{X}$ of the order of 7% when compared with calculations based on a one force constant model. This upward frequency shift in Ni(001) has been attributed to an increase in the force constants between the first and second layer by about 15 to 20% [5.7]. This in turn is explained by the small amount of contraction of the relaxed Ni(001) surface which from ion back scattering experiments is estimated to amount to $\Delta d_{12}/d_{12} = -3.2\%$ [5.116]. The extent of the actual discrepancy depends, of course, on the accuracy of the calculations of the surface phonons for that surface. Fortunately for this system very good agreement is found between calculations of surface modes based on a single force constant and more elaborate calculations based on a 6th nearest neighbor fit of the bulk phonon measurements [5.117]. However, in view of the contradictory evidence against a simple relationship between relaxation and force constant stiffening coming from the Pd(110) helium scattering experiments [5.97], this simple explanation of the Ni(100) measurements may have to be reexamined.

In addition to the Rayleigh mode the longitudinal edge of the bulk bands is seen in Fig. 5.30 along the $\bar{\Gamma}-\bar{X}$ direction and designated by R_3. Also along the $\bar{\Gamma}-\bar{X}$ direction the longitudinally polarized S_6 mode at $\hbar\omega = 30$ meV could be clearly resolved in a number of spectra taken near the zone boundary. Another mode, designated S_2, which appears near the M point is particularly interesting. This mode was already predicted in the early calculations of *de Wette* and coworkers [5.17] and can be attributed to a complex vibrational motion with greatest amplitudes in the second layer. Unfortunately, scattering from this mode is rather weak and buried in the flank of the S_1 mode. It was clearly resolved in only one spectrum.

The resonances R_4, R_5 and R_6 show up strongly in the energy loss spectra already presented in Fig. 5.19. These resonances could not be explained by invoking only surface vibrations and are therefore attributed to vibrations localized within the second layer. R_6 is attributed to transverse motions and R_5 to longitudinal motions. The fact that these resonances, and also the resonances R_1 and R_2 especially, do not show up over the entire region of Q space is attributed to complex interference effects discussed previously in Sect. 5.5.3.

Finally it is worth mentioning that surface phonon effects have also recently been observed in the dipole scattering regime on the (110) surface of the fcc crystals Cu and Ni [5.118, 119]. For this special surface there is a pseudogap in the density of states of the dipole active modes with displacements normal to the surface. Longitudinal modes moving in the bulk with wave vectors normal to the surface have transverse components at the surface. Thus they can excite a surface resonance in the band gap, which in turn can interact with scattered electrons. The effect is so far restricted to the (110) surface and is not expected for (100) and (111) surfaces.

In the impact regime, electron scattering has also been extensively used by Ibach and coworkers to study the adsorbate covered surfaces of Ni(001) [5.120–124] and Cu(001) [5.125]. In all the reported systems involving S, O and C additional optical modes could be found above the maximum frequency of the bulk bands. The effect of these adsorbates on the Rayleigh phonon dispersion curves of the ideal substrate is small. This is surprising since S, O and C present a wide range of different interaction strengths which increases in the order in which they are listed. The interaction in the case of carbon is so strong that it leads to a reconstruction of the surface.

Table 5.5 and the figures presented in Chap. 8 show that the amount of data accumulated by helium scattering and EELS measurements is already very ex-

Table 5.5. Some systems for which surface phonon dispersion curves have been measured

| | Method and References | |
	Helium scattering	EELS
Clean metals		
Al(111)	5.33	–
Al(110)	5.126	–
Ag(111)	5.14, 41	–
Ag(110)	5.127	–
Au(111)	5.41, 95, 128	–
Au(110)	5.129	–
Cu(111)	5.41	5.130
Cu(110)	5.131	–
Cu(100)	–	5.113, 114
Ni(100)	5.114	5.7, 70, 112
Pb(110)	5.132	–
Pd(110)	5.96	–
Pt(111)	5.88, 133, 134	–
W(100)	5.135	–
NbC(100)	–	5.136
TaC(100)	–	5.136, 137
TaS$_2$(0001)	5.138	–
TaSe$_2$(0001)	5.139	–
TiC(100)	–	5.140
ZrC(100)	–	5.141
Adsorbate covered metals		
Ag(110) + O(2 × 1)	5.142	5.142
Cu(100) + Ag(111)	–	5.143
Cu(100) + N c(2 × 2)	–	5.125
Ni(110) (1 × 2) + H	–	5.144
Ni(100) + C $p4g$(2 × 2)	–	5.124
Ni(100) + N c(2 × 2)	–	5.123
Ni(100) + O c(2 × 2)	–	5.7, 121
Ni(100) + O p(2 × 2)	5.114	–
Ni(100) + S c(2 × 2)	–	5.122
Ni(100) + CO	5.114	
Ni(100) + Ag(111)	–	5.143
Pt(111) + H(1 × 1)	5.89	–
Pt(111) + D(1 × 1)	5.89	–
Pt(111) + O(1 × 1)	5.145	–
Pt(111) + O p(2 × 2)	5.146	–
Pt(111) + CO $\sqrt{3} \times \sqrt{3}$	5.110	–
Pt(111) + CO c(4 × 2)	5.110	–
W(100) + H(1 × 1)	5.147	–
Ag(111) + Kr and Xe	5.105–107	–
Pt(111) + Kr and Xe	5.108, 109	–
Semimetals and Semiconductors		
GaAs(110)	5.104, 147	–
GaSe(0001)	5.149	–
Graphite(0001)	5.150	5.151
Si(111) (2 × 1)	5.103	–
Si(111) (1 × 1) + H(1 × 1)	5.152	–
Insulators		
LiF(100)	5.28, 153, 154	5.155
NaF(100)	5.28, 156, 157	5.155
NaCl(100)	5.158	–
KCl(100)	5.28	–
MgO	5.159	–

tensive. As we have seen from the above discussion the two techniques of the He atom scattering and EELS very nicely complement each other so that the information compiled in Table 5.5 has a different emphasis depending on the technique. These differences can be concisely summarized by way of a simple pros and cons comparison listed in Table 5.6. He atoms with the greater energy and angular resolution are ideally suited for studying vibrational modes at low energies ($\hbar\omega < 30\,meV$), whereas at higher vibrational frequencies electron scattering is the only choice.

Table 5.6. Pros and cons of helium atom and electron scattering for studying surface phonons

	Helium scattering	Electron scattering
Advantages	1. High resolution ($\approx 0.5\,meV$)	1. Wide vibrational energy range up to 500 meV
	2. Applicable to all substances (insulators, semiconductors and metals)	2. Sensitive to deeper layers
	3. Simple theoretical interpretation	3. Simple commercially available apparatus
Disadvantages	1. Restricted to low energy transfers ($< 30\,meV$)	1. Low resolution
	2. Low intensity at zone boundary	2. Not universal in application (insulators and semiconductors difficult)
	3. Large apparatus	3. Complicated interpretation

Acknowledgements. The author would like to thank his students and colleagues for their collaboration, which has made it possible to produce, analyze and understand the He scattering results. I am particularly greatful to G. Benedek (Milan) and the Modena group of V. Bortolani, A. Franchini and G. Santoro, as well as Ch. Wöll (Göttingen) for many fruitful discussions. Finally, I am indebted to H. Froitzheim (Erlangen), J. Hinch (Göttingen), A. Lahee (Heidelberg) and D.M. Smilgies (Göttingen) for constructive comments on the manuscript and to M. Quentin, V. Rosenthal and B. Schillings for their careful typing.

References

5.1   B.N. Brockhouse, A.T. Stewart: Phys. Rev. **100**, 756 (1955)
5.2   J.W. White, R.K. Thomas, T. Trewern, I. Marlow, G. Bomchil: Surf. Sci. **76**, 13 (1978)
5.3   K.H. Rieder, E.M. Hörl: Phys. Rev. Lett. **20**, 209 (1968); K.H. Rieder, W. Drexel: Phys. Rev. Lett. **34**, 148 (1975)
5.4   J. Aldag, R.M. Stern: Phys. Rev. Lett. **14**, 857 (1965); R.F. Wallis, A.A. Maradudin: Phys. Rev. **148**, 962 (1966); D.L. Huber: Phys. Rev. **153**, 772 (1967)
5.5   F.M. Probst, T.C. Piper: J. Vac. Sci. Technol. **4**, 53 (1967); H. Ibach: Phys. Rev. Lett. **24**, 1416 (1970); **27**, 253 (1971); J. Vac. Sci. Technol. **9**, 713 (1971)
5.6   C.H. Li, S.Y. Tong, D.L. Mills: Phys. Rev. B **21**, 3057 (1980)
5.7   J.M. Szeftel, S. Lehwald, H. Ibach, T.S. Rahman, J.E. Black, D.L. Mills: Phys. Rev. Lett. **51**, 268 (1983)
5.8   N. Carbrera, V. Celli, R. Manson: Phys. Rev. Lett. **22**, 346 (1969)
5.9   R. Manson, V. Celli: Surf. Sci. **24**, 495 (1971)

5.10   S.S. Fisher, J.R. Bledsoe: J. Vac. Sci. Technol. **9**, 814 (1971); Surf. Sci. **46**, 129 (1974)
5.11   R.E. Subbarao, D.R. Miller: J. Vac. Sci. Technol. **9**, 808 (1972); J.M. Horne, D.R. Miller: Phys. Rev. Lett. **41**, 511 (1978)
5.12   B.E. Mason, B.R. Williams: J. Chem. Phys. **55**, 3220 (1971); **61**, 2765 (1974)
5.13   G. Brusdeylins, R.B. Doak, J.P. Toennies: Phys. Rev. Lett. **44**, 1417 (1980); G. Brusdeylins, R.B. Doak, J.P. Toennies: Phys. Rev. Lett. **46**, 437 (1981)
5.14   R.B. Doak, U. Harten, J.P. Toennies: Phys. Rev. Lett. **51**, 578 (1983)
5.15   J.P. Toennies: J. Vac. Sci. Technol. A **5**(4), 440 (1987)
5.16   H. Ibach: J. Vac. Sci. Technol. A **5**(4), 419 (1987)
5.17   R.E. Allen, G.P. Alldredge, F.W. de Wette: Phys. Rev. B **4**, 1648 (1971); B **4**, 1661 (1971); G.P. Alldredge, R.E. Allen, F.W. de Wette: Phys. Rev. B **4**, 1682 (1971)
5.18   G. Benedek: Surf. Sci. **61**, 603 (1976)
5.19   F.O. Goodman, H.Y. Wachman: *Dynamics of Gas-Surface Scattering* (Academic, New York 1976)
5.20   D. Eichenauer, U. Harten, J.P. Toennies, V. Celli: J. Chem. Phys. **86**, 3693 (1987)
5.21   V. Bortolani, A.C. Levi: Atom Surface Scattering Theory, La Rivista del Nuovo Cimento 9, N.11 (1986)
5.22   A.Z. Msezane, R.J.W. Henry: Phys. Rev. A **33**, 1631 (1986)
5.23   J. Politiek, J.J.M. Schipper, J. Los: Physica **49**, 165 (1970)
5.24   A. Euceda, D.M. Bylander, L. Kleinman: Phys. Rev. B **28**, 528 (1983)
5.25   H. Ibach, D.L. Mills: *Electron Energy Loss Spectroscopy and Surface Vibrations* (Academic, New York 1982) pp. 105–120
5.26   J.P. Toennies: In *Springer Series in Chemical Physics* **21**, ed. by G. Benedek, U. Valbusa (Springer, Berlin, Heidelberg 1982) p. 208
5.27   K.M. Martini: PhD Thesis, Boston University 1986, pp. 86–93
5.28   G. Brusdeylins, R.B. Doak, J.P. Toennies: Phys. Rev. B **27**, 3662 (1983)
5.29   K. Kern, R. David, G. Comsa: Surf. Sci. **164**, L831 (1985)
5.30   G. Benedek: Phys. Rev. Lett. **35**, 234 (1975)
5.31   G. Benedek, G. Brusdeylins, J.P. Toennies, R.B. Doak: Phys. Rev. B **27**, 2488 (1983)
5.32   G. Benedek, G. Brusdeylins, R.B. Doak, J.G. Skofronick, J.P. Toennies: Phys. Rev. B **28**, 2104 (1983)
5.33   A. Lock, J.P. Toennies, Ch. Wöll, V. Bortolani, A. Franchini, G. Santoro: Phys. Rev. B **37**, 7087 (1988-II)
5.34   W.L. Nichols, J.H. Weare: Surf. Sci. **148**, 42 (1984); Phys. Rev. Lett. **56**, 753 (1985)
5.35   R. Avila, M. Lagos: Surf. Sci. Lett. **103**, L104 (1981)
5.36   J.P. Toennies, K. Winkelmann: J. Chem. Phys. **66**, 3965 (1977)
5.37   I. Estermann, R. Frisch, O. Stern: Z. Physik **73**, 348 (1932)
5.38   B.F. Mason, B.R. Williams: Rev. Sci. Instrum. **49**, 897 (1978)
5.39   B.F. Mason, B.R. Williams: Phys. Rev. Lett. **46**, 1138 (1981); J. Chem. Phys. **75**, 2199 (1981)
5.40   G. Lilienkamp, J.P. Toennies: J. Chem. Phys. **78**, 5210 (1983)
5.41   U. Harten, J.P. Toennies, Ch. Wöll: Faraday Discuss. Chem. Soc. **80**, 137 (1985)
5.42   H.D. Meyer: Bericht 137/1972, Max-Planck-Institut für Strömungsforschung, Göttingen 1972
5.43   G. Comsa, R. David, B.J. Schumacher: Rev. Sci. Instrum. **52**, 789 (1981)
5.44   G. Rotzoll: J. Phys. E (Sci. Instrum.) **15**, 708 (1982)
5.45   For a review of nozzle beams see the article by D.R. Miller in *Atomic and Molecular Beam Methods*, ed. by G. Scoles (Oxford University Press, Oxford 1987)
5.46   G. Brusdeylins, H.-D. Meyer, J.P. Toennies, K. Winkelmann: *Progress in Astronautics and Aeronautics* **51**, ed. by J.L. Poetter (AIAA, New York 1977) p. 1047
5.47   K. Kern, R. David, G. Comsa: Rev. Sci. Instrum. **56**, 369 (1985)
5.48   H.C.W. Beijerinck, R.J.F. van Gerwen, E.R.T. Kerstel, J.F.M. Martens, E.J.W. van Vliembergen, M.R.Th. Smits, G.H. Kaashoek: Chem. Phys. **96**, 153 (1985)
5.49   H. Buchenau, R. Götting, A. Scheidemann, J.P. Toennies: Proc. 15th Int. Symposium on Rarefield Gas Dynamics, ed. by V. Boffi, C. Cercignani, Vol. II (1986) p. 197
5.50   H.-D. Meyer, K.T. Tang: Z. Physik A **279**, 349 (1976)
5.51   Y.-H. Uang, W.C. Stwalley: J. Chem. Phys. Ö **76**, 5069 (1982)
5.52   R. Feltgen, H. Kirst, K.A. Köhler, H. Pauly, F. Torello: J. Chem. Phys. **76**, 2360 (1982)
5.53   G. Brusdeylins, J.P. Toennies: unpublished
5.54   G. Brusdeylins, J.P. Toennies: unpublished
5.55   W.R. Gentry, D.F. Giese: Rev. Sci. Instrum. **49**, 595 (1978)

163

5.56   W.R. Gentry: In *Atomic and Molecular Beam Methods*, ed. by G. Scoles (Oxford University Press, Oxford 1987) Chap. 3

5.57   B. Feuerbacher, M.A. Adriaens, H. Thuis: Surf. Sci. **94**, 2171 (1980)

5.58   W. Allison, B. Feuerbacher: Phys. Rev. Lett. **45**, 2040 (1980)

5.59   K.L. Saenger, J.B. Fenn: J. Chem. Phys. **79**, 6043 (1983)

5.60   B.D. Kay, T.D. Raymond, J.K. Rice: Rev. Sci. Instrum. **57**, 2266 (1986)

5.61   W.R. Gentry: private communication

5.62   R. Götting, H.R. Mayne, J.P. Toennies: J. Chem. Phys. **85**, 6396 (1986)

5.63   D. Smilgies, J.P. Toennies: Rev. Sci. Instrum. **59**, 2185 (1988)

5.64   A. Lock: Max-Planck-Institut für Strömungsforschung, Bericht (1987)

5.65   A. Lock, J.P. Toennies, Ch. Wöll, V. Bortolani, A. Franchini, G. Santoro: in preparation

5.66   D. Eichenauer, U. Harten, J.P. Toennies, V. Celli: J. Chem. Phys. **86**, 3693 (1987)

5.67   M. Rocca: Berichte der Kernforschungsanlage Jülich Nr. 2000, 1985

5.68   H. Ibach, T.S. Rahman: In *Chemistry and Physics of Solid Surfaces V*, ed. by R. Vanselow, R. Howe, Springer Ser. Chem. Phys., Vol. 35 (Springer, Berlin, Heidelberg 1984) pp. 455–482

5.69   P.A. Thiry, M. Liehr, J.J. Pireaux, R. Caudano: J. Electr. Spectr. Rel. Phen. **39**, 69 (1986)

5.70   M. Rocca, S. Lehwald, H. Ibach, T.S. Rahman: Surf. Sci. **171**, 632 (1986)

5.71   V. Celli, D. Eichenauer, A. Kaufhold, J.P. Toennies: J. Chem. Phys. **83**, 2504 (1985)

5.72   For a discussion of some corrections to the simple two body potential see Ref. [5.71] and P.W. Fowler, J.H. Hutson: Phys. Rev. B **33**, 3724 (1986)

5.73   R. Smoluchowski: Phys. Rev. **60**, 661 (1941)

5.74   J.M. Jackson, N.F. Mott: Proc. Roy. Soc. A **137**, 703 (1932)

5.75   H.D. Meyer: Surf. Sci. **104**, 117 (1981)

5.76   H. Hoinkes, N. Nahr, H. Wilsch: Surf. Sci. **88**, 221 (1972)

5.77   G. Armand, J. Lapujoulade, Y. Lejay: Surf. Sci. **63**, 143 (1977)

5.78   G. Armand, J.R. Manson, C.S. Jayanthi: Phys. Rev. B **34**, 6627 (1986)

5.79   V. Celli, A.A. Maradudin: Phys. Rev. B **31**, 825 (1985)

5.80   J. Idiodi, V. Bortolani, A. Franchini, G. Santoro, V. Celli: Phys. Rev. B **35**, 6029 (1987)

5.81   For a derivation of the Baule formula, see R.E. Stickney: Advan. At. Mol. Phys. **3**, 143 (1967)

5.82   J.L. Beeby: J. Phys. C (Solid State Phys.) **4**, L359 (1971)

5.83   G. Brusdeylins, R.B. Doak, J.P. Toennies: J. Chem. Phys. **75**, 1784 (1981)

5.84   J.H. Weare: J. Chem. Phys. **61**, 2900 (1974)

5.85   D. Evans, V. Celli, G. Benedek, J.P. Toennies, R.B. Doak: Phys. Rev. Lett. **50**, 1854 (1983)

5.86   D. Eichenauer, J.P. Toennies: J. Chem. Phys. **85**, 532 (1986)

5.87   M.-L. Xu, B.M. Hall, S.Y. Tong, M. Rocca, H. Ibach, S. Lehwald, J.E. Black: Phys. Rev. Lett. **54**, 1171 (1985)

5.88   B.M. Hall, D.L. Mills: Phys. Rev. B **34**, 8318 (1986)

5.89   U. Harten, J.P. Toennies, Ch. Wöll, G. Zhang: Phys. Rev. Lett. **35**, 2308 (1985)

5.90   V. Bortolani, A. Franchini, G. Santoro, J.P. Toennies, Ch. Wöll, G. Zhang: Phys. Rev. B **40**, 3524 (1989)

5.91   G. Armand: Solid State Commun. **48**, 261 (1983)

5.92   V. Bortolani, A. Franchini, F. Nizzoli, G. Santoro: Phys. Rev. Lett. **52**, 429 (1984); V. Bortolani, A. Franchini, G. Santoro: In *Dynamical Phenomena at Surface Interfaces and Superlattices*, ed. by F. Nizzoli, K.H. Rieder, R.F. Willis, Springer Ser. Surf. Sci., Vol. 3 (Springer, Berlin, Heidelberg 1985) p. 92

5.93   V. Bortolani, A. Franchini, G. Santoro: 'Surface Phonon Calculations in Metals and Comparison with Experimental Techniques'. In *Electronic Structure, Dynamics and Quantum Structural Properties of Condensed Matter*, ed. by J.T. Devreese, P. van Lamp (Plenum, New York 1984)

5.94   G.W. Farnell: In *Physical Acoustics*, ed. by W.P. Mason, R.N. Thurston (Academic, New York 1970) Vol. 6, pp. 109–166

5.95   V. Bortolani, A. Franchini, G. Santoro: Private communication

5.96   V. Bortolani, G. Santoro, U. Harten, J.P. Toennies: Surf. Sci. **148**, 82 (1984)

5.97   A. Lahee, J.P. Toennies, Ch. Wöll: Surf. Sci. **177**, 371 (1986)

5.98   V. Heine, L.D. Marks: Surf. Sci. **165**, 66 (1986)

5.99   C.S. Jayanthi, G. Benedek, W. Kress, H. Bilz: Phys. Rev. Lett. **59**, 795 (1987)

5.100   H.J. Ernst, E. Hulpke, J.P. Toennies: Phys. Rev. Lett. **58**, 1941 (1987)

5.101 M.K. Debe, D.A. King: Phys. Rev. Lett. **39**, 708 (1977); R.A. Barker, P.J. Estrup, F. Jona, P.M. Marcus: Solid State Commun. **25**, 375 (1978)

5.102 A. Fasolino, G. Santoro, E. Tosatti: Phys. Rev. Lett. **44**, 1648 (1980); C.L. Fu, A.J. Freeman, E. Wimmer, M. Weinert: Phys. Rev. Lett. **54**, 2261 (1985); L.D. Roelofs, J.F. Wendelken: Phys. Rev. B **34**, 3319 (1986)

5.103 U. Harten, J.P. Toennies, Ch. Wöll: Phys. Rev. Lett. **57**, 2947 (1986)

5.104 U. Harten, J.P. Toennies: Europhys. Lett. **4**, 833 (1987)

5.105 K.D. Gibson, S.J. Sibener, B.M. Hall, D.L. Mills, J.E. Black: J. Chem. Phys. **83**, 4256 (1985)

5.106 K.D. Gibson, S.J. Sibener: Phys. Rev. Lett. **55**, 1514 (1985)

5.107 K.D. Gibson, S.J. Sibener: Faraday Discuss. Chem. Soc. **80**, 203 (1985)

5.108 K. Kern, R. David, R.L. Palmer, G. Comsa: Phys. Rev. Lett. **56**, 2823 (1986)

5.109 K. Kern, P. Zeppenfeld, R. David, G. Comsa: Phys. Rev. B **35**, 886 (1987)

5.110 A.M. Lahee, J.P. Toennies, Ch. Wöll: Surf. Sci. **177**, 371 (1986)

5.111 K.M. Martini, W. Franzen, M. El-Batanouny: Rev. Sci. Instrum. **58**, 1027 (1987)

5.112 J. Szeftel, S. Lehwald: Surf. Sci. **143**, 11 (1984)

5.113 M. Wuttig, R. Franchy, H. Ibach: Solid State Commun. **57**, 445 (1986)

5.114 L.L. Kesmodel, M.L. Xu, S.Y. Tang: Phys. Rev. B **34**, 2010 (1986)

5.115 R. Bernd, J.P. Toennies, Ch. Wöll: J. Electron. Spectr. Relat. Phenom. **44**, 183 (1987)

5.116 J.W.M. Frenken, R.G. Smeenk, J.F. van der Veen: Surf. Sci. **135**, 147 (1983)

5.117 V. Bortolani, A. Franchini, G. Santoro: Private communication

5.118 J.A. Stroscio, M. Persson, S.R. Bare, W. Ho: Phys. Rev. Lett. **54**, 1428 (1985)

5.119 J.A. Stroscio, M. Persson, W. Ho: Phys. Rev. B **33**, 6758 (1986); J.A. Stroscio, M. Persson, C.E. Bartsch, W. Ho: Phys. Rev. B **33**, 2879 (1986)

5.120 S. Lehwald, J.W. Szeftel, H. Ibach, T.S. Rahman, D.L. Mills: Phys. Rev. Lett. **50**, 518 (1983)

5.121 T.S. Rahman, D.L. Mills, J.E. Black, J.M. Szeftel, S. Lehwald, H. Ibach: Phys. Rev. B **30**, 589 (1984)

5.122 S. Lehwald, M. Rocca, H. Ibach, T.S. Rahman: Phys. Rev. B **31**, 3477 (1985)

5.123 W. Daum, S. Lehwald, H. Ibach: Surf. Sci. **178**, 528 (1986)

5.124 S. Lehwald, M. Rocca, H. Ibach, T.S. Rahman: J. Electr. Spectr. Relat. Phenom. **38**, 29 (1986)

5.125 R. Franchy, M. Wuttig, H. Ibach: Z. Phys. B (Condensed Matter) **64**, 453 (1985)

5.126 J.P. Toennies, Ch. Wöll: Phys. Rev. B **36**, 4475 (1987)

5.127 G. Bracco, R. Tatarek, F. Tommasini, U. Linke, M. Persson: Phys. Rev. B **36**, 2928 (1987)

5.128 M. Cates, D.R. Miller: Phys. Rev. B **28**, 3615 (1983)

5.129 A.M. Lahee, J.P. Toennies, Ch. Wöll, K.P. Bohnen, K.M. Ho: Europhys. Lett. **10**, 261 (1989)

5.130 B.M. Hall, D.L. Mills, M.H. Mohamed, L.L. Kesmodel: Phys. Rev. B **58**, 5856 (1988-II)

5.131 B.F. Mason, K. McGreer, B.R. Williams: Surf. Sci. **130**, 282 (1983)

5.132 J.W.M. Frenken, J.P. Toennies, Ch. Wöll: Proc. of ICSOS II, Amsterdam, June 1987, Springer Ser. Surf. Sci., ed. by J.F. van der Veen, M.A. Hove

5.133 D. Neuhaus, F. Joo, B. Feuerbacher: Surf. Sci. Lett **165**, L90 (1986)

5.134 K. Kern, R. David, R.L. Palmer, G. Comsa, T.S. Rahman: Phys. Rev. B **33**, 4334 (1986)

5.135 H.J. Ernst, E. Hulpke, J.P. Toennies: Phys. Rev. Lett. **58**, 1941 (1987)

5.136 C. Oshima, R. Souda, M. Aono, S. Otani, Y. Ishizawa: Phys. Rev. Lett. **56**, 240 (1986)

5.137 C. Oshima, R. Souda, M. Aono, S. Otani, Y. Ishizawa: Solid State Commun. **57**, 283 (1986)

5.138 G. Brusdeylins, C. Heimlich, J. Skofronick, J.P. Toennies XIth Int. Symp. on Molecular Beams, Univ. of Edinburgh, July 1987

5.139 G. Benedek, L. Miglio, J.G. Skofronick, G. Brusdeylins, C. Heimlich, J.P. Toennies: J. Vac. Sci. Technol. **5**, 1093 (1987)

5.140 C. Oshima, R. Souda, M. Aono, S. Otani, Y. Ishizawa: Surf. Sci. **178**, 519 (1986)

5.141 R. Franchy, C. Oshima, T. Aizawa, S. Souda, S. Otani, Y. Ishizawa: J. Electr. Spectr. Relat. Phenom. **44**, 289 (1987)

5.142 G. Bracco, R. Tatarek, S. Terrini, F. Tommasini, U. Linke: J. Electr. Spectr. Relat. Phenom. **44**, 197 (1987)

5.143 W. Daum: J. Electr. Spectr. Relat. Phenom. **44**, 271 (1987)

5.144 H. Ibach, S. Lehwald, B. Voigtländer: J. Electr. Spectr. Relat. Phenom. **44**, 263 (1987)

5.145 D. Neuhaus, F. Joo, B. Feuerbacher: Phys. Rev. Lett. **58**, 694 (1987)

5.146 K. Kern, R. David, R.L. Palmer, G. Comsa, J. He, T.S. Rahman: Phys. Rev. Lett. **56**, 2064 (1986)

5.147 H.J. Ernst, E. Hulpke: J. Vac. Sci. Technol. A **5**(4), 460 (1987)

165

5.148 R.B. Doak, D.B. Nguyen: J. Electr. Spectr. Relat. Phenom. **44**, 205 (1987)

5.149 G. Brusdeylins, R. Rechsteiner, J.G. Skofronick, J.P. Toennies, G. Benedek, L. Miglio: Phys. Rev. B **34**, 902 (1986)

5.150 U. Valbusa, G. Brusdeylins, C. Heimlich, J.P. Toennies, G. Benedek: Surf. Sci. ECOSS-8 (1986)

5.151 J.L. Wilkes, R.E. Palmer, R.F. Willis: J. Electr. Spectr. Relat. Phenom. **44**, 355 (1987)

5.152 U. Harten, J.P. Toennies, Ch. Wöll, L. Miglio, P. Ruggerone, L. Colombo, G. Benedek: Phys. Rev. B **38**, 3305 (1988)

5.153 G. Bracco, R. Tatarek, S. Terreni, F. Tommasini: Phys. Rev. B **34**, 9045 (1986)

5.154 G. Bracco, M. D'Avanzo, C. Salvo, R. Tatarek, S. Terreni, F. Tommasini: Surf. Sci. **189/190**, 2928 (1987)

5.155 P.A. Thiry, J.L. Longueville, C. Bertoluzza, J.J. Pireaux, R. Candono: 5th Int. Conf. on Vibrations at Surfaces, Sept. 1987, Grainau, unpublished

5.156 G. Brusdeylins, R. Rechsteiner, J.G. Skofronick, J.P. Toennies, G. Benedek, L. Miglio: Phys. Rev. Lett. **54**, 466 (1985)

5.157 G. Benedek, L. Miglio, G. Brusdeylins, J.G. Skofronick, J.P. Toennies: Phys. Rev. B **35**, 6593 (1987)

5.158 G. Benedek, G. Brusdeylins, R.B. Doak, J.G. Skofronick, J.P. Toennies: Phys. Rev. B **28**, 2104 (1983)

5.159 G. Brusdeylins, R.B. Doak, J.G. Skofronick, J.P. Toennies: Surf. Sci. **128**, 191 (1983)

6. Theory of Helium Scattering from Surface Phonons

V. Celli

With 5 Figures

This is a review of the application of atom–surface scattering theory to the determination of surface phonon spectra throughout the surface Brillouin zone. It focuses on the interpretation of time-of-flight (TOF) experiments using He beams and on the discussion of the He–surface interaction. Earlier reviews covering some of the same subjects include those by *Goodman* [6.1], *Goodman* and *Wachman* [6.2], *Engel* and *Rieder* [6.3], *Celli* [6.4], *Barker* and *Auerbach* [6.5], and *Bortolani* and *Levi* [6.6], as well as the book edited by *Benedek* and *Valbusa* [6.7]. A fuller review of the experiments is given in Chap. 5.

In addition to He TOF scattering, other atom beam techniques used in the study of surface phonon spectra include He beam scattering with a LiF crystal analyzer [6.8] and Ne beam scattering [6.9]. Most of what follows also applies to the analysis of these experiments, although we do not consider in detail the Ne–surface potential.

6.1 Kinematics

Most of the data analysis requires nothing more than simple kinematics, namely the conservation laws for energy and for the lateral components of momentum in the atom–surface collision. In the notation consistently used in this book, the z axis is perpendicular to the mean plane of the surface and points towards the vacuum; "lateral" vectors are perpendicular to the z axis and are denoted by the subscript $\parallel$, reciprocal lattice vectors of the surface mesh are denoted by $g_\parallel$. The wave vector of the incident atom is denoted by $k_i = (k_{i\parallel}, k_{iz})$, with $k_{iz} < 0$, and the corresponding initial energy is $\varepsilon_i = \hbar^2 k_i^2/2m$; the final wave vector is $k_f = (k_{f\parallel}, k_{fz})$, with $k_{fz} > 0$, and the corresponding final energy is $\varepsilon_f = \hbar^2 k_f^2/2m$. The energy gain $\Delta\omega$ and the lateral momentum transfer $\hbar\Delta k_\parallel$ are given by

$$\hbar\Delta\omega = \varepsilon_f - \varepsilon_i , \tag{6.1}$$

$$\Delta k_\parallel = k_{f\parallel} - k_{i\parallel} . \tag{6.2}$$

Monoenergetic, well collimated He beams have a well-defined k_i; the final angles θ_f and ϕ_f are set by the detector's position, and time-of-flight (TOF) measurements give ε_f. From this information $\Delta\omega$ and $\Delta k_\parallel$ are uniquely inferred, using (6.1) and (6.2).

Springer Series in Surface Sciences, Vol. 21   **Surface Phonons**   167
Editors: W. Kress · F. W. de Wette   © Springer-Verlag Berlin, Heidelberg 1991

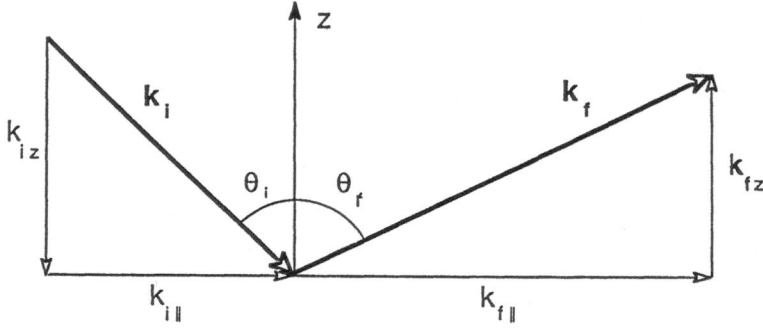

Fig. 6.1. Geometry of an in-plane inelastic scattering process

On a given TOF run, the values of $\Delta\omega$ and $\Delta k_\parallel$ for different arrival times (different k_f) lie on a parabola called the scan curve. In the usual case of in-plane scattering ($\phi_f = 0$), (6.2) reduces to $\Delta k_\parallel = k_f \sin\theta_f - k_i \sin\theta_i$ (see Fig. 6.1) and (6.1) gives the scan curve

$$\hbar\Delta\omega = \varepsilon_i \left[\frac{(\sin\theta_i + \Delta k_\parallel/k_i)^2}{\sin^2\theta_f} - 1 \right] . \tag{6.3}$$

The phonon frequencies of the semi-infinite crystal are denoted by $\omega(\boldsymbol{q}_\parallel, j)$, where $\boldsymbol{q}_\parallel$ is the lateral phonon momentum and j stands for all other phonon quantum numbers. Quite generally, $\omega(\boldsymbol{q}_\parallel + \boldsymbol{g}_\parallel, j) = \omega(\boldsymbol{q}_\parallel, j)$ and $\omega(-\boldsymbol{q}_\parallel, j) = \omega(\boldsymbol{q}_\parallel, j)$ [6.10]. It is convenient to take $\omega(\boldsymbol{q}_\parallel, j)$ to be positive for phonon absorption (annihilation) and negative for phonon emission (creation). It is also understood that the wave vector of an absorbed phonon is $\boldsymbol{q}_\parallel + \boldsymbol{g}_\parallel$, that of an emitted phonon is $-\boldsymbol{q}_\parallel - \boldsymbol{g}_\parallel$. With these conventions both creation and annihilation processes can be displayed on the same graph (see Fig. 6.2), and in every case for one-phonon processes $\Delta k_\parallel = \boldsymbol{q}_\parallel + \boldsymbol{g}_\parallel$ and $\Delta\omega = \omega(\boldsymbol{q}_\parallel, j)$. Thus the dispersion relations of surface phonons are directly determined by the location of peaks in the one-phonon cross sections.

From the point of view of surface phonon spectroscopy, multiphonon processes, which are difficult to disentangle, are in general a nuisance. An exception is the case of a nondispersive surface mode (an Einstein oscillator), where of course multiphonon processes appear simply as harmonics [6.11, 12].

The first question to be answered is then: when are the one-phonon processes clearly visible over the multiphonon background? This is almost the same as the more theoretical question: when is the one-phonon exchange approximation valid? Empirically, a TOF spectrum where one-phonon processes are dominant is often easy to recognize from the fact that the scattered intensity is consistently larger in regions of the $(\omega, \Delta k_\parallel)$ plane where one expects to find the surface-projected frequencies of the bulk phonons, assumed to be known. These are the shaded regions in Fig. 6.2. Outside these regions one may see sharp peaks,

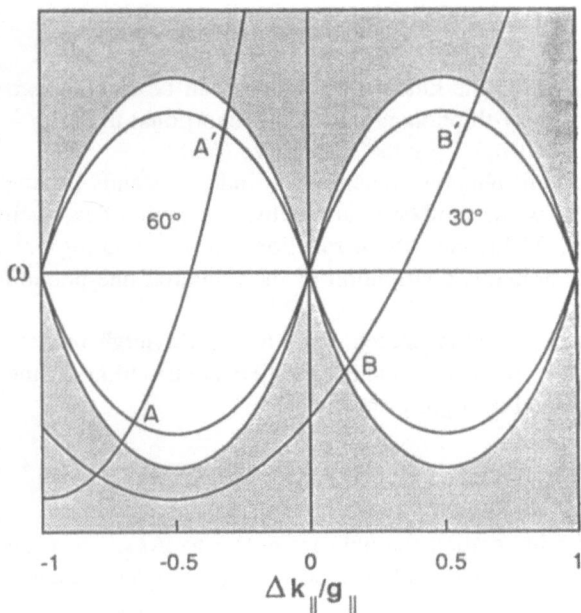

Fig. 6.2. Two scan curves, for incident angles θ_i of 30° and 60°, intersect a Rayleigh phonon branch at A, A' and at B, B'. Sharp peaks of the reflection coefficient, owing to one-phonon exchange, are expected at the corresponding values of $\Delta\omega$ and $\Delta k_{\parallel}$. One-phonon exchange processes also occur where the scan curves cross the bulk phonon continuum (shaded area). The scan curves are shown for $\theta_i + \theta_f = 90°$, hence the 45° curve (not shown) passes through the origin

corresponding to surface phonons, that can usually be associated with an edge of the bulk phonon continuum.

This review discusses the inelastic cross sections in the one-phonon approximation. Multiphonon processes generally cause a reduction of the one-phonon cross sections. This reduction is analogous to the Debye–Waller reduction of (elastic) diffraction due to the competition of inelastic processes (involving one or more phonons). Thus when the Debye–Waller factor e^{-2W} is small, or $2W > 1$, for a typical momentum transfer, most one-phonon processes are swamped out. A rough criterion, suggested by *Weare* [6.13], is obtained by using $2W = \Delta k^2 \langle u^2 \rangle / 3$, where Δk can be approximated by its value for specular scattering, $\Delta k = 2|k_{iz}|$, and $\langle u^2 \rangle$ is an effective mean square thermal displacement of the surface atoms. In the Debye model with characteristic temperature T_D, $\langle u^2 \rangle = 9\hbar^2 T / M k_B T_D^2$, where k_B is Boltzmann's constant, M is an average mass of the surface atoms, and T is the surface temperature; it is supposed that $T > T_D/10$, so that quantum corrections are negligible. As discussed again in Sect. 6.3, T_D is close to the bulk Debye temperature. The criterion for one-phonon processes to be prominent, $W < 1$, can then be written as

$$24(m/M)T\varepsilon_i \cos^2\theta_i < k_B T_D^2 \ . \tag{6.4}$$

169

6.1.1 Kinematic Focusing

For in-plane one-phonon scattering, the kinematic relations can be summarized as follows: the contribution of the j-th phonon branch is proportional to $\delta(\Delta\omega - \omega(\Delta k_\parallel, j))$, where $\Delta\omega$ is related to $\Delta k_\parallel$ by the scan curve (6.3).

For the surface-projected bulk phonon branches, the index j stands for the quasi-continuous perpendicular wave number q_z and a discrete band index, such as longitudinal acoustic (LA). In this case the δ functions for neighboring values of q_z overlap, giving a continuous distribution of the scattered one-phonon intensity.

On the other hand, a surface phonon branch, such as the Rayleigh branch, corresponds to a discrete value of j and makes a δ-like contribution to the scattered intensity, with integrated strength proportional to

$$I_j(\theta_f) = \int \delta(\Delta\omega - \omega(\Delta k_\parallel, j)) d\Delta\omega .\qquad(6.5)$$

It is convenient to change the integration variable from $\Delta\omega$ to $\Delta k_\parallel$. For each intersection of the j-th phonon branch with the scan curve we have then a contribution

$$I_j(\theta_f) = \frac{\partial\Delta\omega/\partial\Delta k_\parallel}{|\partial\Delta\omega/\partial\Delta k_\parallel - \partial\omega(\Delta k_\parallel, j)/\partial\Delta k_\parallel|} ,\qquad(6.6)$$

which diverges when the scan curve osculates the phonon dispersion curve, as shown in Fig. 6.3. This condition is known as kinematic focusing [6.14].

To evaluate the singular behavior of $I_j(\theta_f)$, expand the argument of the δ function in (6.5). For brevity, define $\bar{v} = \partial\Delta\omega/\partial\Delta k_\parallel$, $\bar{w} = \partial^2\Delta\omega/\partial\Delta k_\parallel^2$, and $\bar{w}_j = \partial^2\omega(q_\parallel, j)/\partial\Delta k_\parallel^2$ evaluated at the osculation point $\overline{\Delta k}_\parallel$ (point B in Fig. 6.3). If the experiment is carried out at fixed θ_i, $\partial\Delta\omega/\partial\theta_f = -(\varepsilon_f/\hbar)\cot\theta_f$, and we have

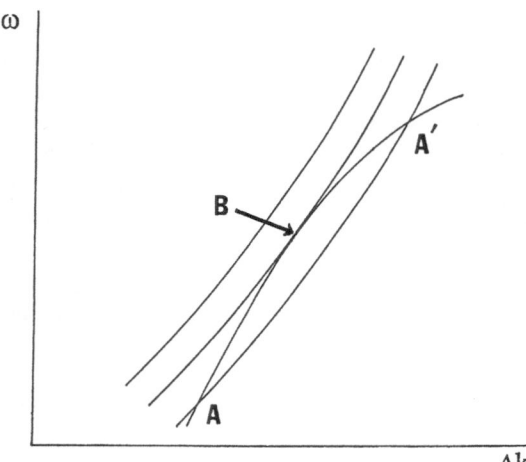

Fig. 6.3. At kinematic focusing, the critical scan curve ($\theta_f = \overline{\theta}_f$) osculates a surface phonon branch at B. Scan curves for $\theta_f < \overline{\theta}_f$ intersect the phonon branch at two points A, A' that coalesce into B

$$I_j(\theta_f) = \bar{v} \int \delta \left[\frac{1}{2}(\bar{w} - \bar{w}_j)(\Delta k_\parallel - \overline{\Delta k_\parallel})^2 - \varepsilon_f \cot \theta_f (\theta_f - \bar{\theta}_f)/\hbar \right] d\Delta k_\parallel$$

$$= \bar{v} \left[(\bar{w} - \bar{w}_j)(\theta_f - \bar{\theta}_f)\varepsilon_f \cot \theta_f / \hbar \right]^{-1/2} \tag{6.7}$$

for $\theta_f < \bar{\theta}_f$, and zero otherwise, in the case of Fig. 6.3. If, as is more usual, the experiment is carried out at fixed $\theta_i + \theta_f = 90°$, the result (6.7) is multiplied by $(1 + (k_f/k_i) \tan \theta_f)^{-1/2}$.

Many observations of kinematic focusing have been reported [6.15, 16]. The effect is not restricted to the surface phonon branches and can lead to enhancements in the continuum spectrum, as pointed out by Nichols and Weare [6.17]. The extension to out-of-plane scattering has been discussed by Avila and Lagos, who have proposed the use of kinematic focusing to measure phonon dispersion relations throughout the Brillouin zone without a TOF apparatus [6.18].

6.2 Dynamical Theory: General Considerations

Explicit formulae for inelastic processes in atom–surface scattering were first obtained by *Jackson* and *Mott* [6.19] and by *Lennard–Jones* et al. [6.20], and have been extended and rewritten by many authors [6.21–31], often without reference to earlier work. Notations and conventions differ, factors of L are sometimes omitted (see Sect. 6.2.1), and the emphasis may be on global energy exchange with the surface, rather than on the individual phonon processes that concern us most. As a result, comparison of different sources can be frustrating. In this section we briefly describe some nomenclature associated with the different approaches.

6.2.1 Box Normalization and Scattering Geometry

In developing the theory, one can work with wave functions normalized in a box, which we shall use here, or with some sort of continuum normalization, such as unit incident amplitude, or unit incident flux.

Box normalization in surface scattering means more than enclosing the whole system in a "quantization box" of volume L_b^3 and letting $L_b \to \infty$ at the end. The sample size must also approach infinity. Atoms scatter (reflect) from an area L_s^2 on the sample's top surface; usually L_s^2 is less than the whole top surface area. The sample thickness, L_z, also appears in the theory through the normalization of bulk phonon modes. The experimental setup corresponds loosely to the scattering theory or Fraunhofer geometry of Fig. 6.4, in which the sample is put at the center of a cubic box of size $L_b \gg L_s$; another convenient theoretical construction, which may be called the diffraction theory or Fresnel geometry, is to erect on top of the surface a box of height L_b and base L_s^2, with $L_s \gg L_b$. See Fig. 6.5. In

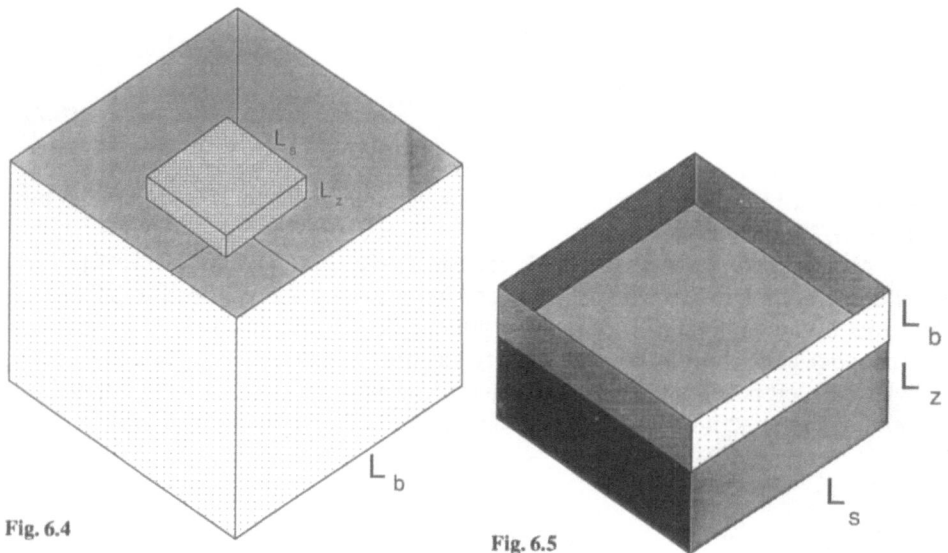

Fig. 6.4

Fig. 6.5

Fig. 6.4. In the scattering-theory or Fraunhofer description of the He–surface interaction, a large quantization box of side L_b encloses the sample

Fig. 6.5. In the diffraction theory or Fresnel description, a quantization box of height L_b is erected on the sample's surface. L_b must be larger than the range of the atom–surface interactions. It is assumed that $L_b \ll L_s$, or else periodic boundary conditions are imposed in the directions parallel to the surface

the final formulae all large lenghts L, be they L_b, L_s, or L_z must independently cancel or be absorbed by the conversion of sums into integrals and of Kronecker deltas into Dirac delta functions, according to the model (for one dimension):

$$\sum_q = \int q/2\pi \,, \tag{6.8}$$

Kronecker $\delta_{qq'} = (2\pi/L)\delta(q - q')$ Dirac . \hfill (6.9)

When the phonon dynamics is computed by the slab method, however, it is convenient to keep L_z finite.

Continuum normalization directly gives the final formulae, but may require tricky manipulations of Dirac delta functions. Hybrid normalization schemes are also possible.

6.2.2 Cross Sections and Reflection Coefficients

In scattering theory, one often computes first a transition rate per unit time, which is then divided by the incident current density (dimensions: $time^{-1}$ $area^{-1}$) to obtain a cross section. Alternatively, one can directly obtain the differential cross sections from the asymptotic behavior of the wave function for large r, where r is the radial distance from the scatterer. (However, this method is usually replaced by the Fresnel method for surface processes.) Either way, all cross sections of the surface scattering problem come out proportional to $L_s^2 \cos \theta_i$, where L_s^2 is the illuminated surface area and θ_i is the incidence angle. This occurs because $L_s^2 \cos \theta_i$ is the geometrical cross section of the incident beam that is intercepted by the surface. Division of the cross section by L_s^2 gives a cross section per unit area, division by $L^2 \cos \theta_i$ gives a reflection coefficient.

The θ_i dependence of the scattered (or reflected) intensity is given by the cross section per unit area when the whole surface is illuminated by the incident beam, but it is given by the reflection coefficient when only a "spot" on the surface is illuminated. It may also happen that the beam is totally intercepted by the surface at small θ_i, but not at near grazing incidence.

Diffraction theory à la Fresnel (Fig. 6.5) naturally gives the reflection coefficients. In any case, edge effects are neglected.

6.2.3 State-to-State Cross Sections vs Differential Cross Sections

The distinction here is the same as in usual scattering theory, and applies equally well to transition rates, cross sections, or reflection coefficients. Box normalization naturally leads to state-to-state transition rates, as we will see in Sect. 6.3; the conversion to reflection coefficients is illustrated there. The quantity of interest is a reflection coefficient that has been averaged over the initial phonon distribution and integrated over all phonon processes that lead to a given k_f (or at least over the processes that are included in the theory). This quantity, denoted by $R(k_f, k_i)$, is state-to-state between the experimentally measurable states of the He atom. By summing $R(k_f, k_i)$ over a small range of k_f, and dividing by that range, we obtain the differential reflection coefficient $dR(k_f, k_i)/d^3 k_f$. (Continuum normalization directly gives this quantity.) In an experiment, $d^3 k_f$ is determined by the final energy range $d\varepsilon_f$ and by the solid angle $d\Omega_f$ subtended by the detector. By using $d^3 k_f = k_f^2 dk_f d\Omega_f$ and $d\varepsilon_f = \hbar^2 k_f dk_f / m$, we obtain

$$\frac{dR}{d\varepsilon_f d\Omega_f} = \frac{m k_f}{\hbar^2} \frac{dR}{d^3 k_f} \ . \tag{6.10}$$

6.3 One-Phonon Exchange Processes

We first derive and discuss the Distorted Wave Born Approximation (DWBA) formula, following closely the papers by *Cabrera* et al. [6.21] suitably updated. Since the extraction of phonon spectra from TOF data relies mostly on one-phonon processes, the first order DWBA is appropriate. Higher order corrections can be included approximately by a Debye–Waller attenuation factor of the form discussed in Sect. 6.1: e^{-2W} with $2W = \langle \Delta \mathbf{k} \cdot \mathbf{u} \rangle^2$. A great deal of effort has been devoted to the justification of this Debye–Waller factor, mostly for the attenuation of the diffraction peaks [6.29, 32, 33, 38, 39, 42–47]. *Armand* et al. [6.28, 42, 44, 45, 47] have attacked the problem by computing higher order terms in the DWBA series. In their conclusions they emphasize deviations from the standard formula, which however are simply the result of stopping the calculation at a lower order. That is, $1-2W$ badly underestimates e^{-2W}, and $1-2W+(2W)^2/2$ overestimates it, when $W \geq 1/2$. The other major approach to multiphonon processes has been by semiclassical approximations [6.29, 32–36]. These naturally lead to a formula of the Debye–Waller type, but with a modified exponent. The limitation of these calculations is that they use highly simplified models of the atom–surface interaction and of the surface phonon spectrum. Detailed calculations of the lowest order correction to elastic scattering, i.e. of $1-2W$, have been carried out for Ag and Pt [6.43, 46]: they support the simple expression with a $\langle u^2 \rangle$ that, by a cancellation of various corrections, is more or less the same as that of a bulk atom (and smaller than that of a surface atom). Although the results of [6.43, 46] have no direct bearing on the attenuation of the one-phonon intensity, they suggest that the standard form for $2W$ is at present still a good guess.

The concentration of this review on one-phonon processes does not mean that multiphonon effects are unimportant. In fact, even when the Weare criterion (6.4) is well satisfied, many phonons with $\hbar\omega < k_B T$ or $\hbar\omega \ll \varepsilon_i$ are exchanged in some of the collisions. The multiphonon background is not necessarily structureless, but the processes involving the exchange of a single higher-energy phonon are often clearly visible until $2W \gg 1$.

6.3.1 The Distorted Wave Born Approximation

We use box normalization with the geometry of Fig. 6.5, which has the advantage that the distorted wave functions are more easily described than for the geometry of Fig. 6.4.

The state-to-state transition rate is

$$\frac{2\pi}{\hbar} |T_{fi}|^2 \delta(E_f - E_i) , \tag{6.11}$$

where E_i and E_f are total energies and T_{fi} is a transition matrix element, to be computed. The shorthand index i stands for the initial wave vector k_i of the He atom and for the initial occupation numbers $n_i(\mathbf{q}_\parallel, j)$ of the normal modes of the crystal, and similarly for f. We are interested in transitions where $n_f(\mathbf{q}_\parallel, j) = n_i(\mathbf{q}_\parallel, j) \pm 1$ for a single phonon mode, and correspondingly

$$E_f - E_i = \varepsilon_f - \varepsilon_i \pm \hbar |\omega(\mathbf{q}_\parallel, j)| . \tag{6.12}$$

The lower sign corresponds to phonon absorption (or annihilation).

We need T_{fi} to first order in the displacements $u_{l\kappa}$ of the crystal atoms from their equilibrium positions $r_{l\parallel} + r_\kappa$. Here $r_{l\parallel}$ denotes a surface net vector and r_κ gives the position of the crystal atoms within a surface unit cell. It must be understood that this cell is large in the z direction, extending from $z = 0$ to $z = -L_z$. In the harmonic approximation, $u_{l\kappa}$ can be expanded in normal modes as

$$u_{l\kappa} = \sum_{q_\parallel j} \left(\frac{\hbar}{2M_\text{c} N_\text{s} |\omega(q_\parallel, j)|} \right)^{1/2} e^{iq_\parallel \cdot r_{l\parallel}} e_\kappa(q_\parallel, j)$$
$$\times \left[a(q_\parallel, j) + a^\dagger(-q_\parallel, j) \right] , \tag{6.13}$$

where $a(q_\parallel, j)$ and $a^\dagger(-q_\parallel, j)$ are the usual annihilation and creation operators, N_s is the number of cells in the surface area L_s^2, and $e_\kappa(q_\parallel, j)$ is a polarization vector, normalized to

$$\sum_\kappa M_\kappa e_\kappa(q_\parallel, j)^* \cdot e_\kappa(q_\parallel, j') = M_\text{c} \delta_{jj'} , \tag{6.14}$$

where M_κ is the mass of the κ-th atom and M_c is the mass in the surface unit cell.

For any atom–surface interaction $V(r, \{u_{l\kappa}\})$, expansion to first order in $u_{l\kappa}$ gives

$$V(r, \{u_{l\kappa}\}) = V_{\text{static}}(r) + \sum_{l\kappa} F_{l\kappa}(r) \cdot u_{l\kappa} , \tag{6.15}$$

where $F_{l\kappa}(r)$ is the force exerted on the He atom by the atom, or ion, at $r_{l\parallel} + r_\kappa$.

To be definite, assume that the atom–surface potential V is a sum of two-body potentials V_κ that depend only on the separation between the position of the He atom, r, and the positions of the atoms in the crystal:

$$V(r, \{u_{l\kappa}\}) = \sum_{l\kappa} V_\kappa(r - r_{l\parallel} - r_\kappa - u_{l\kappa}) . \tag{6.16}$$

Then we have explicitly

$$V_{\text{static}}(r) = \sum_{l\kappa} V_{l\kappa}(r - r_{l\parallel} - r_\kappa) , \tag{6.17}$$

$$F_{l\kappa}(r) = -\nabla V_\kappa(r - r_{l\parallel} - r_\kappa) . \tag{6.18}$$

We assume that the eigenfunctions of V_{static} are known: these will be the distorted waves in our DWBA. In particular, we need the eigenfunction $\chi_\text{i}(r)^{(+)}$ that corresponds to the incoming (incident) wave $\exp(ik_\text{i} \cdot r)$ and the eigenfunction $\chi_\text{f}(r)^{(-)}$ that corresponds to the single outgoing wave $\exp(ik_\text{f} \cdot r)$ with all the incoming waves necessary to make up this outgoing wave. The superscripts $(+)$ and $(-)$ explicitly denote the boundary conditions (given incoming wave and given outgoing wave, respectively).

We are now in a position to compute T_{fi}. According to the DWBA for the potential (6.16), we have

$$T_{\mathrm{fi}} = \sum_{l\kappa} \int d^3 r \chi_{\mathrm{f}}(\boldsymbol{r})^{(-)*} \, F_{l\kappa}(\boldsymbol{r}) \chi_{\mathrm{i}}(\boldsymbol{r})^{(+)} \cdot \langle n_{\mathrm{i}}(\boldsymbol{q}_{\parallel}, j) \pm 1 | u_{l\kappa} | n_{\mathrm{i}}(\boldsymbol{q}_{\parallel}, j) \rangle \quad (6.19)$$

for the transition involving the emission or absorption of a single phonon in the $(\boldsymbol{q}_{\parallel}, j)$ normal mode.

6.3.2 The Phonon Matrix Elements

Using the expansion (6.13) and the well-known properties of the creation and annihilation operators, we find for the phonon absorption process

$$\langle n_{\mathrm{i}}(\boldsymbol{q}_{\parallel}, j) - 1 | u_{l\kappa} | n_{\mathrm{i}}(\boldsymbol{q}_{\parallel}, j) \rangle = \left(\frac{\hbar n_{\mathrm{i}}(\boldsymbol{q}_{\parallel}, j)}{2 M_{\mathrm{c}} N_{\mathrm{s}} |\omega(\boldsymbol{q}_{\parallel}, j)|} \right)^{1/2} e^{i\boldsymbol{q}_{\parallel} \cdot \boldsymbol{r}_{l\parallel}} e_{\kappa}(\boldsymbol{q}_{\parallel}, j). \quad (6.20)$$

Initially the crystal is in thermal equilibrium at temperature T and the average value of $n_{\mathrm{i}}(\boldsymbol{q}_{\parallel}, j)$ is $\langle n_{\mathrm{i}}(\boldsymbol{q}_{\parallel}, j) \rangle = n(|\omega(\boldsymbol{q}_{\parallel}, j)|)$, where $n(\omega)$ is the Bose function

$$n(\omega) = \left[\exp(\hbar\omega / k_{\mathrm{B}} T) - 1 \right]^{-1} . \tag{6.21}$$

The transition rate for all possible one-phonon absorption processes, thermally averaged over initial states, is then, from (6.11, 12, 19)

$$2\pi \sum_{\boldsymbol{q}_{\parallel} j} \frac{n(|\omega(\boldsymbol{q}_{\parallel}, j)|)}{2 M_{\mathrm{c}} N_{\mathrm{s}} |\omega(\boldsymbol{q}_{\parallel}, j)|} |M_{\mathrm{fi}}|^2 \delta(\varepsilon_{\mathrm{f}} - \varepsilon_{\mathrm{i}} - \hbar|\omega(\boldsymbol{q}_{\parallel}, j)|) , \tag{6.22}$$

$$M_{\mathrm{fi}} = \int d^3 r \chi_{\mathrm{f}}(\boldsymbol{r})^{(-)*} \left[\sum_{l\kappa} F_{l\kappa}(\boldsymbol{r}) \cdot e_{\kappa}(\boldsymbol{q}_{\parallel}, j) e^{i\boldsymbol{q}_{\parallel} \cdot \boldsymbol{r}_{l\parallel}} \right] \chi_{\mathrm{i}}(\boldsymbol{r})^{(+)} . \tag{6.23}$$

When the procedure leading to (6.22) is repeated for one-phonon emission processes, we find three differences: $\boldsymbol{q}_{\parallel}$ is replaced by $-\boldsymbol{q}_{\parallel}$, which is irrelevant when we sum over all $\boldsymbol{q}_{\parallel}$; $n(|\omega(\boldsymbol{q}_{\parallel}, j)|)$ is replaced by $n(|\omega(\boldsymbol{q}_{\parallel}, j)|) + 1$; and, in the argument of the δ function, $-\hbar|\omega(\boldsymbol{q}_{\parallel}, j)|$ is replaced by $+\hbar|\omega(\boldsymbol{q}_{\parallel}, j)|$. If we let $\omega(\boldsymbol{q}_{\parallel}, j)$ have either sign, then $\delta(\varepsilon_{\mathrm{f}} - \varepsilon_{\mathrm{i}} - \hbar\omega)$ automatically picks the $+$ sign for absorption and the $-$ sign for emission, and we can write the thermally averaged transition rate for all one-phonon processes in the compact form

$$2\pi \sum_{\boldsymbol{q}_{\parallel} j} \frac{n(\omega(\boldsymbol{q}_{\parallel}, j))}{2 M_{\mathrm{c}} N_{\mathrm{s}} \omega(\boldsymbol{q}_{\parallel}, j)} |M_{\mathrm{fi}}|^2 \delta(\varepsilon_{\mathrm{f}} - \varepsilon_{\mathrm{i}} - \hbar\omega(\boldsymbol{q}_{\parallel}, j)) , \tag{6.24}$$

where, for creation processes, we have used the fact that $n(|\omega|) + 1 = -n(-|\omega|)$.

6.3.3 The Atom–Surface Matrix Elements

The problem is now reduced to the calculation of M_{fi} (6.23). It will be recalled that $\boldsymbol{F}_{l\kappa}(\boldsymbol{r})$ (6.8) is the force acting between the He atom and the atom, or ion, at $\boldsymbol{r}_{l\parallel} + \boldsymbol{r}_{\kappa}$, while $\boldsymbol{e}_{\kappa}(\boldsymbol{q}_{\parallel}, j)\mathrm{e}^{\mathrm{i}\boldsymbol{q}_{\parallel} \cdot \boldsymbol{r}_{l\parallel}}$ are the normalized displacements of the $(\boldsymbol{q}_{\parallel}, j)$ normal mode. Thus the bracket in (6.23) is the effective interaction of the He atom with the $(\boldsymbol{q}_{\parallel}, j)$ normal mode of vibration. It can be written in the form

$$\mathrm{e}^{-\mathrm{i}\boldsymbol{q}_{\parallel} \cdot \boldsymbol{r}} \sum_{l\kappa} \boldsymbol{F}_{l\kappa}(\boldsymbol{r}) \cdot \boldsymbol{e}_{\kappa}(\boldsymbol{q}_{\parallel}, j)\mathrm{e}^{-\mathrm{i}\boldsymbol{q}_{\parallel} \cdot (\boldsymbol{r} - \boldsymbol{r}_{l\parallel})} , \tag{6.25}$$

where the sum is a periodic function of $\boldsymbol{r}_{l\parallel}$.

The calculation is much simpler when the static corrugation can be neglected, as is the case for scattering from the (111) surfaces of fcc metals; therefore we consider this case first. The wave functions have the form

$$\chi_{\mathrm{i}}(\boldsymbol{r})^{(+)} = \chi(k_{\mathrm{i}z}|z)\mathrm{e}^{\mathrm{i}\boldsymbol{k}_{\mathrm{i}\parallel} \cdot \boldsymbol{r}}/(L_{\mathrm{s}}^2 L_{\mathrm{b}})^{1/2} , \tag{6.26}$$

$$\chi_{\mathrm{f}}(\boldsymbol{r})^{(-)} = \chi(k_{\mathrm{f}z}|z)\mathrm{e}^{\mathrm{i}\boldsymbol{k}_{\mathrm{f}\parallel} \cdot \boldsymbol{r}}/(L_{\mathrm{s}}^2 L_{\mathrm{b}})^{1/2} . \tag{6.27}$$

The sign of the perpendicular momentum (negative for $k_{\mathrm{i}z}$, positive for $k_{\mathrm{f}z}$) is sufficient to distinguish outgoing from incoming wave functions; anyhow, for a planar surface, the difference between $\chi^{(+)}$ and $\chi^{(-)}$ is an irrelevant phase factor; we may choose $\chi(k_z|z)$ to be real for convenience. Using (6.25–27) in the expression for M_{fi}, the integral over $\boldsymbol{r}_{\parallel}$ is seen to vanish unless $\boldsymbol{k}_{\mathrm{f}\parallel} = \boldsymbol{k}_{\mathrm{i}\parallel} + \boldsymbol{q}_{\parallel} + \boldsymbol{g}_{\parallel}$. This result expresses the conservation of lateral momentum, as anticipated in (6.2). We obtain

$$M_{fi} = \frac{1}{L_{\mathrm{b}} A_{\mathrm{c}}} \sum_{\kappa \boldsymbol{g}_{\parallel}} \delta_{\boldsymbol{k}_{\mathrm{f}\parallel} - \boldsymbol{k}_{\mathrm{i}\parallel} - \boldsymbol{q}_{\parallel} - \boldsymbol{g}_{\parallel}}$$

$$\times \int \chi(k_{\mathrm{f}z}|z)^* \boldsymbol{e}_{\kappa}(\boldsymbol{q}_{\parallel}, j) \cdot \boldsymbol{F}_{\kappa}(\boldsymbol{q}_{\parallel} + \boldsymbol{g}_{\parallel}|z)\chi(k_{\mathrm{i}z}|z)dz , \tag{6.28}$$

where $A_{\mathrm{c}} = L_{\mathrm{s}}^2/N_{\mathrm{s}}$ is the area of the surface unit cell. For a sum of two-body potentials

$$\boldsymbol{F}_{\kappa}(\boldsymbol{q}_{\parallel} + \boldsymbol{g}_{\parallel}|z) = -(\mathrm{i}(\boldsymbol{q}_{\parallel} + \boldsymbol{g}_{\parallel}), \partial/\partial z)V_{\kappa}(\boldsymbol{q}_{\parallel} + \boldsymbol{g}_{\parallel}|z - z_{\kappa})\mathrm{e}^{\mathrm{i}(\boldsymbol{q}_{\parallel} + \boldsymbol{g}_{\parallel}) \cdot \boldsymbol{r}_{\kappa}} , \tag{6.29}$$

where we have introduced the lateral Fourier transform of the two-body potential in the usual way:

$$V_{\kappa}(\boldsymbol{r}) = \int \mathrm{e}^{\mathrm{i}\boldsymbol{p}_{\parallel} \cdot \boldsymbol{r}_{\parallel}} V_{\kappa}(\boldsymbol{p}_{\parallel}|z)d^2 p_{\parallel}/4\pi^2 , \tag{6.30}$$

$$V_{\kappa}(\boldsymbol{p}_{\parallel}|z) = \int \mathrm{e}^{-\mathrm{i}\boldsymbol{p}_{\parallel} \cdot \boldsymbol{r}_{\parallel}} V_{\kappa}(\boldsymbol{r})d^2 r_{\parallel} . \tag{6.31}$$

More generally, $\boldsymbol{F}_{\kappa}(\boldsymbol{q}_{\parallel} + \boldsymbol{g}_{\parallel}|z)$ is the lateral Fourier transform of $\boldsymbol{F}_{\kappa}(\boldsymbol{r})$. The dominant contribution to M_{fi} comes usually from the top atom in each surface unit cell, which can be taken at $\boldsymbol{r}_{\kappa} = 0$.

Turning now to the more general case of a corrugated surface, we write the eigenfunctions of the static potential in the Bloch form:

$$\chi_i(\mathbf{r})^{(+)} = \sum_{\mathbf{g}_\parallel'} \chi(\mathbf{k}_{i\parallel} + \mathbf{g}_\parallel', k_{iz}|z) e^{i(\mathbf{k}_{i\parallel} + \mathbf{g}_\parallel') \cdot \mathbf{r}_\parallel} / (L_s^2 L_b)^{1/2} , \qquad (6.32)$$

$$\chi_f(\mathbf{r})^{(-)} = \sum_{\mathbf{g}_\parallel''} \chi(\mathbf{k}_{f\parallel} + \mathbf{g}_\parallel'', k_{fz}|z) e^{i(\mathbf{k}_{f\parallel} + \mathbf{g}_\parallel'') \cdot \mathbf{r}_\parallel} / (L_s^2 L_b)^{1/2} . \qquad (6.33)$$

Solution of the eigenvalue problem for V_{static} by the method of coupled channels, for instance, will automatically give the He wave functions in this form. Time-reversal invariance implies that

$$\chi(\mathbf{k}_\parallel + \mathbf{g}_\parallel, k_z|z) = \chi(-\mathbf{k}_\parallel - \mathbf{g}_\parallel, -k_z|z)^* . \qquad (6.34)$$

Further, if the surface has inversion symmetry, there is no distinction between $\mathbf{k}_\parallel + \mathbf{g}_\parallel$ and $-\mathbf{k}_\parallel - \mathbf{g}_\parallel$. The final result for M_{fi} is a generalization of (6.28) and (6.29): there is a triple sum over $\mathbf{g}_\parallel, \mathbf{g}_\parallel'$ and $\mathbf{g}_\parallel''$ that accounts for all possible umklapp processes, $\mathbf{g}_\parallel$ is replaced by $\mathbf{g}_\parallel + \mathbf{g}_\parallel' + \mathbf{g}_\parallel''$, $\chi(k_{iz}|z)$ is replaced by $\chi(\mathbf{k}_\parallel + \mathbf{g}_\parallel', k_{iz}|z)$, and similarly for $\chi(k_{fz}|z)$.

Calculations using these formulae have been carried out by *Eichenauer* et al. [6.48–50].

6.3.4 The Differential Reflection Coefficient

From the transition rate (6.24) upon dividing by the incident flux $\hbar k_i \cos \theta_i$, we obtain $R(\mathbf{k}_f, \mathbf{k}_i)$. This is the state-to-state reflection coefficient for the He atom, averaged over intial phonon states and summed over all one-phonon processes. Using the matrix element (6.28), we obtain

$$R(\mathbf{k}_f, \mathbf{k}_i) = \frac{\pi m}{L_b A_c^2 \hbar N_s M_c k_i \cos \theta_i} \sum_{\mathbf{q}_\parallel \mathbf{g}_\parallel j} \frac{n(\omega(\mathbf{q}_\parallel, j))}{\omega(\mathbf{q}_\parallel, j)} \delta_{\mathbf{k}_{f\parallel} - \mathbf{k}_{i\parallel} - \mathbf{q}_\parallel - \mathbf{g}_\parallel}$$

$$\times \delta(\varepsilon_f - \varepsilon_i - \hbar\omega(\mathbf{q}_\parallel, j)) \left| \sum_\kappa \mathbf{e}_\kappa(\mathbf{q}_\parallel, j) \cdot \mathbf{F}_\kappa(\mathbf{k}_f, \mathbf{k}_i) \right|^2 , \qquad (6.35)$$

where, with $\mathbf{F}_k(\mathbf{q}_\parallel + \mathbf{g}_\parallel|z)$ given by (6.29),

$$\mathbf{F}_\kappa(\mathbf{k}_f - \mathbf{k}_i) = \int dz \chi(k_{fz}|z)^* \mathbf{F}_\kappa(\mathbf{q}_\parallel + \mathbf{g}_\parallel|z) \chi(k_{iz}|z) . \qquad (6.36)$$

This is the matrix element of the force exerted by the κ-th atom on the impinging He.

To obtain the differential reflection coefficient $dR(\mathbf{k}_f, \mathbf{k}_i)/d^3 k_f$, we multiply $R(\mathbf{k}_f, \mathbf{k}_i)$ by the density of states $L_s^2 L_b/(2\pi)^3$. Finally, we use (6.10) to obtain the quantity $dR(k_f, k_i)/d\varepsilon_f d\Omega_f$ that is most easily compared with experiment. In the general case of a corrugated surface, the result is

$$\frac{dR(k_f, k_i)}{d\varepsilon_f d\Omega_f} = \frac{m^2 k_f}{8\pi^2 \hbar^3 M_c A_c k_i \cos\theta_i} \sum_j \frac{n(\omega(q_{\parallel}, j))}{\omega(q_{\parallel}, j)} \delta(\varepsilon_f - \varepsilon_i - \hbar\omega(q_{\parallel}, j))$$

$$\times \left| \sum_\kappa e_\kappa(q_{\parallel}, j) \cdot F_\kappa(k_f, k_i) \right|^2 , \tag{6.37}$$

where $q_{\parallel}$ differs from $k_{f\parallel} - k_{i\parallel}$ by a reciprocal lattice vector and, with the definitions of (6.32) and (6.33),

$$F_\kappa(k_f, k_i) = \sum_{g'_{\parallel}, g''_{\parallel}} \int dz \chi(g''_{\parallel}, k_{fz}|z)^* \sum_\kappa e_\kappa(q_{\parallel}, j)$$

$$\cdot F_\kappa(k_{f\parallel} - k_{i\parallel} - g'_{\parallel} + g''_{\parallel}|z) \chi(g'_{\parallel}, k_{iz}|z) . \tag{6.38}$$

As expected, all quantization lengths cancel out (recall that the χ eigenfunctions that appear in (6.36) and (6.38) are normalized to unit incoming wave $\exp(ik_{iz}z)$, according to (6.26)). Although M_c is proportional to the sample thickness L_z, the result (6.37) is also independent of L_z, as we now show. For the extended phonon modes, j consists of a discrete polarization index and of a quasi-continuum perpendicular wave vector index q_z, and $e_\kappa(q_{\parallel}, j)$ does not depend on L_z, according to the normalization condition (6.4). However, when the sum over q_z is converted to an integral according to (6.8), a factor $L_z/2\pi$ is introduced. For the surface phonon modes, the index j is discrete and $e_\kappa(q_{\parallel}, j)$ is proportional to $L_z^{1/2}$, decreasing exponentially with $|z_\kappa|$. Therefore in either case the r.h.s. of (6.36) is proportional to L_z/M_c, which is size-independent.

6.3.5 Relation to Phonon Density of States and Correlation Functions

It is often useful, at least conceptually, to write the reflection coefficient in terms of the local phonon density of states for lateral momentum $\hbar q_{\parallel}$, which is actually a matrix in the layer indices $\kappa\kappa'$ and in the Cartesian components $\alpha\beta$:

$$D_{\kappa\kappa'}^{\alpha\beta}(q_{\parallel}, \omega) = \sum_j \frac{\hbar}{2M_c A_c \omega} e_\kappa^\alpha(q_{\parallel}, j) e_\kappa^\beta(q_{\parallel}, j)^* \delta(\omega - \omega(q_{\parallel}, j)) . \tag{6.39}$$

We have

$$\frac{dR(k_f, k_i)}{d\varepsilon_f d\Omega_f} = \frac{m^2 k_f}{4\pi^2 \hbar^4 k_i \cos\theta_i} \sum_{\kappa\kappa'} \sum_{\alpha\beta}$$

$$\times D_{\kappa\kappa'}^{\alpha\beta}(\Delta k_{\parallel}, \Delta\omega) n(\Delta\omega) F_\kappa^\alpha(k_f, k_i) F_{\kappa'}^\beta(k_f, k_i)^* , \tag{6.40}$$

where $\Delta k_{\parallel} = k_{f\parallel} - k_{i\parallel}$ and $\hbar\Delta\omega = \varepsilon_f - \varepsilon_i$. If only the atoms in the top layer ($\kappa = 0$) contribute appreciably to the scattering, the reflection coefficient is proportional to the surface-projected phonon density of states $D_{00}^{\alpha\beta}(\Delta k_{\parallel}, \Delta\omega)$.

179

Another physically meaningful quantity is the displacement correlation function

$$
\begin{aligned}
\langle u_{l\kappa}^{\alpha}(t) u_{l'\kappa'}^{\beta}(0) \rangle &= \sum_{\boldsymbol{q}_{\parallel} j} \frac{\hbar n(\omega(\boldsymbol{q}_{\parallel}, j))}{2 M_{\mathrm{c}} N_{\mathrm{s}} \omega(\boldsymbol{q}_{\parallel}, j)} \\
&\quad \times \mathrm{e}^{\mathrm{i}\boldsymbol{q}_{\parallel} \cdot (\boldsymbol{r}_{l\parallel} - \boldsymbol{r}'_{l\parallel})} \mathrm{e}^{-\mathrm{i}\omega(\boldsymbol{q}_{\parallel}, j) t} e_{\kappa}^{\alpha}(\boldsymbol{q}_{\parallel}, j) e_{\kappa}^{\beta}(\boldsymbol{q}_{\parallel}, j)^{*} \\
&= \frac{A_{\mathrm{c}}}{N_{\mathrm{s}}} \sum_{\boldsymbol{q}_{\parallel}} \int d\omega \, n(\omega) \mathrm{e}^{\mathrm{i}\boldsymbol{q}_{\parallel} \cdot (\boldsymbol{r}_{l\parallel} - \boldsymbol{r}'_{l\parallel})} \mathrm{e}^{-\mathrm{i}\omega t} D_{\kappa\kappa'}^{\alpha\beta}(\boldsymbol{q}_{\parallel}, \omega) . \quad (6.41)
\end{aligned}
$$

The He reflection coefficient is related to this correlation function in the same way as the neutron cross section, for instance, is related to the correlation function in the bulk. The connection is established by the usual Glauber–Van Hove trick of writing from the start of the theory

$$
2\pi\delta(\varepsilon_{\mathrm{f}} - \varepsilon_{\mathrm{i}} - \hbar\omega(\boldsymbol{q}_{\parallel}, j)) = \int dt \, \mathrm{e}^{\mathrm{i}(\varepsilon_{\mathrm{f}} - \varepsilon_{\mathrm{i}} - \hbar\omega(\boldsymbol{q}_{\parallel}, j))t/\hbar} \qquad (6.42)
$$

or, more directly at this point, by inverting (6.41) and inserting (6.40). The result is

$$
\begin{aligned}
\frac{dR(\boldsymbol{k}_{\mathrm{f}}, \boldsymbol{k}_{\mathrm{i}})}{d\varepsilon_{\mathrm{f}} d\Omega_{\mathrm{f}}} &= \frac{m^2 k_{\mathrm{f}}}{4\pi^2 \hbar^4 M_{\mathrm{c}} A_{\mathrm{c}} k_{\mathrm{i}} \cos\theta_{\mathrm{i}}} \sum_{\boldsymbol{r}_{l\parallel}} \int dt \, \mathrm{e}^{-\mathrm{i}\boldsymbol{q}_{\parallel} \cdot \boldsymbol{r}_{l\parallel}} \mathrm{e}^{\mathrm{i}(\varepsilon_{\mathrm{f}} - \varepsilon_{\mathrm{i}})t/\hbar} \\
&\quad \times \sum_{\kappa\kappa'} \sum_{\alpha\beta} \langle u_{l\kappa}^{\alpha}(t) u_{0\kappa'}^{\beta}(0) \rangle F_{\kappa}^{\alpha}(\boldsymbol{k}_{\mathrm{f}}, \boldsymbol{k}_{\mathrm{i}}) F_{\kappa'}^{\beta}(\boldsymbol{k}_{\mathrm{f}}, \boldsymbol{k}_{\mathrm{i}})^{*} . \qquad (6.43)
\end{aligned}
$$

This expression, like the earlier ones, simplifies if only the surface atoms contribute ($\kappa = \kappa' = 0$), if there is enough symmetry that correlations between different components vanish (only $\beta = \alpha$ remains), and even more for Einstein oscillators (only the $\boldsymbol{r}_{l\parallel} = 0$ term contributes and the scattered distribution is that due to independent oscillators in the presence of the average surface potential).

6.4 The Inelastic Atom–Surface Interaction

In the previous section, we have expressed all the one-phonon DWBA results in terms of the eigenfunctions χ of the static potential V_{static} and of the force $\boldsymbol{F}_{l\kappa}$ between the He atom and the surface atom at $\boldsymbol{r}_{l\kappa} = \boldsymbol{r}_{l\parallel} + \boldsymbol{r}_{\kappa'}$. V_{static} and $\boldsymbol{F}_{l\kappa}$ should be derived from the same model of the atom–surface interaction. A sum of rigid pairwise potentials provides such a model and leads to (6.17) and (6.18). Each of these expressions is the first term in an expansion that may include significant three-body and four-body terms. In the case of simple insulators one knows, at least in principle, how to compute the various terms; in the case of metals there are still many open questions. We discuss very briefly the general theory of the atom–surface interaction and focus on the semiempirical formulae that have been used extensively for the calculation of scattered inelastic intensities by *Bortolani* et al. [6.51–64].

It is common (but not universal) practice to divide the interaction into an attractive part V_{att} due to Van der Waals forces and a repulsive part V_{rep} due to the overlap of the surface wave functions with the closed-shell He orbitals. The theory has been developed for the static interaction, but the results can be taken over directly to the dynamic interaction, within the usual Born–Oppenheimer approximation.

6.4.1 The Static Repulsive Potential

It has been shown by several methods that to a good approximation

$$V_{rep}(r) = A\varrho(r) , \tag{6.44}$$

where $\varrho(r)$ is the unperturbed electronic density of the surface and the constant A depends only weakly on the nature of the surface. Unfortunately in the important case of metal surfaces different approaches yield different values of A. *Esbjerg* and *Norskov* [6.65] and *Stott* and *Zaremba* [6.66] were the first to obtain (6.44) for a He atom immersed in a homogeneous electron gas with a positive uniform charge background (the jellium model). The values of A calculated from this effective medium approach range from 305 to 329 eV a_{Bohr}^3, with the latter value presumed to be the most accurate [6.67]. However, the He atom lies outside the jellium background in a region of varying $\varrho(r)$. When the appropriate corrections are made, (6.44) still holds, but the effective A changes from 329 to 255 eV a_{Bohr}^3. Hartree–Fock calculations for He interacting with a metal atom cluster can also be fitted to (6.44) with $A = 373$ eV a_{Bohr}^3 [6.68].

A different approach starts with the He atom well outside the surface and treats it as a perturbation on the surface electrons. This approach was successfully used by *Zaremba* and *Kohn* [6.69] to predict the laterally averaged interaction with metal surfaces and was perfected by *Harris* and *Liebsch* [6.70, 71]. By carrying out the calculation to second order in the He–surface overlap integral, *Nordlander* and *Harris* [6.72, 73] arrive at (6.44) again, but with $A \cong 500$ eV a_{Bohr}^3. When the He is well outside the surface, it is actually possible to infer V_{rep} from e–He scattering data. In the simplest approximation, which goes back to Fermi [6.74], the He atom is represented by a pseudopotential $(2\pi\hbar^2 a_s/m)\delta(r_e - r)$, where a_s is the s-wave scattering length at zero energy. This gives (6.44) with $A = 2\pi\hbar^2 a_s/m \cong 200$ eV a_{Bohr}^3. However, one should exclude the contribution to a_s coming from the long-range polarization force, because this force cannot be represented by a δ-like pseudopotential and in fact gives rise to the Van der Waals attraction [6.75]. Further, the e–He scattering data must be continued to negative energies of a few eV, because the surface electrons are bound. The resulting best estimate of A [6.76, 77] is about 500 eV a_{Bohr}^3, in agreement with the *Nordlander* and *Harris* [6.73] calculations, but about twice the value given by the effective medium theory [6.67]. However, these refinements of the pseudopotential method show that it is subject to considerable uncertainty, so that values of A ranging from 300 to 600 eV a_{Bohr}^3 are obtained, depending also on the surface work function [6.77].

For insulators, one can start directly from pairwise repulsive interactions between the He atom and the surface constituents, which in first approximation can be taken to be undistorted atoms in the case of graphite and of noble gas overlayers [6.78–82], and undistorted ions in the case of LiF(100) [6.83, 84]. In these cases $\varrho(r)$ is by construction the sum of the densities of the constituents and (6.44) turns out to be valid within the uncertainties of the calculation. The fact that one can take undistorted ions for LiF(100), however, results from accidental cancellation of several corrections; in the other ionic crystal surface that has been studied in detail, NaCl(100), ion distortion is important, but good agreement with He diffraction data is obtained also in this case when all the corrections to the simple model are carefully evaluated [6.85].

The main trouble with the attractively simple formula (6.44) is that, while it works well for the laterally averaged atom–surface potential, it generally overestimates the effective corrugation for the metal surfaces that have been carefully studied, such as Cu, Ni and Ag(111). Actually, the correct corrugation is obtained if V_{att} is neglected altogether [6.86], but this is of course inconsistent. The effect of V_{att} is to move the classical turning point of the collision closer to the surface, where the variation of the density is greater [6.87]. *Takada* and *Kohn* [6.76] have pointed out that in the pseudopotential approach the discrepancy could in principle be removed by the contribution of p-wave scattering; however, this correction turns out to be small for Cu, Ni and Ag(111) [6.88, 89]. Corrections due to the presence of empty states above the Fermi level have been proposed by *Annett* and *Haydock* [6.90–92] and are still a subject of controversy: Annett finds that they are sufficient to remove the discrepancy with experiment, while *Harris* and *Zaremba* [6.93] find them negligible. It is possible that the error comes from uncertainties in the separation of V_{rep} from the attractive part of the potential, to be discussed below. It could also come from a failure of the methods used to compute $\varrho(r)$ in the region of interest for the scattering of 20–60 meV He atoms ($z = 3$–4 Å). Surprisingly, the full calculation of $\varrho(r)$ gives a result that, in the region of interest, is practically indistinguishable [6.94] from a simple superposition of atomic Cu densities, as tabulated [6.95] by *Herman* et al.

As a practical matter the laterally averaged He–surface repulsion is generally well approximated by

$$V_{rep}(z) = U_{rep} e^{-\beta z} , \tag{6.45}$$

which is consistent with a pairwise sum of Yukawa potentials $U_\kappa e^{-\beta r}/\beta r$ over a monatomic top layer, with $U_{rep} = 2\pi U_\kappa/\beta^2 A_c$. However, it is more customary to approximate the pairwise repulsion by the Born–Mayer form $U_\kappa e^{-\beta r}$, which then gives $(2\pi U_\kappa/\beta^2 A_c)(1 + \beta z)e^{-\beta z}$ instead of (6.45). If V_{att} is neglected, the potential (6.45) defines the simple soft-wall model, and for any exponential-type potential β is called the softness parameter. The hard-wall model is recovered in the limit $\beta \to \infty$.

6.4.2 The Static Attractive Potential

V_{att} is usually dominated by the Van der Waals dispersion force. At large distances z from the surface (but not large enough that retardation effects come into play), $V_{att} = -C_3/z^3$. The exact value of C_3 is given by the Lifshitz formula

$$C_3 = \frac{\hbar}{4\pi} \int_0^\infty du \frac{\varepsilon(iu) - 1}{\varepsilon(iu) + 1} \alpha_{He}(iu) , \qquad (6.46)$$

where ε is the dielectric function of the surface material and α_{He} is the He atom polarizability. The extension of ε and α_{He} to imaginary frequencies iu is easily performed if these functions are known for real frequencies; further, the integral is numerically convenient because $\varepsilon(iu)$ and $\alpha(iu)$ are real, non-negative, and monotonically decreasing functions of u. Thus C_3 can often be accurately evaluated, but the corrections to $-C_3/z^3$ are much less certain.

For a crystal consisting of localized electronic units (atoms, molecules, or ions) having polarizability α, the crudest way to compute V_{att} is to sum pairwise interactions $-c_6/|\boldsymbol{r} - \boldsymbol{r}_l|^6$, where

$$c_6 = \frac{3\hbar}{\pi} \int_0^\infty \alpha(iu)\alpha_{He}(iu)du . \qquad (6.47)$$

The sum of $-c_6/|\boldsymbol{r} - \boldsymbol{r}_l|^6$ over a monatomic plane gives a laterally averaged interaction $-C_4/(z-z_l)^4$, with $C_4 = (\pi/2A_c)c_6$. A_c is the area per atom. The sum over equally spaced planes of $-C_4/(z-z_l)^4$ gives asymptotically $-C_3(z-d/2)^3$, where the origin of z is on the top atom plane and d is the interplanar spacing. The value of C_3 is $C_4/3d$, or $(\pi/6A_cd)c_6$, in agreement with (6.46) in the low-density limit where $\varepsilon = 1 + 4\pi\alpha/A_cd$. However, $-C_3/(z - d/2)^3$ gives a poor approximation to the sum over planes in the region of interest ($z \cong 3$ Å); a better approximation is

$$V_{att}(z) = -C_3/(z + d/2)^3 - C_4/z^4 , \qquad (6.48)$$

which treats the top atom layer separately.

A systematic improvement of the pairwise sum formula can be obtained by including n-body forces, beginning with the Axilrod–Teller three body interaction [6.96–98]. The result is approximately of the form (6.48), with C_3 now given by (6.46) and C_4 in general different from $3dC_3$.

A very general expression for V_{att} can be written down, assuming only that the He charge density does not overlap the surface charge density and that the response to electrical fluctuations is linear. In the limit when retardation effects can be neglected, V_{att} involves only the surface density response function $\chi_{g_\parallel}(z_e, z'_e, \boldsymbol{q}_\parallel, \omega)$ and the 2^l-pole He polarizabilities $\alpha_{He}^{(l)}(\omega)$. The surface density response to an external electric potential $\Phi(z'_e, \boldsymbol{q}_\parallel, \omega)$ is in general given by

$$e\varrho(z_e, \boldsymbol{q}_\parallel, \omega) = \sum_{\boldsymbol{g}_\parallel} \int dz'_e \chi_{g_\parallel}(z_e, z'_e, \boldsymbol{q}_\parallel, \omega)\Phi(z'_e, \boldsymbol{q}_\parallel - \boldsymbol{g}_\parallel, \omega) . \qquad (6.49)$$

When the surface corrugation is negligible, one finds [6.99]

$$V_{att}(z) = - \sum_{l=1}^{\infty} \frac{2^l}{2\pi l!(2l-1)!!} \int_0^{\infty} dq_{\parallel} \int_0^{\infty} du \, q_{\parallel}^{2l}$$
$$\times \chi(q_{\parallel}, iu)\alpha_{He}^{(l)}(iu)e^{-2q_{\parallel}z} , \tag{6.50}$$

where (with the medium extending to $z_e = -\infty$)

$$\chi(q_{\parallel}, \omega) = \frac{2\pi}{q_{\parallel}} \int_{-\infty}^{\infty} dz_e \int_{-\infty}^{\infty} dz'_e \chi_0(z_e, z'_e, q_{\parallel}, \omega)e^{q_{\parallel}(z_e + z'_e)} . \tag{6.51}$$

The difficulty of course is in the evaluation of $\chi(q_{\parallel}, \omega)$. For large z, V_{att} is obtained by expanding

$$\chi(q_{\parallel}, \omega) = \chi(0, \omega) + q_{\parallel}\chi'(0, \omega) + \tfrac{1}{2}q_{\parallel}^2\chi''(q, \omega) . \tag{6.52}$$

Since $\chi(0, \omega) = [\varepsilon(\omega) - 1]/[\varepsilon(\omega) + 1]$, the leading term in (6.50) (from $l = 1$) is $-C_3/z^3$, with C_3 given by (6.46). The next term, coming from $q\chi'(0, \omega)$, is proportional to z^{-4}. The result can then be written in the form (6.48) with an appropriate value of C_4. However, it is possible to make $\chi'(q, \omega)$ vanish by shifting the origin of z_e and z'_e by an appropriate amount z_{vW}; the origin of the He coordinate z must of course be shifted by z_{vW} too. One can then write, up to terms of order z^{-5},

$$V_{att}(z) = -C_3/(z - z_{vW})^3 . \tag{6.53}$$

The value of z_{vW} for a variety of surfaces (jellium model and noble metals) was first given by *Zaremba* and *Kohn* [6.100]. Later papers by *Liebsch* [6.101] suggested that the exact values of z_{vW} are uniformly less – typically about 20% closer to the surface than the ZK values. Based on these values, the currently accepted z_{vW} for the noble metals is close to $d/2$, which, in the jellium model, is the jellium edge. The results are summarized in Table I of Ref. 101.

The form (6.53) has become customary for metal surfaces, although the pairwise sum approach suggests that (6.48) is a better approximation. Comparison of (6.48) and (6.53) shows that z_{vW} differs from $d/2$ if C_4 for the topmost atom plane differs from $3dC_3$.

Corrections to (6.48) or (6.53) are of three kinds: (i) multipolar interactions, (ii) charge overlap, and (iii) lateral variation.

(i) Multipolar corrections from (6.50) begin with $-C_5/z^5$ and consist of two terms, one coming from $\alpha_{He}^{(2)}$ and $\chi(0, \omega)$ (He quadrupole–surface dipole), the other from α_{He} and $q^2\chi''(0, \omega)/2$ (He dipole-surface quadrupole). The latter is generally more important and hard to estimate. Here again, the pairwise summation model gives explicit results. Pairwise multipolar terms c_8, c_{10}, and c_{12} are routinely included in interatomic potentials: they are of the form (6.47) with α and/or α_{He} replaced by $\alpha^{(l)}$ and $\alpha_{He}^{(l)}$. The pairwise sum of $-c_{2n}/|r - r_l|^{2n}$ gives asymptotically $-C_{2n-3}/z^{2n-3}$ with $C_{2n-3} = [\pi/2n(2n-1)](N_a/d)c_{2n}$.

(ii) The effect of charge overlap is to reduce the attraction at close range. In pairwise interactions it is found semiempirically that the reduction of $-c_n/|r - r_l|^n$ is often well described by the Tang–Toennies "damping factor" [6.102]

$$f_n(x) = 1 - e^{-x} \sum_{k=1}^{n} \frac{x^k}{k!} , \tag{6.54}$$

where $x = \beta|r - r_l|$ and β is approximately the same parameter that describes the fall-off of V_{rep}. Other "damping factors" have been proposed, but (6.54) works as well as any and does not introduce new parameters. A pairwise sum of Tang–Toennies interactions gives the following modification of (6.48):

$$V_{att}(z) = -C_3 F_3[\beta(z + d/2)]/(z + d/2)^3 - C_4 F_4(\beta z)/z^4 \tag{6.55}$$

with

$$F_3(x) = 1 - e^{-x} \left[1 + x + \frac{x^2}{2} + \frac{2x^3}{45} + \frac{x^4}{180} \right] , \tag{6.56}$$

$$F_4(x) = 1 - e^{-x} \left[1 + x + \frac{x^2}{2} + \frac{x^3}{6} + \frac{7x^4}{180} + \frac{x^5}{180} \right] . \tag{6.57}$$

In practice, one can probably just as well use f_3 and f_4 instead of F_3 and F_4.

Another popular form of "damping" is obtained [6.73] by cutting off the integral in (6.50) at $q = \beta$. To leading order, this prescription gives

$$V_{att}(z) = -C_3 f_2(\beta|z - z_{vW}|)/|z - z_{vW}|^3 . \tag{6.58}$$

For the same β (6.58) gives a stronger attraction than (6.54).

(iii) The lateral variation of V_{att} for a sum of pairwise interactions is caused almost entirely by the atoms in the top surface layer and falls off like $e^{-g_\| z}$. The dipole–dipole term takes the form [6.103]

$$-(C_4/2) \sum_{g_\| \neq 0} (g_\|/2z)^2 (e^{ig_\| \cdot r_\|} K_2(g_\| z) , \tag{6.59}$$

where K_2 is a modified Bessel function and $C_4 = \pi c_6/A_c$. A formula of the same type, with $C_4/2$ replaced by an appropriate $C_{g_\|}$, is obtained from the generalization of (6.50) to a corrugated surface [6.104, 105]. For a material consisting of point polarizabilities, $C_{g_\|} = C_4/2 = 3dC_3/2$, with C_3 given by the Lifshitz formula (6.46); this is an upper limit for the true $C_{g_\|}$ [6.104].

6.4.3 The Total Static Potential

In order to obtain relatively small effects such as the corrugation of metal surfaces, it may not be a good idea to separate V_{static} into V_{att} and V_{rep}. Around the classical turning point of a typical He–metal collision there is a large cancellation of V_{att} and V_{rep}, which tends to amplify the uncertainties.

A direct calculation of the total V_{static} is feasible, using density functional theory in the local density approximation, for a closed shell atom interacting with a jellium surface. *Lang* and *Norskov* [6.106] find by this method very reasonable shapes of V_{static} for $z \leq z_0$, where z_0 is the bottom of the attractive well. The values of the well depth D are also in good agreement with the values obtained by *Zaremba* and *Kohn* [6.100], and by others, using the separate evaluation of V_{att} and V_{rep}, as discussed above. Of course for $z > z_0$ the *Lang–Norskov* potential decays exponentially, rather than as z^{-3}; thus it will not give the correct energies of the surface-bound states. However, it is presumably accurate around the classical turning point; hence a calculation of He interacting with a corrugated metal surface would be very useful.

It seems prudent at present to regard the corrugations of metal surfaces as adjustable parameters to be determined experimentally. If one wants to preserve the relation $V_{\text{rep}}(r) = A\varrho(r)$ with $\varrho(r)$ a superposition of atomic charge densities, one must assume nonspherical atoms without theoretical justification. Thus *Dondi* et al. [6.107] are able to fit the corrugation of Ag(110) by superimposing atomic potentials of the form $U_\kappa \exp[-\beta_x^2 x^2 + \beta_y^2 y^2 + \beta_z^2 z^2)^{1/2}]$ with $\beta_x = 2.03\,\text{Å}^{-1}$, $\beta_y = 2.17\,\text{Å}^{-1}$, $\beta_z = 2.78\,\text{Å}^{-1}$. Here the x direction is along the close-packed (110) atom rows. For comparison, the *Herman–Skillman* tables [6.95] give $\beta = 2.42\,\text{Å}^{-1}$.

6.4.4 The Dynamic Repulsion and the Cutoff Factor

In most of the inelastic scattering calculations performed to date, V_{rep} has been approximated by a pairwise sum of Born–Mayer or Yukawa potentials, or even more drastically by a vibrating wall, soft or hard. The dynamical effects of V_{att} can be fully included in the calculation if V_{att} is represented by a pairwise sum, but it is usually assumed with some justification (see below) that V_{att} does not contribute appreciably to the inelastic matrix elements, except at small lateral momentum transfer $\Delta k_{\parallel}$. Since the prime interest is in the study of surface phonon spectra throughout the surface Brillouin zone, the dynamical V_{att} can then be neglected altogether. We therefore proceed by considering at first V_{rep} alone.

According to the discussion of Sect. 6.3, what is needed are the matrix elements of the lateral Fourier transform of the pair potential (6.31). For a Yukawa potential $V_\kappa(r) = U_\kappa e^{-\beta r}$

$$V_\kappa(p_{\parallel}|z) = \frac{2\pi}{\beta(p_{\parallel})} U_\kappa e^{-\beta(p_{\parallel})z} , \qquad (6.60)$$

where A_c, as before, is the area of a surface unit cell and

$$\beta(p_{\parallel}) = \sqrt{\beta^2 + p_{\parallel}^2} . \qquad (6.61)$$

For a Born–Mayer potential $V_\kappa(r) = U_\kappa e^{-\beta r}$

$$V_\kappa(p_\parallel|z) = \frac{2\pi\beta}{\beta(p_\parallel)^3}[1 + \beta(p_\parallel)z]U_\kappa e^{-\beta(p_\parallel)z} . \tag{6.62}$$

For $p_\parallel \ll \beta$ one has $\beta(p_\parallel) \cong \beta + p_\parallel^2/2\beta$ and (6.60) or (6.62), neglecting the $p_\parallel$ dependence of the prefactor, give

$$V_\kappa(p_\parallel|z) = V_\kappa(0|z)e^{-p_\parallel^2 z/2\beta} , \tag{6.63}$$

where $V_\kappa(0|z)/A_c$ is in fact the same as the laterally averaged $V_{\rm rep}(z)$ (compare (6.45)). The total interaction, to first order in the phonon displacement u_{lz} and neglecting $u_{l\parallel}$, is then approximately the same as

$$V_{\rm rep}\left(z - \sum_l \exp[-\beta(r_\parallel - r_l)^2/2z]u_{lz}\right) . \tag{6.64}$$

This formula is valid as long as $z \gg |r - r_l|$ for the atoms that contribute appreciably to the pairwise sum over l. Physically this means that the classical turning point z_t of the He–surface collision is not smaller than the interatomic distance on the surface and that β is also large enough that the He atom interacts at most with four surface atoms in a square surface array, for instance. The equivalence of (6.62) and (6.63) hinges on the fact that $\exp(-\beta r_\parallel^2/2z)$ and $(2\pi z/\beta)\exp(-p_\parallel^2 z/2\beta)$ are the Fourier transform of each other. It is possible to obtain (6.64) directly from the sum of pairwise exponential potentials by expanding

$$e^{-\beta|r-r_l|} \cong e^{-\beta z}e^{-\beta(r_\parallel - r_l)^2/2z} . \tag{6.65}$$

If z is replaced by z_t in the "influence factor" $\exp[-\beta(r_\parallel - r_l)^2/2z]$, (6.64) describes a "rippling soft wall", which in the limit $\beta \to \infty$ becomes a "rippling hard wall". The influence factor describes how the displacement of the turning point is affected by the displacement of the underlying surface atoms. The corresponding factor in momentum space is the "cutoff factor" $\exp(-p_\parallel^2 z_t/2\beta)$, according to (6.63). The influence factor in coordinate space averages over the displacements of neighboring surface atoms; the cutoff factor in momentum space reduces the contribution to the scattering of short-wavelength phonons. The underlying physical effect was recognized and discussed qualitatively by *Hoinkes* [6.108], *Armand* [6.109], and *Levi* [6.23, 26]. The fall-off of the scattering one-phonon intensity as $\exp(-q_\parallel^2/q_c^2)$, with $q_c^2 \cong \beta/z_t$, describes rather well the decay of the intensity with momentum transfer for scattering from a surface with negligible static corrugation.

6.4.5 Dynamical Effects of the Attractive Potential

The dynamical effects of the static part of $V_{\rm att}$ are fully included in calculations using (6.37) and (6.38) with the correct wave functions χ of $V_{\rm static}$. It is however common practice to keep only $V_{\rm rep}$ throughout the calculation and to account for

V_{att} approximately by applying the "Beeby correction" [6.110], i.e. by increasing the perpendicular He momentum from k_z to $\sqrt{k_z^2 + 2mD/\hbar^2}$. Here D is supposed to be the surface well depth, but comparison with full DWBA calculations shows that it can be considerably larger, especially at low incident He energies [6.111]. This is a reasonable approximation for weakly corrugated surfaces, but does not account for the occurrence of inelastic resonant scattering.

Three types of resonant one-phonon scattering processes can occur on a corrugated surface. In a process of the first type the He atom is elastically scattered (diffracted) from the initial state into a surface-bound state, and then inelastically scattered out of it. This process occurs when the initial energy and angles are close to the resonance condition

$$\varepsilon_i = \varepsilon_b + (\hbar^2/2m)(k_{i\parallel} + g_\parallel)^2 . \tag{6.66}$$

Here $\varepsilon_b (< 0)$ is the binding energy of the n-th surface-bound state and $g_\parallel$ is a surface reciprocal lattice vector. In a process of the second type the He atom is inelastically scattered from the initial state into a surface-bound state, and then elastically scattered out of it. This process occurs when the final energy and angles are close to the resonance condition

$$\varepsilon_f = \varepsilon_{b'} + (\hbar^2/2m)(k_{f\parallel} + g'_\parallel)^2 . \tag{6.67}$$

In a process of the third type both (6.66) and (6.67) are approximately satisfied and an inelastic one-phonon transition occurs between the surface-bound states of binding energies ε_b and $\varepsilon_{b'}$. This process usually causes a large increase of the inelastic cross section.

It is possible to use (6.67) to determine the surface phonon dispersion relation without a TOF apparatus, since the resonance condition itself acts as a final energy analyzer [6.112]. The resonance manifests itself as a sharp variation of the total inelastic intensity. With TOF analysis, the effect of inelastic resonances in LiF, for instance, is very sharp [6.113]. The theory of the inelastic resonance line shape and of its magnitude has been worked out by *Cantini* and *Tatarek* [6.114] in the model of a vibrating hard wall with a stationary attractive potential in front of it. Detailed applications of this model have been made to the analysis of inelastic resonances in LiF [6.115, 116]. Resonances are useful to enhance the scattering from surface phonon branches that are weakly coupled to the He atom; however, one must be careful not to mistake a bound-state resonance for a peak in the surface phonon density of states.

We have discussed so far dynamical effects that arise from the presence of a static attraction. There are also direct dynamical effects of V_{att}. They can be evaluated in the pairwise summation model: the pairwise force has the form (6.29), where $V_\kappa(g_\parallel + q_\parallel | z - z_\kappa)$ is of the form (6.59) with $g_\parallel \to g_\parallel + q_\parallel$. Because this force decays rapidly with $q_\parallel$ it is ignored in the calculations by *Bortolani* et al. [6.51–64]. It was however included in the calculations by *Eichenauer* et al. [6.48–50, 83].

Acknowledgement. This article was written with the support of NATO Grant 33-0673/88 and NSF Grant INT-8913115. I am indebted to P. Tran and T.T. Ong for help in the preparation of this review and to D. Himes for a critical reading of the manuscript.

References

6.1 F.O. Goodman: Surf. Sci. **24**, 667 (1971)
6.2 F.O. Goodman, H.Y. Wachman: *Dynamics of Gas-Surface Scattering* (Academic, New York 1976)
6.3 T. Engel, K.H. Rieder: In *Structural Studies of Surfaces*, ed. by. G. Hohler (Springer, Berlin, Heidelberg 1982) p.55
6.4 V. Celli: In *Many-Body Phenomena at Surfaces*, ed. by D. Langreth, H. Suhl (Academic, Orlando 1984) p.315
6.5 J.A. Barker, D.J. Auerbach: Surf. Sci. Rep. **4**, 1 (1985)
6.6 V. Bortolani, A.C. Levi: Riv. Nuovo Cimento **9**, 1 (1986)
6.7 *Dynamics of Gas-Surface Interaction*, ed. by G. Benedek, U. Valbusa (Springer, Berlin, Heidelberg 1982)
6.8 B.R. Williams: J. Chem. Phys. **55**, 3220 (1971).
     B.F. Mason, B.R. Williams: ibid., **61**, 2765 (1974); **75**, 2199 (1981).
     B.F. Mason, B.R. Williams: Phys. Rev. Lett. **46**, 1138 (1981).
     B.F. Mason, K. McGreer, B.R. Williams: Surf. Sci. **130**, 282 (1983)
6.9 E. Semerad, E.M. Horl: Surf. Sci. **115**, 346 (1982) and **126**, 661 (1983)
6.10 A.A. Maradudin, E.W. Montroll, C.H. Weiss, I.P. Ipatova: *Theory of Lattice Dynamics in the Harmonic Approximation*, 2nd ed. (Academic, New York 1971)
6.11 K.D. Gibson, S.J. Sibener: Phys. Rev. Lett. **55**, 1514 (1985).
     K.D. Gibson, S.J. Sibener, D.L. Mills, B.M. Hall, J.E. Black: J. Chem. Phys. **83**, 4256 (1985)
6.12 K. Kern, P. Zeppenfeld, R. David, G. Comsa: Phys. Rev. B **35**, 886 (1987)
6.13 J.H. Weare: J. Chem. Phys. **61**, 2900 (1974)
6.14 G. Benedek: Phys. Rev. Lett. **35**, 234 (1975)
6.15 G. Benedek, G. Brusdeylins, J.P. Toennies: Phys. Rev. B **27**, 2488 (1983)
6.16 G. Benedek, G. Brusdeylins, R.B. Doak, J.G. Skofronick, J.P. Toennies: Phys. Rev. B **28**, 2104 (1983)
6.17 W.L. Nichols, J.H. Weare: Phys. Rev. Lett. **56**, 753 (1985)
6.18 R. Avila, M. Lagos: Surf. Sci. **103**, L104 (1981)
6.19 J.M. Jackson, N.F. Mott. Proc. Roy. Soc. A **137**, 703 (1932)
6.20 J.E. Lennard-Jones, C. Strachan: Proc. Roy. Soc. A **150**, 442 (1935).
     A.F. Devonshire: ibid., **158**, 269 (1937); C. Strachan: ibid., **158**, 591 (1937)
6.21 N. Cabrera, V. Celli, R. Manson: Phys. Rev. Lett. **22**, 346 (1969).
     R. Manson, V. Celli: Surf. Sci. **24**, 495 (1971)
6.22 F.O. Goodman: Surf. Sci. **30**, 1 (1972)
6.23 U. Garibaldi, A.C. Levi, R. Spadacini, G.E. Tommei: Surf. Sci. **38**, 269 (1973)
6.24 J.L. Beeby: J. Phys. C **5**, 3438 and 3457 (1972)
6.25 M. Lagos, L. Birstein: Surf. Sci. **52**, 391 (1975).
     M. Lagos: ibid., **65**, 124 (1977)
6.26 A.A. Marvin, F. Toigo: Nuovo Cimento B **53**, 25 (1979)
6.27 A.C. Levi: Nuovo Cimento B **54**, 357 (1979)
6.28 G. Armand, J.R. Manson: Surf. Sci. **80**, 532 (1979)
6.29 G. Benedek, N. Garcia: Surf. Sci. **80**, 543 (1979)
6.30 A.C. Levi, H. Suhl: Surf. Sci. **88**, 221 (1979)
6.31 W. Brenig: Z. Phys. B **36**, 81 (1979).
     W. Brenig, R. Sedlmeir: ibid., **36**, 245 (1980)
6.32 G. Benedek, N. Garcia: Surf. Sci. **103**, L143 (1981)
6.33 H.-D. Meyer: Surf. Sci. **104**, 117 (1981).
     H.-D. Meyer, R.D. Levine: Chem. Phys. **85**, 189 (1984)
6.34 J. Stutzki, W. Brenig: Z. Phys. B **45**, 49 (1981)

6.35 R. Brako, D.M. Newns: Phys. Rev. Lett. **48**, 1859 (1982); Surf. Sci. **117**, 42 (1982)
6.36 R. Brako: Surf. Sci. **123**, 439 (1982)
6.37 V. Celli, G. Benedek, U. Harten: Surf. Sci. **143**, L376 (1984)
6.38 V. Celli, A.A. Maradudin: Phys. Rev. B **31**, 825 (1985)
6.39 A. Marvin, V. Celli, F. Toigo: Surf. Sci. **154**, 121 (1985)
6.40 M.D. Stiles, J.W. Wilkins, M. Persson: Phys. Rev. B **34**, 4490 (1986)
6.41 M.D. Stiles, J.W. Wilkins: Phys. Rev. B **37**, 7306 (1988)
6.42 G. Armand, J.R. Manson, C.S. Jayanthi: Phys. Rev. B **34**, 6627 (1986)
6.43 J. Idiodi, V. Bortolani, A. Franchini, G. Santoro, V. Celli: Phys. Rev. B **35**, 6029 (1987)
6.44 J.R. Manson, G. Armand: J. Vac. Sci. Technol. A **5**, 448 (1987); Surf. Sci. **195**, 513 (1988)
6.45 G. Armand, J.R. Manson: Phys. Rev. B **37**, 4363 (1988)
6.46 V. Bortolani, V. Celli, A. Franchini, J. Idiodi, G. Santoro, K. Kern, B. Poelsema, G. Comsa: Surf. Sci. **208**, 1 (1989)
6.47 G. Armand: J. Phys. France **50**, 1493 (1989)
6.48 D. Eichenauer, J.P. Toennies: In *Dynamics at Surfaces*, ed. by B. Pullman et al. (Reidel, Dordrecht 1984) p.1
6.49 D. Eichenauer, J.P. Toennies: J. Chem. Phys. **85**, 532 (1987)
6.50 D. Eichenauer, U. Harten, J.P. Toennies, V. Celli: J. Chem. Phys. **86**, 3693 (1987)
6.51 V. Bortolani, A. Franchini, F. Nizzoli, G. Santoro, G. Benedek, V. Celli: Surf. Sci. **128**, 249 (1983)
6.52 V. Bortolani, A. Franchini, F. Nizzoli, G. Santoro, G. Benedek, V. Celli, N. Garcia: Solid State Commun. **48**, 1045 (1983)
6.53 V. Bortolani, A. Franchini, N. Garcia, F. Nizzoli, G. Santoro: Phys. Rev. B **28**, 7358 (1983)
6.54 V. Bortolani, A. Franchini, F. Nizzoli, G. Santoro: Phys. Rev. Lett. **52**, 429 (1984)
6.55 V. Celli, G. Benedek, U. Harten, J.P. Toennies, R.B. Doak, V. Bortolani: Surf. Sci. **143**, L376 (1984)
6.56 V. Bortolani, A. Franchini, G. Santoro, U. Harten, J.P. Toennies: Surf. Sci. **148**, 82 (1984)
6.57 V. Bortolani, A. Franchini, F. Nizzoli, G. Santoro: In *Electronic Structure, Dynamics, and Quantum Structural Properties of Condensed Matter*, ed. by J.T. Devreese, P. Van Camp (Plenum, New York 1985) p.401
6.58 V. Bortolani, G. Santoro: In *Phonon Physics*, ed. by J. Kollar, N. Kroo, M. Menyhard, T. Siklos (World Scientific, Singapore 1985) p.566
6.59 V. Bortolani, A. Franchini, G. Santoro: In *Dynamical Phenomena at Surfaces, Interfaces, and Superlattices*, ed. by F. Nizzoli, K.H. Rieder, R.F. Willis (Springer, Berlin, Heidelberg 1985) p.92
6.60 V. Bortolani, A. Franchini, F. Nizzoli, G. Santoro: Surf. Sci. **152**, 811 (1985)
6.61 J.E. Black, A. Franchini, V. Bortolani, G. Santoro, R.F. Wallis: Phys. Rev. B **36**, 2996 (1987)
6.62 G. Santoro, A Franchini, V. Bortolani, U. Harten, J.P. Toennies, Ch. Woll: Surf. Sci. **183**, 180 (1987)
6.63 V. Bortolani, A. Franchini, G. Santoro: Surf. Sci. **189/90**, 675 (1987)
6.64 A. Lock, J.P. Toennies, Ch. Woll, V. Bortolani, A. Franchini, G. Santoro: Phys. Rev. B **37**, 7087 (1988)
6.65 N. Esbjerg, J.K. Norskov: Phys. Rev. Lett. **45**, 807 (1980)
6.66 M.J. Stott, E. Zaremba: Phys. Rev. B **22**, 1564 (1980)
6.67 M. Manninen, J.K. Norskov, M.J. Puska, C. Umrigar: Phys. Rev. B **29**, 2314 (1984)
6.68 I.P. Batra: Surf. Sci. **148**, 1 (1984).
     I.P. Batra, P.S. Bagus, J.A. Barker: Phys. Rev. B **31**, 1737 (1985)
6.69 E. Zaremba, W. Kohn: Phys. Rev. B **15**, 1769 (1977)
6.70 J. Harris, A. Liebsch: Phys. Rev. Lett. **49**, 341 (1982)
6.71 J. Harris, A. Liebsch: J. Phys. C **15**, 2275 (1982)
6.72 P. Nordlander: Surf. Sci. **126**, 675 (1983)
6.73 P. Nordlander, J. Harris: J. Phys. C **17**, 1141 (1984)
6.74 E. Fermi: Nuovo Cimento **11**, 157 (1934)
6.75 M.W. Cole, F. Toigo: Phys. Rev. B **31**, 727 (1985)
6.76 Y. Takada, W. Kohn: Phys. Rev. Lett. **54**, 470 (1985)
6.77 Y. Takada, W. Kohn: Phys. Rev. B **27**, 826 (1987)
6.78 W.E. Carlos, M.W. Cole: Surf. Sci. **77**, L173 (1978); Phys. Rev. Lett. **43**, 697 (1979); Surf. Sci. **91**, 339 (1980)
6.79 M.W. Cole, D.R. Frankl, D.L. Goodstein: Rev. Mod. Phys. **53**, 199 (1981)

6.80 G. Vidali, M.W. Cole: Phys. Rev. B **22**, 4661 (1980); Surf. Sci. **110**, 10 (1981).
      G. Vidali, M.W. Cole, C. Schwartz: Surf. Sci. **87**, L273 (1979)
6.81 H. Jonsson, J. Weare: Phys. Rev. Lett. **57**, 412 (1986)
6.82 H. Jonsson, J. Weare: J. Chem. Phys. **86**, 3711 (1987)
6.83 V. Celli, D. Eichenauer, A. Kaufhold, J.P. Toennies: J. Chem. Phys. **83**, 2504 (1985)
6.84 P.W. Fowler, J.M. Hutson: Phys. Rev. B **33**, 3724 (1986)
6.85 J.M. Hutson, P.W. Fowler: Surf. Sci. **173**, 337 (1986)
6.86 D.R. Hamann: Phys. Rev. Lett. **46**, 1227 (1981)
6.87 N. Garcia, J.A. Barker, K.H. Rieder: Solid State Commun. **45**, 567 (1983)
6.88 J. Tersoff: Phys. Rev. Lett. **55**, 140C (1985)
6.89 Y. Takada, W. Kohn: Phys. Rev. Lett. **55**, 141C (1985)
6.90 J.F. Annett, R. Haydock: Phys. Rev. Lett. **53**, 838 (1984); Phys. Rev. Lett. **57**, 1382 (1986)
6.91 J.F. Annett, R. Haydock: Phys. Rev. B **29**, 3773 (1984); ibid., **34**, 6860 (1986)
6.92 J.F. Annett: Phys. Rev. B **35**, 7826 (1987)
6.93 J. Harris, E. Zaremba: Phys. Rev. Lett. **55**, 1940C (1985)
6.94 J.A. Barker, N. Garcia, I.P. Batra, M. Baumberger: Surf. Sci. **141**, L317 (1984)
6.95 F. Herman, S. Skillman: *Atomic Structure Calculations* (Prentice-Hall, Englewood Cliffs 1963).
      E. Clementi, C. Roetti: At. Data Nucl. Data Tables **14**, 177 (1974)
6.96 H.-Y. Kim, M.W. Cole: Phys. Rev. B **35**, 3990 (1987)
6.97 B.R.A. Nijboer, M.J. Renne: Chem. Phys. Lett. **2**, 35 (1969)
6.98 E. Cheng, M.W. Cole: Phys. Rev. B **38**, 987 (1988)
6.99 J.M. Hutson, P.W. Fowler, E. Zaremba: Surf. Sci. **175**, L775 (1986)
6.100 E. Zaremba, W. Kohn: Phys. Rev. B **13**, 2270 (1976)
6.101 A. Liebsch: Phys. Rev. B **33**, 7249 (1986)
6.102 K.T. Tang, J.P. Toennies: J. Chem. Phys. **80**, 3726 (1984)
6.103 W.A. Steele: *The Interaction of Gases with Solid Surfaces* (Pergamon, New York 1974) p.22
6.104 N.R. Hill, M. Haller, V. Celli: Chem. Phys. **73**, 363 (1982)
6.105 A.M. Lahee, W. Allison, R.F. Willis: Surf. Sci. **147**, L630 (1984)
6.106 N.D. Lang, J.K: Norskov: Phys. Rev. B **27**, 4612 (1983)
6.107 M.G. Dondi, S. Terreni, F. Tommasini, U. Linke: Phys. Rev. B **37**, 8034 (1988)
6.108 H. Hoinkes, H. Nahr, H. Wilsch: Surf. Sci. **33**, 516 (1972)
6.109 G. Armand, J. Lapujoulade, Y. Lejay: Surf. Sci. **63**, 143 (1977)
6.110 J.L. Beeby: J. Phys. C **4**, L359 (1971)
6.111 G. Armand, J.R. Manson: J. Phys. France **44**, 473 (1983)
6.112 P. Cantini, G.P. Felcher, R. Tatarek: Phys. Rev. Lett. **37**, 606 (1976); P. Cantini, G.P. Felcher, R. Tatarek: In *Proceedings of the Seventh International Vacuum Congress and the Third International Conference on Solid Surfaces*, ed. by R. Dobrozemsky et al. (Vienna 1977) p.1357
6.113 G. Brusdeylins, R.B. Doak, J.P. Toennies: J. Chem. Phys. **75**, 1784 (1981).
      G. Lilienkamp, J.P. Toennies: Phys. Rev. B **26**, 4752 (1982); J. Chem. Phys. **78**, 5210 (1983)
6.114 P. Cantini, R. Tatarek: Phys. Rev. B **23**, 3030 (1981)
6.115 D. Evans, V. Celli, G. Benedek, J.P. Toennies, R.B. Doak: Phys. Rev. Lett. **50**, 1854 (1983)
6.116 W.L. Nichols, J.H. Weare: Phys. Rev. Lett. **56**, 753 (1985)

7. The Study of Surface Phonons by Electron Energy Loss Spectroscopy: Theoretical and Experimental Considerations

D.L. Mills, S.Y. Tong, J.E. Black

With 2 Figures

The method of electron energy loss spectroscopy recently has been applied to the study of surface phonon dispersion relations on clean and adsorbate covered surfaces. Here we discuss the experimental arrangement used in the new studies, and contrast it with the conventional scheme employed in earlier studies of dipole scattering. We also review the Green's function approach we have used to study surface phonons and surface resonances in systems explored by the method, and then turn to calculations of surface phonon excitation cross sections, with emphasis on their use to guide the experimentalists, and their sensitivity to surface structural parameters.

Since the pioneering experiments carried out by *Ibach* [7.1, 2] twenty years ago, electron energy loss spectroscopy has evolved into a widely used probe of vibrational modes of molecules or atoms adsorbed on crystal surfaces, and of the clean crystal surface. While the energy resolution of the method (2–3 meV in near specular studies of dipole spectra, 4 meV in the off specular impact regime) is inferior to that of infrared spectroscopy and offered by highly monoergetic helium beams, the spectral range of the electron energy loss method is very large. Normal modes with frequency as small as $30 \, cm^{-1}$ have been detected [7.3] , and the loss spectrum also may be scanned continuously from these low freqencies into the electron volt range. This allows the study of overtones and double losses associated with high frequency intra-molecular vibration modes [7.4], and in fact one may scan the spectrum continuously, up to the range where electronic degrees of freedom may be explored [7.5].

In the past few years, the technique has been employed to measure the dispersion curves of surface phonons (and surface resonance modes) throughout the two dimensional Brillouin zone [7.3, 6, 7]. From the experimental point of view, this requires the use of primary beams with a kinetic energy much higher than previously employed, along with an off specular scattering geometry. An experimental challenge was to develop the required high energy beams (in the 100 eV–300 eV range), maintaining an energy resolution sufficient to resolve vibrational losses, and then one must detect the weak signals associated with the required off specular geometry.

Theoretical developments have also allowed the extraction of a substantial amount of information from the data. We have developed and implemented a Green's function approach to surface lattice dynamics that allows one to generate quickly model loss spectra from the theory, including features from both surface phonons and surface resonance modes which lie within the bulk phonon bands. We have found the latter difficult to reproduce with the more conventional slab method on occasion. Also, under conditions appropriate to the surface phonon studies, one may develop a quantitative theory of the energy and angle variation of the loss cross section [7.8]; one finds new selection rules that may be exploited in the experiments [7.9]. The theory has now been applied to the data with impressive success [7.10]. It has proved possible to predict in advance of experiment the energy range within which the excitation cross section of a previously unobserved surface phonon is large enough for the mode to be seen. Also, detailed structural information is contained in data on the cross section [7.11], so from this we have a new probe of surface structure.

This chapter reviews these recent developments. We begin in Sect. 7.1 with a brief review of the conventional mode of electron energy loss spectroscopy, and contrast it with the new work. Section 7.2 discusses the excitation cross section, and the ingredients required for its calculation. Section 7.3 is a summary of the Green's function approach to surface lattice dynamics we have employed in our work, and Sect. 7.4 summarizes comparisons between theory and experiment.

7.1 A Brief Review

An electron energy loss experiment is illustrated in Fig. 7.1 for the case where the electron creates a surface phonon of wave vector $Q_\parallel$, Which propagates along a perfect surface, or possibly one with a periodic overlayer present. Direct measurement of the energy loss provides one with the frequency of the mode created in the scattering process. Wave vector components parallel to the surface are conserved to within a reciprocal lattice vector, so from the angle of reflection θ_s, one may calculate its wave vector. In this manner, one may construct surface phonon dispersion curves from the data. To an excellent approximation, electrons of various energies which emerge with trajectories that make a particular angle θ_s with the normal to the surface have created phonons with the *same* wave vector $Q_\parallel$. This follows simply from the fact that the kinetic energy of the electrons is very much larger than that of the vibrational quanta created in the scattering process [7.12], and this feature of the data which emerges from electron energy loss studies simplifies its interpretation. One does not encounter the awkward scan curves necessary for analysis of inelastic He scattering data.

The angular distribution of electrons scattered inelastically from phonons consists of two distinct regions, within which different physical pictures are invoked, in the analysis of data. Quite typically, in the angular distribution one finds an intense, very narrow peak centered around the specular direction of a

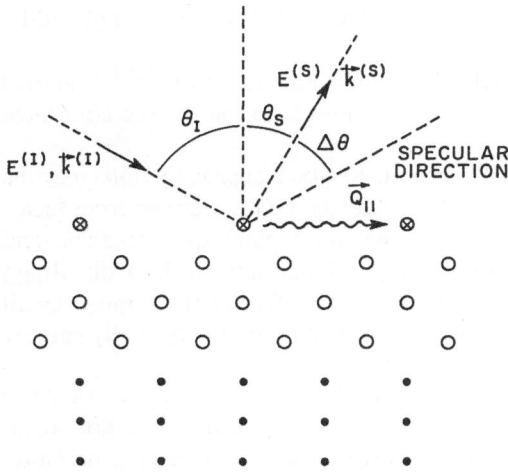

Fig. 7.1. A schematic diagram of an electron energy loss experiment. An electron of energy $E^{(I)}$ and wave vector $k^{(I)}$ strikes the surface, and the electron emerges with energy E^S and wave vector k^S, after creating a vibrational quantum of wave vector $Q_{||}$ and frequency $\omega(Q_{||})$. Wave vector components parallel to the surface are conserved (to within a reciprocal lattice vector), so the electron emerges at an angle θ_s controlled by the value of $Q_{||}$

Bragg beam (angular width less than one degree). This has a wing which extends to large scattering angles and within which the scattering intensity is very much weaker than the near specular lobe. The wing covers the entire 2π solid angle above the surface. We thus have intense, small angle inelastic scattering events, while the large angle deflections produced by scattering off short wavelength phonons have very much smaller cross sections. The wings far from the specular peak in fact form the thermal diffuse background observed in low energy electron diffraction; here we are concerned with an analysis of the energy spectrum of electrons which contribute to the thermal diffuse background [7.12].

When an atom within or adsorbed on the crystal surface vibrates, necessarily there is an oscillatory electric dipole moment, and this produces a long ranged electric field in the vacuum above the crystal. This long ranged field gives rise to the intense small angle scattering. The reason is that a phonon of wave vector $Q_{||}$ produces a field with the spatial variation $\exp(-Q_{||}z)$, with z the distance above the surface, so only long wavelength phonons, which produce small angle scatterings as a consequence of wave vector conservation, contribute to the dipole fields. The theory of dipole scattering was developed some years ago [7.13–15], and the phenomenon appears well understood at this point. The main characteristics of dipole scattering are as follows.

(a) Typical deflection angles are $\Delta\Theta = \hbar\omega_0/2E^{(I)}$, with ω_0 the frequency of the vibrational mode responsible for the loss peak, and $E^{(I)}$ the electron beam energy. One estimates $\Delta\Theta \cong 0.1°$, under typical conditions.

(b) The wave vector of the surface phonon excited by the dipole mechanism is, on the average, $Q_{||} \cong k^{(I)}\Delta\Theta \cong 10^6 \text{cm}^{-1}$, where $k^{(I)}$ is the electron wave vector. Thus, only the vicinity of the origin of the Brillouin zone is explored by the method.

(c) There is a selection rule [7.13, 15] which asserts that only vibrational modes which create a dynamic dipole moment normal to the surface contribute to the near specular peak.

(d) The inelastic excitation cross section, to good approximation, may be written in the form [7.13] $f_{dip}(E)I_{LEED}(E)$, where $f_{dip}(E)$ is a dipole form factor which decreases with energy above 2 or 3 eV for typical spectrometers, and $I_{LEED}(E)$ is the intensity of elastic scattering off the surface, into the Bragg beam about which the "dipole lobe" is centered. Both factors favor rather small beam energies, and as a consequence studies of dipole scattering typically employ rather small beam energies, in the range of 5 eV

Because the "dipole lobe" is intense, nearly all spectrometers are arranged to study dipole losses. It is clear from statement (b) above that one does not obtain information on surface phonon dispersion curves from this data. Furthermore, the dipole selection rule [statement (c)] limits the number of modes one may observe in this scattering geometry.

To measure surface phonon dispersion curves out to the Brillouin zone boundary, one must probe the large angle "wing" of the dipole loss, where the phonon wave vector $Q_\parallel$ is the order of 10^8 cm^{-1}. Long ranged fields, again which exhibit the spatial variation $\exp(-Q_\parallel z)$ in the vacuum above the crystal, play only a minor role in the excitation process here. One must recognize instead that the electron penetrates into the crystal to sample its outermost three or four atomic layers before reemerging into the vacuum. While within the crystal, it interacts with surface phonons (and also bulk phonons [7.12]) by virtue of modulations in the crystal potential provided by the thermal motions of the atoms. It has become customary to refer to the regime in which this picture applies as the impact regime.

The first off specular experiments were studies of the vibrations of hydrogen adsorbed on W(100), reported by *Ho* et al [7.16]. Modest beam energies were employed in this work, and no information on dispersion of the highly localized hydrogen vibrations is evident in the data. These studies were the first to study nondipole active parallel motions at the surface, and clearly demonstrated the power of combining information obtained from off specular studies, in combination with the dipole spectrum.

Ibach and collaborators [7.3], were the first to extract the complete dispersion curve of a surface phonon by the electron energy loss method. A key feature of their work is the use of beams with large kinetic energies (180 eV and 320 eV, in the initial experiments). Earlier theoretical studies [7.8] had illustrated that, for a variety of reasons, the use of impact energies in the low energy electron diffraction range offer very substantial advantages. The experimental studies, along with theoretical analyses, have been extended to the study of surface phonon dispersion curves, and surface resonances of surfaces covered with ordered adsorbate layers [3.6, 7]. At the time of writing, data is emerging from other experimental laboratories.

7.2 The Surface Phonon Excitation Mechanism in the Impact Regime

We have seen that, for small angle inelastic scattering events, long ranged fields seen by the electrons above the crystal couple the electron to the surface vibrational quanta. Estimates show that the electron is quite far above the crystal when it suffers its energy loss (100 Å typically), and a consequence is that a phenomenological approach which ignores structural details provides a very good description of the excitation process [7.15].

At large momentum transfers, as discussed above, a microscopic theory of the excitation cross section is required. One may proceed to construct such a theory, as follows.

Consider a semi–infinite crystal, with the atom at site l displaced from its equilibrium position $R_l^{(0)}$, to the point $R_l = R_l^{(0)} + U_l$. In the crystal, with atoms vibrating by virtue of their thermal motions, in fact all atoms are off–site at any instant of time. Let us imagine, in the spirit of the adiabatic approximation, that we freeze the atoms in place at some instant, then calculate the scattering amplitude $f(k^{(I)}, k^{(S)}; \{R_l\})$ for elastic scattering of an electron from an initial state $k^{(I)}$ to a final state $k^{(S)}$. In principle, within a muffin tin approximation, through the use of multiple scattering theory, one may express this scattering amplitude in terms of the phase shifts for scattering off the various atoms in the crystal and Green's functions which describe propagation of the electron through the crystalline medium, between the muffin tins [7.17].

For small amplitude motions, one may develop the scattering amplitude in a Taylor series:

$$f(k^{(I)}, k^{(S)}; \{R_l\}) = f(k^{(I)}, k^{(S)}; \{R_l^{(0)}\})$$
$$+ \sum_l U_l \cdot (\nabla_{R_l} f)_{\{R_l^{(0)}\}} + \dots . \tag{7.1}$$

The first term in (7.1) is the amplitude for elastic scattering off the perfectly ordered crystal (the LEED amplitude), and from the second, the cross section for scattering from surface phonons and surface resonances may be synthesized [7.8].

It is possible to express the gradient of the scattering amplitude, ∇f, which appears in (7.1), in terms of the same scattering phase shifts and electron propagation matrices that enter the theory of low energy electron diffraction [7.8]. One new ingredient is required. We must calculate matrix elements of the change δV in crystal potential produced by displacing an atom from its equilibrium position. If we represent δV by rigidly shifting a muffin tin potential, then the required matrix elements may be expressed in terms of the phase shifts for *elastic* scattering of the electron from the muffin tin [7.18]. Thus, in this picture, we may calculate $\nabla_{R_l} f$ from information available from probe electrons which elastically scatter from the surface region, as in low energy electron diffraction and angle resolved photoemission studies. Of course, calculation of the scattering amplitude derivatives requires substantially more computer time [7.19].

When the atoms are displaced from their equilibrium positions, the displacement U_l may be expressed in terms of the eigenvectors $e_s(l)$ of the vibrational normal modes. Here s stands for the set of quantum numbers required to designate a particular normal modes. We specify this more completely below. We may write, if $l = l_\parallel + \hat{z}l_z$, and z is normal to the surface,

$$u_l = \sum_s \left(\frac{\hbar}{2\omega_s M(l_z)}\right)^{1/2} e_s(l)(a_s + a_s^+) \tag{7.2}$$

so if we let δf be that part of the scattering amplitude which describes surface phonon excitation, we have

$$\delta f = \sum_s \sum_{l_z} \left(\frac{\hbar}{2\omega_s M(l_z)}\right)^{1/2} (a_s + a_s^+) \sum_{l_\parallel} e_s(l) \cdot (\nabla_{R_l} f)_{\{R_l^{(0)}\}} . \tag{7.3}$$

Our interest is in the perfect crystal, or possibly a crystal with a perfectly periodic overlayer. Such structures then have perfect periodicity in the two directions parallel to th surface. This may be exploited to simplify (7.3) greatly. As we have seen elsewhere in this volume, each normal mode of such a crystal may be labeled by a wave vector $Q_\parallel$, which lies in the appropriate two dimensional Brillouin zone, and a second index α, which labels the various modes of wave vector $Q_\parallel$. The index α may refer to a surface phonon or a bulk phonon which propagates up to and reflects off the surface. Then the dependence of $e_s(l)$ on $l_\parallel$, the component of l parallel to the surface, is fixed. We have

$$e_s(l) = \xi(Q_\parallel \alpha; l_z) e^{iQ_\parallel \cdot l_\parallel} . \tag{7.4}$$

Next let

$$\Lambda(l_\parallel, l_z) = (\nabla_{R_l} f)_{\{R_l^{(0)}\}} . \tag{7.5}$$

One may prove that, for general $l_\parallel$, $\Lambda(l_\parallel, l_z)$ may be related to the scattering amplitude derivative $\Lambda(0, l_z)$ for an atom in plane l_z, located in a unit cell at the origin of plane l_z:

$$\Lambda(l_\parallel, l_z) = \exp\left[i(k^{(S)} - k^{(I)}) \cdot l_\parallel\right] \Lambda(0, l_z) , \tag{7.6}$$

so that

$$\delta f = \sum_{Q_\parallel \alpha} \left(\frac{\hbar}{2\omega(Q_\parallel \alpha)}\right)^{1/2} \left[a_{Q_\parallel \alpha} + a_{-Q_\parallel \alpha}^+\right] \Delta f(Q_\parallel \alpha) , \tag{7.7}$$

where

$$\Delta f(Q_\parallel \alpha) = \sum_{l_z} \frac{\xi(Q_\parallel \alpha; l_z)}{[M(l_z)]^{1/2}} \cdot \Lambda(0, l_z) \sum_{l_\parallel} e^{i(k^{(S)} - k^{(I)} + Q_\parallel) \cdot l_\parallel} . \tag{7.8}$$

Or, with N_s the number of the unit cells in a plane parallel to the surface, after

198

carrying out the sum on $l_\parallel$ in (7.8), we have

$$\Delta f(\boldsymbol{Q}_\parallel \alpha) = \left\{ N_s \sum_{\boldsymbol{G}_\parallel} \delta_{\boldsymbol{k}_\parallel^{(S)};\, \boldsymbol{k}_\parallel (I) - \boldsymbol{Q}_\parallel + \boldsymbol{G}_\parallel} \right\}$$
$$\times \left\{ \sum_{l_z} \frac{\boldsymbol{\xi}(\boldsymbol{Q}_\parallel \alpha;\, l_z) \cdot \boldsymbol{\Lambda}(0, l_z)}{\sqrt{M(l_z)}} \right\} . \tag{7.9}$$

We have $\delta_{\boldsymbol{k}_\parallel;\, \boldsymbol{k}'_\parallel} = 1$ if $\boldsymbol{k}_\parallel = \boldsymbol{k}'_\parallel$ and zero otherwise, so the first factor in (7.9) simply expresses wave vector conservation parallel to the surface. The quantity $\boldsymbol{\Lambda}(0, l_z)$ depends on both $\boldsymbol{k}^{(I)}$ and $\boldsymbol{k}^S$, but is independent of which normal mode is considered.

The scattering efficiency for exciting the mode $(\boldsymbol{Q}_\parallel \alpha)$ is readily related to $\Delta f(\boldsymbol{Q}_\parallel \alpha)$. To calculate it we require the phonon eigenvector $\boldsymbol{\xi}(\boldsymbol{Q}_\parallel \alpha; l_z)$, so we need a model of the surface lattice dynamics of the material, and we also require the scattering amplitude derivatives $\boldsymbol{\Lambda}(0, l_z)$ for each layer within which the phonon eigenvectors and electron wave functions have appreciable amplitude.

As we have remarked in Sect. 7.1, in inelastic electron scattering studies of surface vibrations, when the incident and scattered angle are both fixed, and the loss spectrum is scanned, one samples the frequency spectrum of those normal modes with fixed wave vector $\boldsymbol{Q}_\parallel$. Quite clearly, upon squaring the scattering amplitude in (7.9), and forming the cross section, one sees that the cross section may be expressed in terms of the *spectral density function*

$$\varrho_{\mu\nu}(\boldsymbol{Q}_\parallel \Omega;\, l_z l'_z) = \sum_\alpha \xi_\nu^*(\boldsymbol{Q}_\parallel \alpha;\, l_z) \xi_\mu(\boldsymbol{Q}_\parallel \alpha;\, l'_z) \delta(\omega(\boldsymbol{Q}_\parallel \alpha) - \Omega) . \tag{7.10}$$

The number of electrons which suffer an energy loss between Ω and $\Omega + d\Omega$ is then proportional to the quantity

$$S(\boldsymbol{Q}_\parallel, \Omega) = \sum_{l_z l'_z} \sum_{\nu\mu} \frac{\Lambda_\nu^*(0 l_z) \Lambda_\mu(0 l'_z)}{(M(l_z) M(l'_z))^{1/2}} \int_\Omega^{\Omega + d\Omega} d\omega' \varrho_{\nu\mu}(\boldsymbol{Q}_\parallel \omega';\, l_z l'_z) .$$

In Sect. 7.3, we discuss out method for calculating the spectral density function defined in (7.10) and the reasons this approach is very useful, in our view. Section 7.4 discusses the implications of the calculations of surface phonon excitation cross sections, which as we see from this section, require knowledge of the lattice dynamics of the surface.

7.3 The Green's Function Approach to Spectral Density Calculations

Elsewhere in this volume, a discussion is presented of the calculation of surface phonon dispersion relations and eigenvectors by the slab method, which has proved useful for many years. Here, the semi-infinite crystal is approximated by a slab with a finite number of atomic layers, N. One chooses N sufficiently large that the surface phonons of interest have amplitudes which are very small in the center of the slab, and one can be assured that when this condition is satisfied accurate results are obtained. This method was introduced many years ago [7.20, 21], and has been used by many groups.

In recent years, we have used a Green's function method which provides exact results for the spectral density for a semi-infinite geometry. We sketch this method here and, as we shall discuss, we shall see that it has substantial advantages for models where it may be implemented. We first begin with a very brief conceptual review of lattice dynamics in the harmonic approximation.

Consider a semi-infinite crystal, considered monatomic or semiplicity. If $U_\alpha(l_\parallel l_z)$ is the displacement of an atom at site $(l_\parallel l_z)$, then the normal mode frequencies are found by solving

$$\omega^2 U_\mu(l_\parallel l_z) - \sum_\beta D_{\mu\nu}(l_\parallel l_z; l'_\parallel l'_z) U_\nu(l'_\parallel l'_z) = 0 \ , \tag{7.11}$$

where the dynamical matrix $D_{\alpha\beta}(l_\parallel l_z; l'_\parallel l'_z)$ is a function of only $l_\parallel - l'_\parallel$ for a semi-infinite crystal perfectly periodic in the two directions parallel to the surface. These are standard prescriptions for constructing D, given an assumed force law between atoms in the crystal.

For a semi-infinite crystal, we may search for eigensolutions for which

$$U(l_\parallel l_z) = \xi(Q_\parallel \alpha; l_z) e^{iQ_\parallel \cdot l_\parallel} \ , \tag{7.12}$$

where, as in Sect. 3.2, the index α labels the various modes with wave vector $Q_\parallel$. One readily sees that $\xi_\nu(Q_\parallel \alpha; l_z)$ obeys a simple eigenvalue equation

$$\omega^2 \xi_\nu(Q_\parallel \alpha; l_z) - \sum_\nu \sum_{l'_z} d_{\mu\nu}(Q_\parallel; l_z l'_z) \xi_\nu(Q_\parallel \alpha : l'_z) = 0 \ , \tag{7.13}$$

with $d_{\mu\nu}(Q_\parallel; l_z l'_z) = \sum_{l_\parallel} D_{\mu\nu}(l_\parallel l_z; l'_\parallel l'_z) \exp[iQ'_\parallel(l_\parallel - l'_\parallel)]$.

The slab method applies (7.13) to a finite number of atomic layers of material. The spectral densities may then be constructed from (7.10) by performing the sum over α directly, with the delta function replaced by a Lorentzian or Gaussian, to generate a smooth, continuous approximation for the spectral densities of interest.

The Green's function approach introduces an object $U_{\mu\nu}(Q_\parallel z; l_z l'_z)$ which is a function of the complex variable z. This is

$$U_{\mu\nu}(Q_{\|}z; l_z l'_z) = \sum_{\alpha} \frac{\xi_\mu(Q_{\|}\alpha; l_z)\xi_\nu^*(Q_{\|}\alpha; l'_z)}{z^2 - \omega^2(Q_{\|}\alpha)} . \tag{7.14}$$

Once this function has been found, the spectral densities may be determined:

$$\varrho_{\mu\nu}(Q_{\|}\Omega; l_z l'_z) = \frac{1}{2\pi i}[U_{\mu\nu}(Q_{\|}, \Omega - i\eta; l_z l'_z) - U_{\mu\nu}(Q_{\|}, \Omega + i\eta; l_z l'_z)] . \tag{7.15}$$

Here η is a positive, infinitesimal number.

While the Green's function may be constructed, in principle, from the eigenvectors and eigenvalues provided by a slab calculation, in fact in a number of cases one may determine it by semi-analytic methods.

One may show that the Green's function obeys an inhomogeneous version of (7.13):

$$z^2 U_{\mu\nu}(Q_{\|}z; l_z l'_z) - \sum_{\lambda}\sum_{l''_z} d_{\mu\lambda}(Q_{\|}; l_z l''_z)U_{\lambda\nu}(Q_{\|}z; l''_z l'_z) = \delta_{\mu\nu}\delta_{l_z l'_z} . \tag{7.16}$$

For the case where the interplanar couplings are of finite range, it is possible to devise a solution to (7.14) that is exact for the semi-infinite crystal [7.22–24]. One proceeds as follows. One chooses a value of $l_z = l_z^{(M)}$ with the property that for $l_z \geq l_z^{(M)}$ and $l'_z \geq l_z^{(M)}$, $d_{\mu\nu}(Q_{\|}l_z, l'_z)$ has the bulk form. In practice, $l_z^{(M)}$ must be deep enough in the crystal that $d_{\mu\nu}$ is unaffected by changes in force constants near the surface, and by couplings disturbed by the presence of broken bonds severed in the process of forming the surface, say, by cutting a large crystal and discarding one half. For $l_z > l_z^{(M)}$, $U_{\mu\nu}(Q_{\|}z; l_z l'_z)$ may be represented as a superposition of exponentials of the form $\exp(ik_\perp^{(i)}l_z)$, where $k_\perp^{(i)}$, may be close to the real axis, pure imaginary, or a complex number always chosen with $\mathrm{Im}(k_\perp^{(i)}) > 0$. Typically one solves for the $k_\perp^{(i)}$ by finding the roots of a polynomial whose form is controlled by the dynamical matrix of the bulk crystal. The exponential form just described is then joined onto the surface region, $0 < l_z < l_z^{(M)}$ and $0 < l'_z < l_z^{(M)}$. This leads to the requirement that a matrix, generally rather small, must be inverted. The above procedure assumes we are interested in calculating $U_{\mu\nu}(Q_{\|}z; l_z l'_z)$ for l_z and l'_z in the regime $0 < l_z < l_z^{(M)}$, $0 < l'_z < l_z^{(M)}$, which is usually sufficient. In the computation, the complex variable z is replaced by $\omega \pm i\eta$, where η is taken small but finite. The spectral densities are calculated directly from the definition in (7.15). We have found that this procedure requires only a very modest expenditure of computer time and, as remarked earlier, it provides us with an exact solution of the surface lattice dynamics for a semi-infinite model crystal rather than a slab. Very shortly, we shall appreciate why this is useful.

In Fig. 7.2, we show results obtained for selected spectral densities through the Green's function method. Figure 7.2a shows the spectral density associated with the vertical motion of oxygen atoms at $Q_{\|} = 0$, for the $p(2 \times 2)$ oxygen overlayer on Ni(100). This is the spectral density probed in the small angle regime, where the dipole mechanism couples the electron to surface phonons.

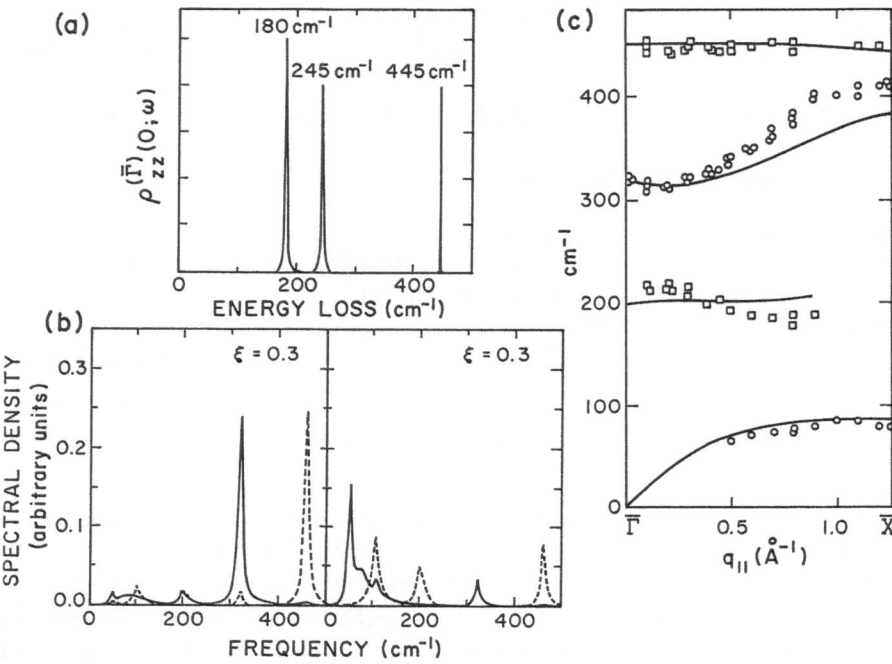

Fig. 7.2. Calculated spectral density functions, illustrating both surface phonon and surface resonance mode features, for (a) vertical motion of oxygen atoms, at $Q_\parallel = 0$, for the $p(2 \times 2)$ oxygen overlayer on Ni(100), (b) (left figure) perpendicular (solid line) and parallel (dashed line) oxygen motions for $Q_\parallel$ 30% of the way $\bar{\Gamma}$ to $\bar{X}$, for the $c(2 \times 2)$ oxygen overlayer on Ni(100), and (right figure) the same for Ni atoms in the first atomic layer. In (c) we show calculated dispersion for the surface phonons and the surface resonance mode, for $c(2 \times 2)$ oxygen on Ni(100). The data is that reported by the group of Ibach in [7.6]

The high frequency feature at $445 \, \text{cm}^{-1}$ is the $Q_\parallel = 0$ surface phonon, induced by the presence of the light adsorbates. The maximum vibration frequency of the Ni substrate is $300 \, \text{cm}^{-1}$, so this mode lies well above the substrate phonon band.

In contrast, the two features at $180 \, \text{cm}^{-1}$ and $245 \, \text{cm}^{-1}$ are embedded *within* the substrate phonon bands. These are examples of surface resonance modes, which are a surface analogue of the well known impurity induced phonon modes induced by impurities in the bulk of crystals [7.25]. We emphasize that surface resonance modes are found even on pure crystals, in the absence of adsorbates. In essence, the "broken bonds" at the surface turn the surface into an extended two dimensional defect, from the lattice dynamical point of view.

While the surface resonances provide a signature in the spectral density very similar to that from surface phonons, they are not normal modes of the system in the harmonic approximation, as is the case for surface phonons. There is a narrow range of frequencies within which a bulk phonon excites a resonant response from the surface region, upon striking the surface. Conversely, if the surface layers are excited by an external probe at the surface resonance frequency, the surface "rings" for an extended period of time, but the energy stored in the resonant response is radiated off into the bulk of the crystal, in the form of phonons. The width of the $180 \, \text{cm}^{-1}$ and $245 \, \text{cm}^{-1}$ features in Fig. 7.2a are finite, in the harmonic approximation, and the inverse of this width is the lifetime of the resonance. (The $445 \, \text{cm}^{-1}$ surface phonon has zero width in the harmonic approximation used to generate Fig. 7.2a. This feature has been broadened artificially.)

In nearly all of our studies of surface lattice dynamics, we find surface resonances in our calculated spectral densities. Figure 7.2b shows spectral density calculations for the Ni(100) surface, with the $c(2 \times 2)$ oxygen overlayer present [7.24]. These are for a wave vector $Q_\parallel$ 30% of the way from the center $\bar{\Gamma}$ of the two dimensional Brillouin zone to the zone boundary point $\bar{X}$. The left figure gives the spectral densities of vertical motion of the oxygen atoms (solid line) and parallel motion of the oxygen atom (dashed line), and the right figure is a similar plot for the outermost Ni layer. Note the prominent resonances in the parallel Ni motions at $100 \, \text{cm}^{-1}$ and near $200 \, \text{cm}^{-1}$.

We have found it difficult to reproduce these narrow resonances reliably with the slab method, though they appear fully resolved in the Green's function description of the full semi-infinite crystal. One may readily appreciate the problem from Fig. 7.2a. The $180 \, \text{cm}^{-1}$ and $240 \, \text{cm}^{-1}$ resonances are roughly $10 \, \text{cm}^{-1}$ wide. Suppose we search for these in a slab calculation with a 30 layer slab. There are four Ni atoms in each unit cell, for Ni(100) with the $p(2 \times 2)$ oxygen overlayer present, so this requires diagonalization and with three degrees of freedom associated with each atom, a matrix larger than 360×360 to find all the eigenfrequencies and eigenvectors. One third of these modes excites vertical motion of the oxygen atom, so within the Ni phonon bands the average spacing between adjacent modes is roughly $3 \, \text{cm}^{-1}$. One will encounter only a small number of modes in the narrow frequency bands where the resonances reside, and even for such a large computation the statistics will be very poor in these regions in spectral densities constructed directly from (7.10) by the slab method. Some time after our work was published, *Andersson* et al. [7.26] calculated the same spectral density function as in Fig. 7.2a, evidently by the slab method. This was done for the model introduced in our earlier work, with parameters chosen in the same manner. They show a third resonance, not evident in our full solution of the same model. We believe the additional feature is an artifact produced by the poor statistics inherent in the slab description of such sharp resonances within the bulk phonon bands. The solution of the hierarchy of equations generated by the Green's function method requires only very modest investments of computer time, so long as the model explored contains interactions between atomic constituents whose range is confined to only a few neighbors, and one then has the exact solution for a semi-infinite crystal in hand, with the bulk phonon bands a true continuum.

Through spectral density plots such as those displayed in Fig. 7.2b, generated for wave vectors throughout the Brillouin zone, one may plot the dispersion of both surface resonances and surface phonons. We show an example in Fig. 7.2c, along a high symmetry line in the Brillouin zone, for the $c(2\times2)$ oxygen overlayer on Ni(100) [7.6]. Other examples of comparison between theory and experiment, for both surface resonances and surface phonons, may be found in papers cited earlier.

7.4 Calculations of the Cross Section for Surface Phonon Excitation

As we have seen, once we know the eigenvectors associated with a given surface phonon, through the use of the formalism discussed in Sect. 7.2, we may calculate the cross section for excitation of the various surface phonons. The first such calculations were directed toward the clean Ni(100) surface [7.10], the Ni surface with a $c(2 \times 2)$ sulfur overlayer [7.11], and the clean Cu(100) surface [7.11]. The results are in excellent accord with the data. For example, calculations of the cross section for exciting the Rayleigh surface phonon at the Brillouin zone boundary (the S_4 mode at the $\bar{X}$ point) agree impressively with data, for impact energies from 50 eV to 270 eV [7.10].

Calculations of the excitation cross section are extremely valuable for several reasons. One is that optimal scattering geometries and impact energies may be selected, through the use of the theoretical cross sections. We have an example in hand. In the first electron energy loss study of surface phonon dispersion [7.3], the Rayleigh wave dispersion relation was obtained experimentally on the clean Ni(100) surface, from the Brillouin center $\bar{\Gamma}$ to the zone boundary point $\bar{X}$. This was the only feature evident in the data. But analysis of the lattice dynamics of the Ni(100) surface leads to the prediction that a second surface phonon, referred to as the S_6 mode, should exist near $\bar{X}$. At $\bar{X}$, in the outermost atomic layer, the displacements associated with the Rayleigh surface phonon are *normal* to the surface, while those associated with the S_6 mode are *parallel* to the surface. A simple kinematic estimate which assumes that the electron scatters off only atomic motions in the outermost layer suggests the cross section for exciting S_6 should be smaller than the S_4 cross section by roughly two orders of magnitude.

The calculations carried out by *Xu* et al. [7.10] showed that indeed, over a wide range of energy, the S_6 cross section is very much smaller than that of S_4, in approximate accord with the simple estimate. However, the calculation predicted the existence of three "energy windows" between 50 eV and 250 eV within which the S_6 cross section becomes comparable to or even larger than that for exciting S_4. The origin of the large enhancement of the S_6 cross section is in a multiple scattering resonance. Experiments carried out by the Ibach group in response to the prediction produced spectra in which the S_6 mode was indeed found only at the energies prediced by theory [7.10]. Thus, one may use the theory to guide the experimentalist in the choice of a scattering geometry designed to enhance excitation of particular modes.

The energy variation of the excitation cross section is sensitive to the surface geometry. The electron engages in multiple scattering from the crystal as it approaches a site where a phonon is excited, and also as it exits the crystal. One finds a nonmonotonic variation of the cross section with energy, reminiscent of low electron diffraction intensities. In electron energy loss studies, when several surface phonons or resonances are probed, the amount of information in hand is substantial. As we shall see shortly, the combined analysis of the dispersion curves and cross section can place strong constraints on both surface geometry and the nature of the force constants which enter the lattice dynamical model.

There have been two applications of electron energy loss cross sections to the determination of surface geometries. While both of these confirm conclusions reached earlier by other methods, they illustrate the potential power of these studies. First of all, as mentioned previously, the calculations of Xu et al. [7.10] provide an excellent account of the energy variation of the S_4 and S_6 mode cross sections, as measured by $Ibach$ and co-workers on Ni(100) [7.10]. In the calculation, the sensitivity of the ratio of the two cross sections to surface relaxation has been explored at the beam energy of 155 eV where both S_4 and S_6 are observed. In the calculation, the ratio is very sensitive to small changes in the interlayer spacing d_{12}. A change of d_{12} of one percent produces a forty percent change in the cross section ratio. The authors of [7.10] have formed a reliability factor R, and find that, to obtain agreement with the data, the spacing between the first and second layers must be contracted by an amount between 1.7% and 3.3%, a result in accord with earlier ion scattering measurements [7.27].

In their analysis of the surface phonon dispersion curves on Ni(100) with the $c(2 \times 2)$ sulfur overlayer present, Lehwald et al. [7.37] fitted the dispersion curves with a simple nearest neighbor central force model. The model provides an excellent account of the data, but to fit the high frequency surface phonon branches associated with the sulfur overlayer, they had to place the overlayer 1.45 Å above the surface, a distance greater than the 1.35 Å distance obtained in earlier LEED and EXAFS studies. While the two vertical distances differ by less than eight percent, the discrepancy lies well outside the uncertainty of the earlier measurements. A detailed series of theoretical studies of the cross section [7.11] show that in fact the 1.45 Å distance is incompatible with the measured energy variation of the various cross sections, while placing the sulfur at the preferred 1.35 Å height produces excellent accord between theory and experiment. With the sulfur placed 1.35 Å above the surface, the introduction of a small noncentral component of the S-Ni coupling constant produces excellent agreement between theory and the measured dispersion curves. This example is an illustration of the fact, well known from bulk lattice dynamics, that more than one empirical force constant model can fit a given set of data. A consequence is that one must be cautious in forming conclusions from non-symmetry-controlled features of phonon dispersion relations calculated from such models. The combined analysis of the dispersion curves and the energy variation of the cross sections places strong constraints on the lattice dynamical models one may use.

7.5 Concluding Remarks

Electron energy loss spectroscopy is a powerful means of probing the dispersion relations for surface phonons on clean and adsorbate covered surfaces. It is complementary to inelastic He scattering, in that all frequencies above $100\,\mathrm{cm}^{-1}$ may be scanned, readily, with no limitations on the upper end of the spectral range accessible experimentally. The limited resolution available to date renders the study of modes below $100\,\mathrm{cm}^{-1}$ difficult, but here the helium beam method works very well. We thus have in hand, with these two techniques, a complete spectroscopy of surface phonons and surface resonance modes.

In our view, in electron energy loss spectroscopy, a key element is the use of incident beams in the energy range of 50–300 eV, rather than the very low energies 1 eV–10 eV used in the early studies of dipole scattering. We have seen that in the high energy range accurate and quantitative calculations of the excitation cross sections may be carried out. These provide guidance for the epxerimentalist and allow quantitative conclusions on surface structure to be obtained from cross section data.

References

7.1   H. Ibach: Phys. Rev. Lett. **24**, 1416 (1970)
7.2   H. Ibach: Phys. Rev. Lett. **27**, 253 (1971)
7.3   See, for example, Fig. 1 of S. Lehwald, J.M. Szeftel, H. Ibach, T.S. Rahman, D.L. Mills: Phys. Rev. Lett. **50**, 518 (1983)
7.4   Numerous examples of the use of electron loss spectroscopy to study high frequency intra-molecular vibrations of adsorbed species are given in H. Ibach, D.L. Mills: *Electron Energy Loss Spectroscopy and Surface Vibrations* (Academic, San Francisco 1982) Chap. 6
7.5   H. Froitzheim, H. Ibach, D.L. Mills: Phys. Rev. B **11**, 4980 (1975)
7.6   J.M. Szeftel, S. Lehwald, H. Ibach, T.S. Rahman, J.E. Black, D.L. Mills: Phys. Rev. Lett. **51**, 268 (1983)
7.7   S. Lehwald, M. Rocca, H. Ibach, S. Rahman: Phys. Rev. B **31**, 3477 (1983). For an erratum, see Phys. Rev. B **32**, 1354 (1983)
7.8   S.Y. Tong, C.H. Li, D.L. Mills: Phys. Rev. Lett. **44**, 407 (1980); C.H. Li, S.Y. Tong, D.L. Mills: Phys. Rev. B **21**, 3057 (1980); S.Y. Tong, C.H. Li, D.L. Mills: Phys. Rev. B **24**, 806 (1981)
7.9   B.M. Hall, S.Y. Tong, D.L. Mills: Phys. Rev. Lett. **50**, 1277 (1983)
7.10   M.L. Xu, B.M. Hall, S.Y. Tong, M. Rocca, H. Ibach, S. Lehwald, J.E. Black: Phys. Rev. Lett. **54**, 1171 (1985)
7.11   Z.Q. Wu, Y. Chen, M.L. Xu, S.Y. Tong, S. Lehwald, M. Rocca, H. Ibach: Phys. Rev. B **39**, 3116 (1989)
7.12   V. Roundy, D.L. Mills: Phys. Rev. B **5**, 1347 (1972)
7.13   E. Evans, D.L. Mills: Phys. Rev. B **5**, 4126 (1972)
7.14   D.L. Mills. Surf. Sci. **48**, 59 (1975)
7.15   See the discussion in H. Ibach, D.L. Mills: *Electron Energy Loss Spectroscopy and Surface Vibrations* (Academic, San Francisco 1982) Chap. 3
7.16   W. Ho, R.F. Willis, E.W. Plummer: Phys. Rev. Lett. **40**, 1463 (1978)
7.17   For a complete discussion, see S.Y. Tong: Prog. Surf. Sci. **7**, 1 (1975)
7.18   G.D. Gaspari, G.L. Gyorffy: Phys. Rev. Lett. **28**, 801 (1972)

7.19 An expanded description of the basic theory and the computational procedures may be found in Burl M. Hall: Thesis, University of Wisconsin, Milwaukee, 1983 (unpublished)

7.20 An early application of the slab method is found in R.F. Wallis, D.L. Mills, A.A. Maradudin: In *Localized Excitations in Solids*, ed. by R.F. Wallis (Plenum, New York 1968) p.403

7.21 Extensive use of the slab method has been made by F.W. de Wette and colleagues. An early study is found in R.E. Allen, C.P. Alldredge, F.W. de Wette: Phys. Rev. Lett. **24**, 301 (1970)

7.22 T.S. Rahman, J.E. Black, D.L. Mills: Phys. Rev. B **25**, 883 (1982)

7.23 T.S. Rahman, J.E. Black, D.L. Mills: Phys. Rev. B **27**, 4059 (1983); J. E. Black T.S. Rahman, D.L. Mills: Phys. Rev. B **27**, 4059 (1983)

7.24 T.S. Rahman, J.E. Black, J.M. Szeftel, S. Lehwald, D.L. Mills: Phys. Rev. B **30**, 589 (1984)

7.25 See the discussion in A.A. Maradudin, E.W. Montroll, G.H. Weiss, I.P. Ipatova: *Theory of Lattice Dynamics in the Harmonic Approximation* (Academic, New York 1971)

7.26 S. Andersson, P.A. Karlsson, M. Persson: Phys. Rev. Lett. **51**, 2378 (1983)

7.27 J.W.M. Frenken, J.F. van der Veen, G. Allan: Phys. Rev. Lett. **51**, 1876 (1983)

8. Vibrational Properties of Clean Surfaces: Survey of Recent Theoretical and Experimental Results

W. Kress

With 82 Figures

Current methods for the experimental investigation of surface vibrations and for the calculation of surface phonon dispersion curves have been reviewed in the preceding contributions. For each method some characteristic results were shown. Our present knowledge of surface vibrations is, however, not restricted to the few examples presented in the preceding chapters. The surface vibrations of a large variety of solids have been studied, both experimentally and theoretically, during recent years and a large amount of data has been accumulated. In this last chapter the attempt is made to provide a systematic and complete survey of the surface phonon dispersion curves of clean surfaces.

For each solid a set of the most reliable and complete data is presented. Reference to earlier work is only made when still required for the current understanding. However, references to most of the pioneering work, which in many cases has stimulated the investigations presented here, can be found in the preceding chapters and in many of the papers cited here. In this compilation the surface phonon dispersion curves are grouped according to the symmetry of the underlying bulk material, since this symmetry is a consequence of the interparticle interaction potentials and imposes many relations between materials belonging to the same space group. Within each space group, the compounds are ordered according to the position of their constituents in the periodic table. In this way similarities and dissimilarities, due to the nature of the interaction potentials, are taken into account in the ordering scheme. An alphabetic index at the end of this book facilitates the search for a particular element or compound.

It has been shown in the preceding chapters that theoretical predictions of the surface phonon dispersion curves based on the detailed knowledge of the bulk properties can, in many cases, be quite reliable and thus may serve as a guideline for future experiments. The present compilation, therefore, not only includes calculated surface-phonon dispersion curves reproducing the experimental data and giving a deeper insight into the underlying physics, but also contains calculated surface phonon dispersion curves of those compounds for which experimental data about the surface vibrations are not yet available. Purely theoretical surface phonon dispersion curves are, however, only presented for those compounds for which reliable predictions can be made from first principles or on the basis of bulk data. The extent to which the interaction potentials governing the bulk phonons determine also the surface vibrations depends strongly on the electronic configuration and is quite different for ionic crystals, covalent crystals, and metals.

The most reliable predictions can be made for those insulators which consist of ions with electronic closed-shell configurations. The interactions between these ions are to a good approximation governed by two-body central potentials. These potentials are, to a large extent, independent of the geometrical arrangement of the interacting ions and remain unaffected by the perturbed geometry at the surface. Thus the ions at and close to the surface relax or reconstruct according to these configuration-independent *potentials*, which can be determined from the knowledge of the bulk vibrations. The *force constants*, which govern the surface vibrations, are the derivatives of these potentials at the new equilibrium positions. Thus quite reliable predictions of the surface vibrations can be made for ionic crystals, in particular for alkali halides. In some cases, however, small deviations between the calculated and the measured surface vibrations occur which are due to charge transfer and many-body corrections at the surface (see LiF). Moreover, the comparison of measured and calculated surface phonon dispersion curves provides a stringent test for the bulk models, which might not always be based on optimal sets of potential parameters.

The situation for metals is quite different from that of ionic crystals. Here, the electronic ground state is already strongly affected by the presence of a surface. The interparticle interaction potential is determined by displacement induced variation of the electronic ground state energy. Thus the interaction *potential* for particles at and close to the surface is quite different from that of particles in the bulk. Even a proper relaxation in the bulk potentials does not, in most cases, lead to the correct force constants at and close to the surface. These force constants may either be obtained from a fit to the measured surface phonon dispersion curves, or they may be obtained from first principle calculations, which first treat the surface problem for the electronic system and then derive the appropriate coupling constants. It should be pointed out that frozen-phonon type calculations are extremely difficult, since the way the displacement vectors of a surface mode decay into the bulk is not governed by symmetry but by the solution of the eigenvalue problem itself. Thus the numerical effort for the surface problem is much higher than for the bulk problem.

The problems in calculating the surface properties of homopolar crystals are very similar to those already discussed for metals. Since the interaction potential for covalent crystals depends strongly on the local cluster structure, drastic changes occur at the surface and lead, as in metals, even in the static case, in many cases to interesting effects such as spontaneous surface reconstruction on cleaving.

8.1 Ionic Crystals

8.1.1 Alkali Halides (Rock Salt Structure)

The alkali halide family is one of the best investigated crystal groups. The dynamical properties of the bulk have been studied using various techniques, such as light scattering, ultrasound propagation, calorimetry, and neutron scattering. The phonon dispersion curves of most alkali halides have been mapped extensively by coherent inelastic neutron scattering measurements [8.1]. Both model theory (e.g. models including displacement induced deformations of the electronic charge density, for instance shell models, breathing shell models, etc.) and first principles theories (which derive lattice vibrations directly from electronic properties) are highly developed [8.1]. In particular, shell models are able to reproduce well the phonon dispersion curves of the bulk material with few adjustable, but physically meaningful, parameters.

Alkali halides have closed-shell electronic configurations. Therefore, major electronic rearrangements do not occur at and close to the surface. The forces which govern the bulk dynamics determine the surface vibrations as well, since relaxation effects are small [8.2] and may be neglected in a first approximation. Thus both the slab [8.3] and the Green's function method [8.4] provide reliable surface phonon dispersion curves on the basis of model parameters which have been determined from the dynamical bulk properties.

Almost all experimental investigations of the surface dynamics of alkali halides have been carried out by atom scattering experiments [8.5] since alkali halides are good insulators and therefore less well suited for electron scattering experiments like EELS [8.6] than metals for which both atom and electron scattering techniques can be used.

LiH(001)

The bulk dynamics of ^{7}LiD has been studied by inelastic neutron scattering [8.7]. The measured phonon dispersion curves are well reproduced by shell-model calculations [8.7] and other theoretical approaches [8.8–11]. Measurements of the surface phonon dispersion curves are still missing.

LiF(001)

The phonon dispersion curves in the bulk material have been determined by inelastic neutron scattering measurements [8.12]. The measured data are well reproduced by various model calculations [8.12–17].

The surface phonon dispersion curves have been calculated using both the slab and the Green's function method. Figure 8.1 shows the surface phonon dispersion curves of an unrelaxed 15-layer slab with free (001) surfaces [8.18]. The calculations are based on the parameters of shell model I of [8.12]. The frequencies of the surface modes S_1–S_7 at the high symmetry points $\overline{\Gamma}$, $\overline{X}$, and $\overline{M}$ of the surface Brillouin zone of a relaxed 15-layer slab with free (001) surfaces

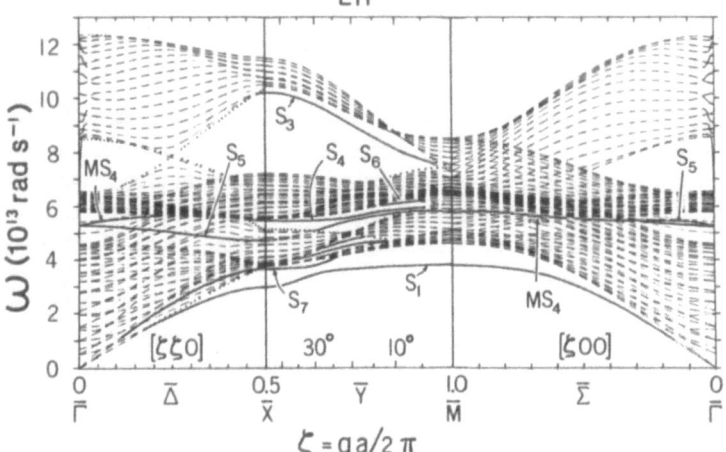

Fig. 8.1. LiF(001): Surface phonon dispersion curves (after [8.18]). Calculations [8.18]: slab method (15 unrelaxed layers, shell model based on bulk data); *dashed lines*: bulk-like modes; *solid lines*: surface localized modes; S_i and MS_i label surface localized modes and resonances, respectively (notation as in [8.3])

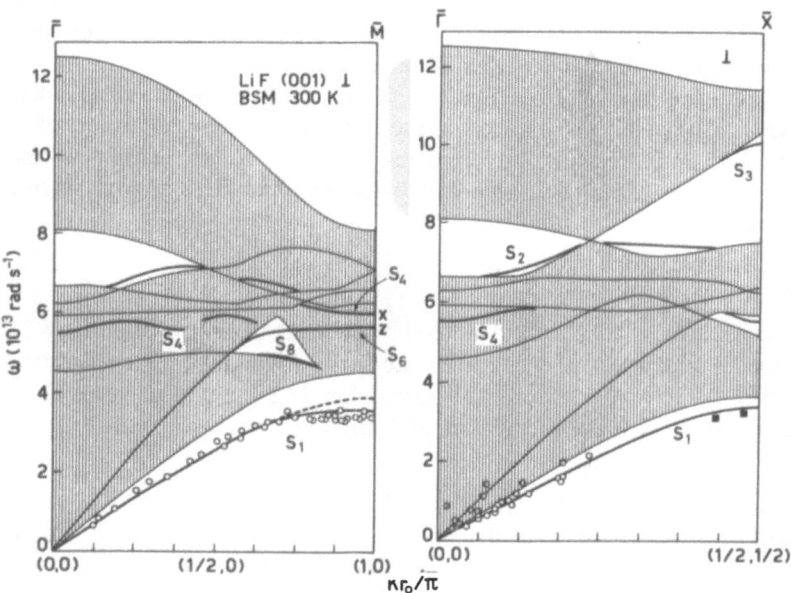

Fig. 8.2. LiF(001): Surface phonon dispersion curves (after [8.27]). Calculations [8.20]: Green's function method (unrelaxed surface, breathing shell model based on macroscopic bulk data); *bold lines*: surface localized modes and resonances; *shaded*: surface projected bulk modes; S_i: labels surface localized modes and resonances (notation as in [8.4]). Panels which show modes with shear horizontal or sagittal polarizations are labeled by $\parallel$ or $\perp$, respectively. Experimental data: surface modes as determined by He-atom scattering measurements [8.21–23] (*open circles*) and [8.42] (*filled squares*)

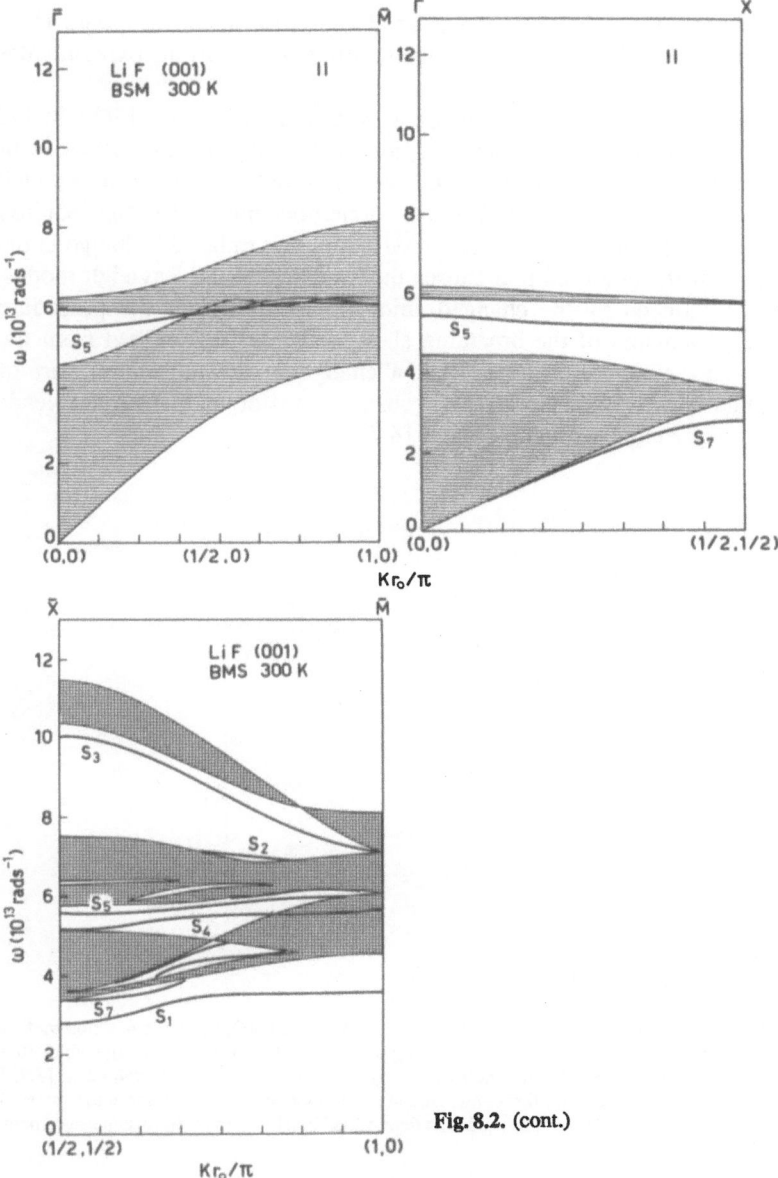

Fig. 8.2. (cont.)

are reported in [8.19]. The frequency shifts due to relaxation are in general small ($\leq 2\%$) except for the S_4-mode which is shifted upwards by 9% at the $\overline{M}$ point.

Green's function calculations [8.20] based on breathing-shell-model parameters [8.13] obtained from the macroscopic bulk data yield results similar to those obtained for the unrelaxed slab. Figure 8.2 shows the results of the Green's function method [8.20] together with recent He-atom scattering data [8.21–23]. The agreement between the calculated surface modes and the inelastic He-scattering data is good except for the Rayleigh waves at and close to the $\overline{M}$ point. Here the calculated frequencies (dashed line) are slightly higher than the experimental data.

It is claimed in [8.20] that an increase of the anion polarizability at the surface by about 20 % yields the curve represented by the full line which is in excellent agreement with the experimental data. It should, however, be pointed out that a change of the anion polarizability alone does not lead to the required lowering of the Rayleigh mode close to the $\overline{M}$ point. It is rather the change in the noncentral part of the potential that causes the lowering of the Rayleigh modes. This change is induced by the changed anion polarizability via the procedure by which the parameters of the breathing shell model are determined from the macroscopic data. Figure 8.3 shows that a change in the noncentral part of the short range interaction as well as a change transfer at the surface leads to satisfactory agreement with experiment [8.75].

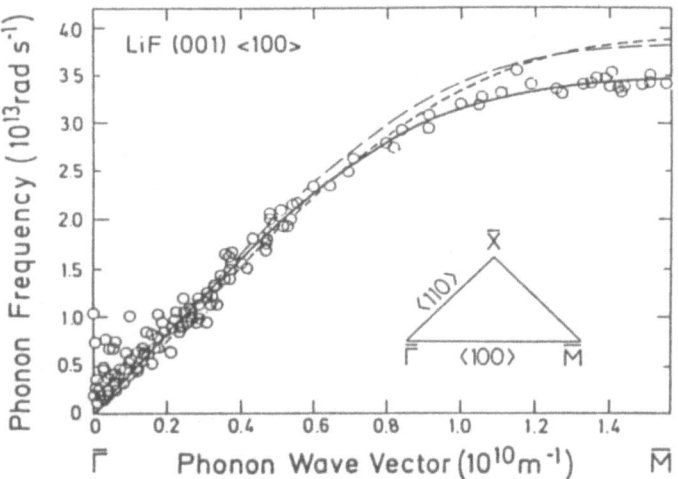

Fig. 8.3. LiF(001): Dispersion of Rayleigh waves (after [8.75]). Calculations: Green's function method (*short-dashed lines* [8.254]: unrelaxed surface, breathing shell model based on bulk data); slab method (*long-dashed lines* [8.18]: unrelaxed slab, shell model based on bulk parameters; *solid line* [8.75]: unrelaxed slab, shell model, noncentral forces changed at the surface or charge transfer at the surface). Experimental data [8.23, 255, 256]: surface modes as determined by He-atom scattering measurements (*open circles*)

Inelastic He-atom scattering measurements have not only been carried out for the Rayleigh waves but have also been extended recently to the optic modes. The results [8.24] are shown in Fig. 8.4. The agreement between the experimental data for the optic surface modes and the results of both the Green's function method and the slab method is rather promising. There remain, however, some deviations which need further investigation. Surface relaxation effects are too small to account for these discrepancies.

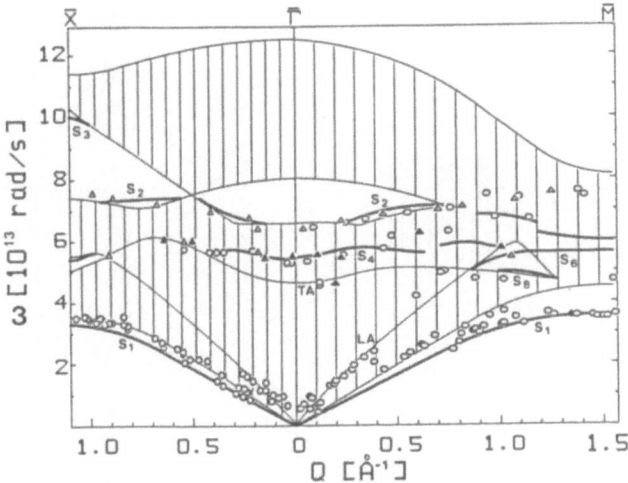

Fig. 8.4. LiF(001): Surface phonon dispersion curves of sagittal modes (after [8.24]). Calculations [8.20]: Green's function method (unrelaxed surface, breathing shell model based on macroscopic bulk data); *bold lines*: surface localized modes and resonances; *vertically hatched*: surface projected bulk modes. Experimental data [8.24]: surface modes as determined by He-atom scattering measurements (*circles*: nonresonant scattering conditions; *triangles*: resonant scattering conditions). The data for the Rayleigh modes are in good agreement with the He-scattering data of [8.21–23]

LiCl(001), LiBr(001), LiI(001)

Inelastic neutron scattering measurements of the phonon dispersion curves in the bulk of these compounds have not been available until now. Nonetheless the phonon dispersion curves of the bulk material have been calculated [8.8, 15, 17]. So far, these calculations have not served as a basis for an investigation of the surface vibrations.

NaH(001)

The only reliable data about the lattice vibrations are incoherent inelastic neutron scattering measurements of the amplitude-weighted bulk phonon density of states [8.25].

NaF(001)

The phonon dispersion curves in the bulk have been determined by inelastic neutron scattering measurements [8.26] as well as by diffuse X-ray scattering [8.257] and are well reproduced by various model calculations [8.26, 29–35].

The surface phonon dispersion curves have been calculated using both the slab and the Green's function method. Figure 8.5 shows the surface phonon dispersion curves of an unrelaxed 15-layer slab with free (001) surfaces [8.18]. The calculations are based on the parameters of shell model VI of [8.26]. The frequencies of the surface modes S_1–S_7 at the high symmetry points $\overline{\Gamma}$, $\overline{X}$,

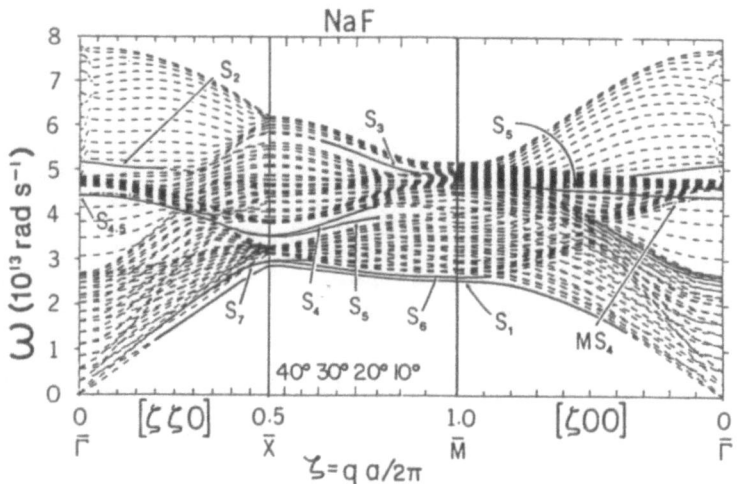

Fig. 8.5. NaF(001): Surface phonon dispersion curves (after [8.18]). Calculations [8.18]: Slab method (15 unrelaxed layers, shell model based on bulk data); *dashed lines*: bulk-like modes; *solid lines*: surface localized modes; S_i and MS_i label surface localized modes and resonances, respectively (notation as in [8.3])

and $\overline{M}$ of the surface Brillouin zone of a relaxed 15-layer slab with free (001) surfaces are reported in [8.19]. The frequency shift due to relaxation is small ($\leq 1\%$) in general, except for the S_4 mode, which shows an upward shift of about 2% at the $\overline{\Gamma}$-point.

Results quite similar to those obtained by the slab method for the unrelaxed NaF(001) slab [8.18] are also obtained by Green's function calculations [8.20, 36] which are based on breathing-shell-model parameters derived from macroscopic bulk data. Figure 8.6 shows the results of the Green's function method together with data obtained from inelastic He-atom scattering experiments [8.23, 36, 37]. The agreement between the experimental data and the calculated surface modes is very convincing. It should be pointed out here that no adjustment of model parameters was made in order to fit the inelastic He-atom scattering data.

Two different interpretations of the mode S_8 of Fig. 8.6 and the mode S_6 in the $\overline{X}-\overline{M}$ direction (which in fact is a sagittal plane mode in which predominantly the light ion vibrates and which might be therefore better called a S_1' mode) can be found in the literature. *Benedek* et al. [8.20, 36, 38] attribute this mode to the folding of a hypothetical monoatomic lattice of the same structure, while *de Wette* et al. [8.39] trace the origin of this mode back to the bulk dynamics and argue that the appearance of this mode is due to the fact that for nearly equal anion and cation masses the TA_1 and TO_1 branches of the bulk dispersion curves in the (110) direction come close together and exchange their vibrational character roughly in the middle of the Σ direction. Both arguments have in common that S_8 and S_1' modes appear only if anion and cation masses are nearly equal. For compounds with a large difference between anion and cation masses such effects are not to be expected.

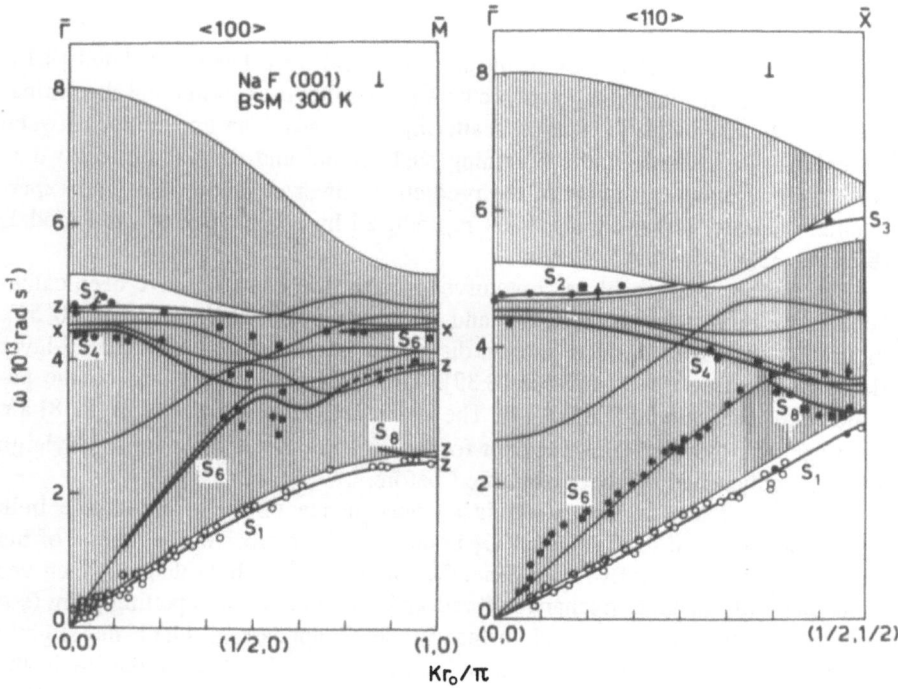

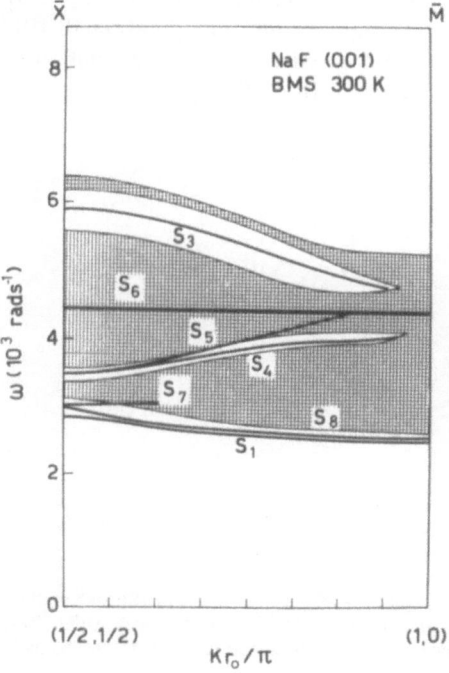

Fig. 8.6. NaF(001): Surface phonon dispersion curves of sagittal modes (after [8.75]). Calculations [8.36]: Green's function method (unrelaxed surface, breathing shell model based on macroscopic bulk data); *solid lines*: surface localized modes and resonances; S_i: labels surface localized modes and resonances (notation as in [8.4]; *vertically hatched*: surface projected bulk modes. Experimental data [8.36]: surface modes as determined by He-atom scattering measurements; *open circles*: Rayleigh modes [8.22, 23]; *filled symbols*: S_2, S_4, S_6, and S_8 modes [8.36, 37]. (The last panel shows all modes in the $\overline{X}$-$\overline{M}$ direction)

NaCl(001)

The phonon dispersion curves in the bulk material have been calculated [8.13] on the basis of available macroscopic data prior to their experimental determination by coherent inelastic neutron scattering [8.40, 44]. The agreement between the calculations based on the breathing shell model and the experimental data is a strong argument in favor of the predictive power of this model. The experimental data are, however, also well reproduced by other types of shell models [8.14, 30, 35, 41, 44–46].

The surface phonon dispersion curves for the (001) surface have been calculated using both the slab [8.18, 49] and the Green's function method [8.50, 51]. Figure 8.7 shows the surface phonon dispersion curves of an unrelaxed 15-layer slab with the free (001) surfaces [8.39]. The calculations are based on the parameters of shell model II of [8.44]. The slab calculations presented in [8.18] are identical with those of [8.39] except for the appearance of the second Rayleigh mode S_1' which had not been identified before.

The appearance of this mode is a consequence of the fact that the bulk dispersion curves of the TA_1 and TO_1 branches almost cross in the middle of the Σ direction of the bulk Brillouin zone. Calculations of the bulk dispersion curves show that both branches exchange vibrational character at the repelling point (see also NaF, KBr). Thus a second surface mode S_1' appears in which the slightly lighter cation is displaced predominantly, whereas the Rayleigh mode S_1 in the $\overline{X}-\overline{M}$ direction is predominantly due to displacements of the heavier anion. The frequencies of the surface modes of this model [8.19] change very little ($\leq 1\%$) when the surface is allowed to relax into its equilibrium configuration [8.19]. Relaxation effects of up to 7% are obtained [8.52], when the slab calculation is based on the shell-model parameters of model 2 of [8.41, 53].

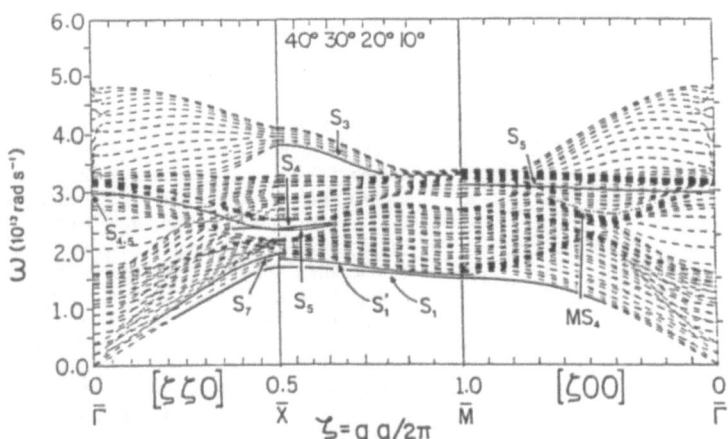

Fig. 8.7. NaCl(001): Surface phonon dispersion curves (after [8.39]). Calculations [8.18]: slab method (15 unrelaxed layers, shell model based on bulk data); *dashed lines*: bulk-like modes; *solid lines*: surface localized modes; S_i and MS_i label surface localized modes and resonances, respectively (notation as in [8.3])

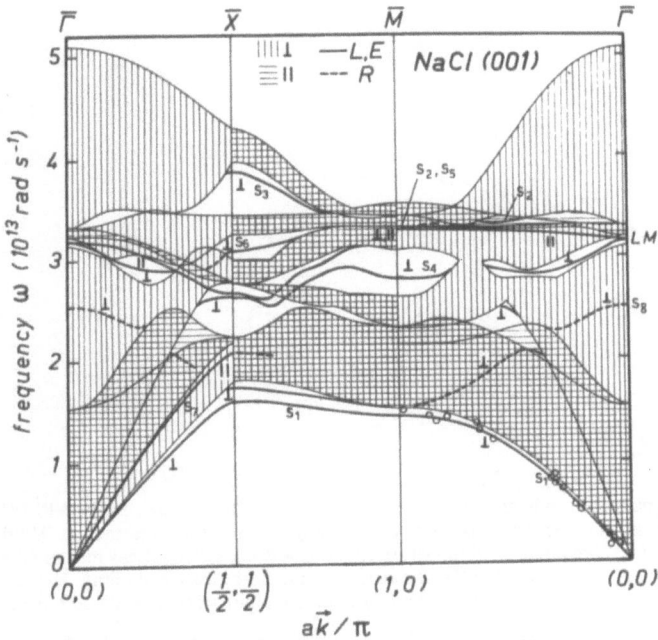

Fig. 8.8. NaCl(001): Surface phonon dispersion curves (after [8.50]). Calculations [8.50]: Green's function method (unrelaxed surface, breathing shell model based on macroscopic bulk data); *solid lines*: surface localized modes with shear horizontal ($\parallel$) and sagittal ($\perp$) polarization; *dashed lines*: resonant modes; S_i: labels surface localized modes and resonances (notation as in [8.4]); *horizontally hatched*: surface projected bulk modes with shear-horizontal polarization; *vertically hatched*: surface projected bulk modes with sagittal polarization. Experimental data [8.21]: surface modes as determined by He-atom scattering measurements (*open* and *filled circles*)

Results similar to those of the slab method are obtained by the Green's function calculations [8.50] based on breathing-shell-model parameters derived from macroscopic bulk data. Figure 8.8 shows the results of the Green's function method together with the results of inelastic He-atom scattering measurements of the Rayleigh waves [8.54]. In Fig. 8.9 the complete He-atom scattering data are compared with the results obtained from calculations using the slab and the Green's function method.

NaBr(001)

The phonon dispersion curves in the bulk material have been determined by inelastic neutron scattering measurements [8.55]. The experimental data are well reproduced by numerous shell-model calculations [8.28, 30, 31, 34, 35, 55, 56].

The surface phonon dispersion curves of a relaxed 15-layer slab with free (001) surfaces have been calculated [8.19] on the basis of a shell model (model V of [8.55]), which reproduces the bulk data satisfactorily. The results are shown in Fig. 8.10. The relaxed equilibrium positions for this slab are given in [8.2]. Although experimental He-scattering data are not yet available for NaBr, there

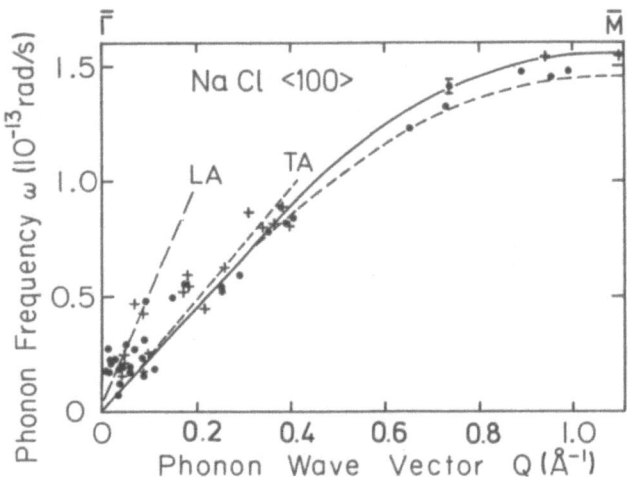

Fig. 8.9. NaCl(001): Dispersion of Rayleigh waves (after [8.54]). Calculations: slab method [8.18] (*solid lines*: unrelaxed slab, shell model based on bulk parameters); Green's function method [8.50] (*dashed lines*: unrelaxed surface, breathing shell model based on bulk data); straight lines indicate the slopes of the longitudinal (LA) and transverse (TA) acoustic bulk modes in the elastic limit. Experimental data [8.54]: Rayleigh modes as determined from inelastic He-atom scattering measurements (*circles*: $T = 300$ K; *crosses*: $T = 220$ K)

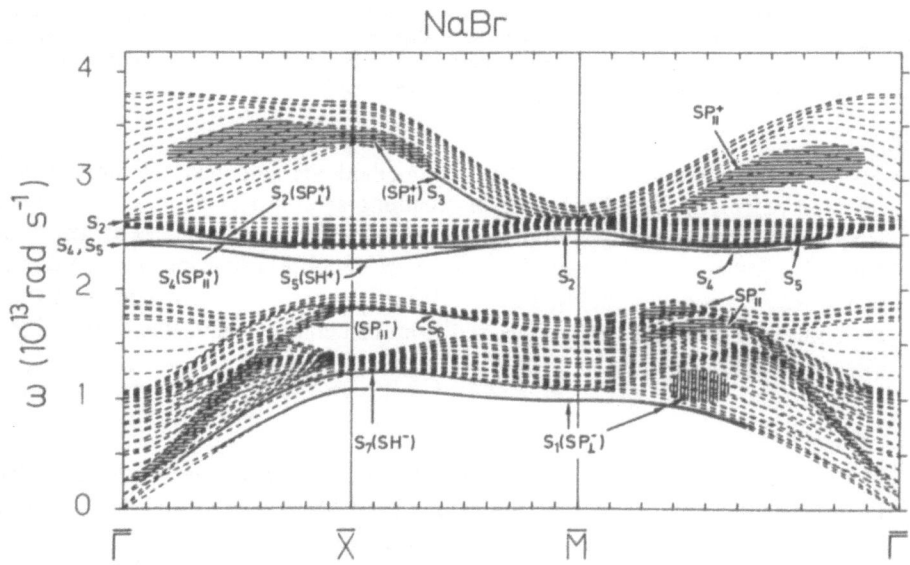

Fig. 8.10. NaBr(001): Surface phonon dispersion curves (after [8.19]). Calculations [8.19]: slab method (15 relaxed layers, shell model based on bulk data); *dashed lines*: bulk-like modes; *solid lines*: surface localized modes; S_i labels surface localized modes; *hatching*: surface resonances (notation as in [8.3]); $SP_\perp$ and $SP_\parallel$: sagittal-plane vibrations perpendicular and parallel to the surface, respectively; SH: shear horizontal modes; superscripts + and −: cation and anion displacements in the top layer, respectively

are good reasons to believe that the calculated dispersion curves are as reliable as those which have been already verified experimentally.

NaI(001)

NaI is one of the compounds for which reliable measurements of the phonon dispersion curves by coherent inelastic neutron scattering measurements [8.57, 58] became available first. Thus much of the early theoretical work has focused on that compound. The measured phonon dispersion curves in the bulk are well reproduced by various types of shell-model calculations [8.28, 30, 34, 35, 57, 59–62]. It should be added here that diffuse X-ray scattering measurements of the phonon dispersion curves [8.257] have also been carried out. The results of these measurements are in agreement with the neutron scattering data.

Both the shell model (model VI of [8.59]) and the breathing shell model [8.60] have served as a basis for calculations of the surface phonon dispersion curves by the slab [8.18] and the Green's function method [8.50], respectively. The results obtained for an unrelaxed 15-layer slab with free (001) surfaces [8.18] are shown in Fig. 8.11. Acoustic and optic bands are well separated due to the large difference between anion and cation mass. Thus the S_4 and S_5 modes, which are peeled off the surface projected transverse optic bulk bands, become nicely visible in the whole surface Brillouin zone.

The Green's function method yields quite similar results for the unrelaxed (001) surface [8.50]. These results are shown in Fig. 8.12. An experimental verification of the calculated surface phonon dispersion curves is still missing, but there are no reasons to believe that major deviations from the calculated curves will result from inelastic He-atom scattering experiments. The relaxed positions for a 15-layer slab have been calculated [8.2] and the effects of the relaxation on

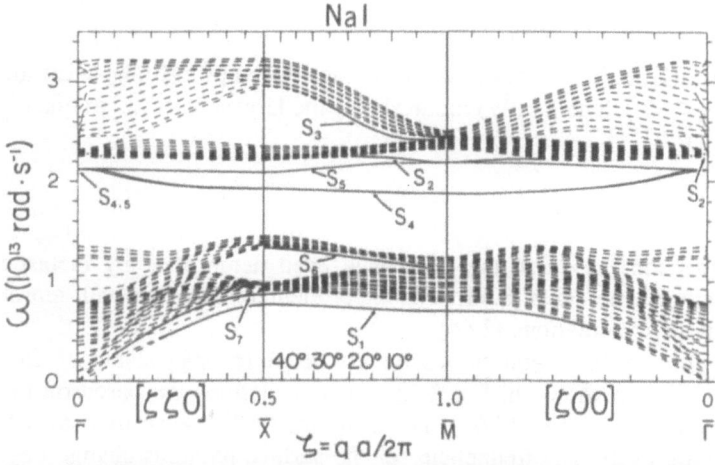

Fig. 8.11. NaI(001): Surface phonon dispersion curves (after [8.18]). Calculations [8.18]: slab method (15 unrelaxed layers, shell model based on bulk data); *dashed lines*: bulk-like modes; *solid lines*: surface localized modes; S_i: labels surface localized modes (notation as in [8.3])

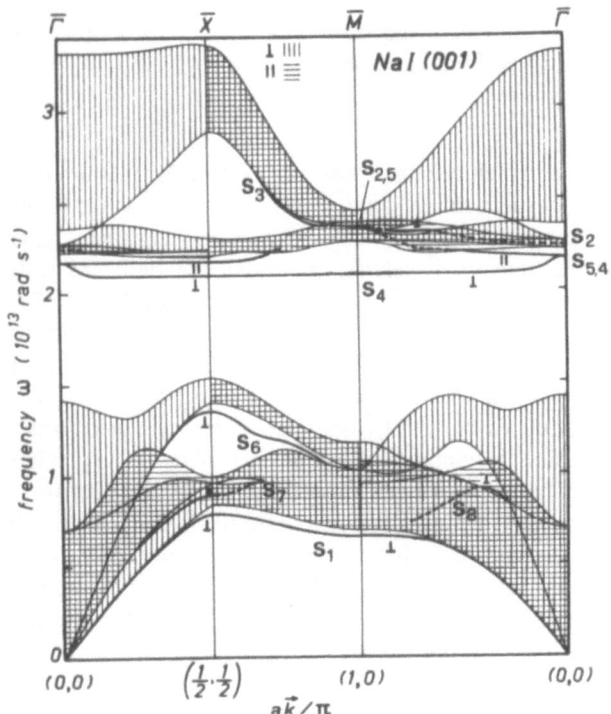

Fig. 8.12. NaI(001): Surface phonon dispersion curves (after [8.50]). Calculations [8.50]: Green's function method (unrelaxed surface, breathing shell model based on macroscopic bulk data); *solid lines*: surface localized modes with shear horizontal (∥) and sagittal (⊥) polarization; *dashed lines*: resonant modes; S_i: labels surface localized modes and resonances (notation as in [8.4]); *horizontally hatched*: surface projected bulk modes with shear horizontal polarization; *vertically hatched*: surface projected bulk modes with sagittal polarization

the frequencies of the surface modes have been investigated [8.19]. The shifts are in general small (≤ 3%). The S_4 ($\overline{M}$) mode shows the largest shift. Its frequency increases by about 7%.

KF(001)

The phonon dispersion curves in the bulk have been determined by inelastic neutron scattering measurements [8.64]. The experimental results are well reproduced by shell-model calculations [8.64].

The surface phonon dispersion curves for a relaxed 15-layer slab with free (001) surfaces [8.19] are shown in Fig. 8.13. The calculations are based on the parameters of shell model IV of [8.64]. The same model has been used to calculate the relaxation [8.2]. The frequencies of the surface phonons change only slightly (≤ 1%) upon relaxation, except for the $S_4(\overline{X})$ and $S_4(\overline{M})$ modes which increase by 8 and 12%, respectively, when the slab relaxes [8.19].

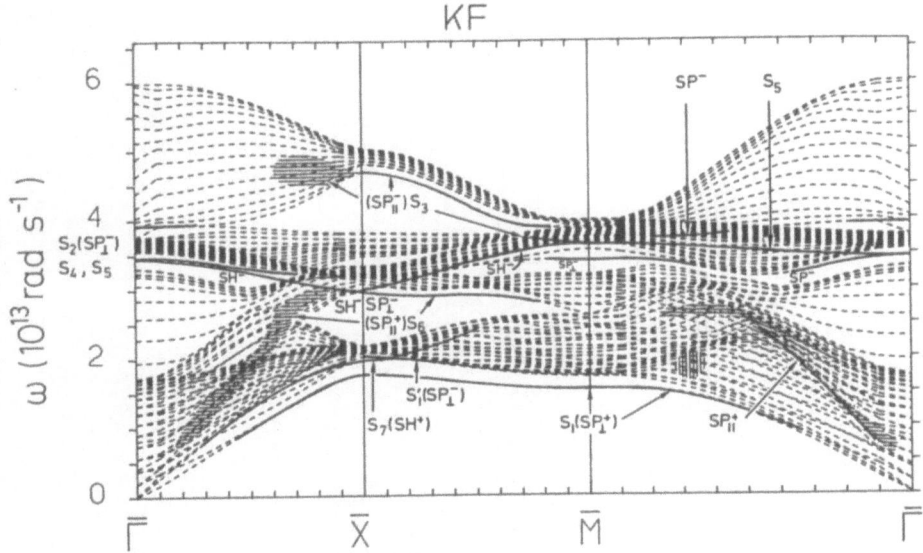

Fig. 8.13. KF(001): Surface phonon dispersion curves (after [8.19]). Calculations [8.19]: slab method (15 relaxed layers, shell model based on bulk data); *dashed lines*: bulk-like modes; *solid lines*: surface localized modes; *hatching*: surface resonances (notation as in [8.3]); S_i: labels surface localized modes; $SP_\perp$ and $SP_\parallel$: sagittal-plane vibrations perpendicular and parallel to the surface, respectively; SH: shear horizontal modes; superscripts $+$ and $-$: cation and anion displacements in the top layer, respectively

The Green's function method has been used to calculate the surface phonon dispersion curves of the unrelaxed (001) surface [8.50]. The computations are based on a breathing shell model. The parameters of the model have been determined from the macroscopic data of the bulk material. The calculated surface phonon dispersion curves are shown in Fig. 8.14. Comparison between Fig. 8.13 and Fig. 8.14 shows that both the slab and the Green's function method yield very similar results. Experimental measurements of the surface phonons are at present not yet available.

KCl(001)

The phonon dispersion curves in the bulk have been determined by inelastic neutron scattering measurements [8.65, 66]. The experimental results are well reproduced by various shell-model calculations [8.14, 41, 43, 47, 67].

The surface phonon dispersion curves for a relaxed 15-layer slab with free (001) surfaces [8.19] are shown in Fig. 8.15. The calculations are based on the shell model parameters (model 2 of [8.41, 53]), which have been determined from a fit to the measured phonon dispersion curves of the bulk. The same model has been used to calculate the relaxation [8.2]. The frequencies of the surface phonons change only slightly ($\leq 2\%$).

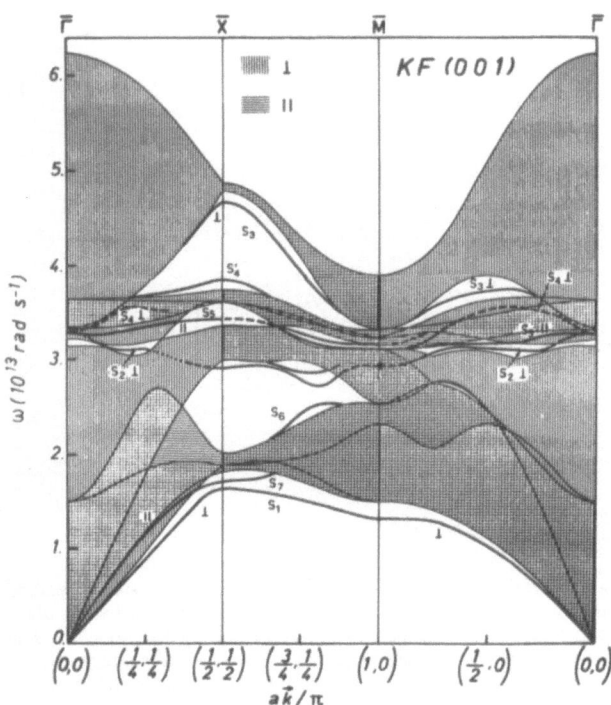

Fig. 8.14. KF(001): Surface phonon dispersion curves (after [8.50]). Calculation [8.50]: Green's function method (unrelaxed surface, breathing shell model based on macroscopic bulk data); *solid lines*: surface localized modes with shear horizontal (∥) and sagittal (⊥) polarization; *dashed lines*: resonant modes; S_i: labels surface localized modes and resonances (notation as in [8.4]); *horizontally hatched*: surface projected bulk modes with shear horizontal polarization; *vertically hatched*: surface projected bulk modes with sagittal polarization

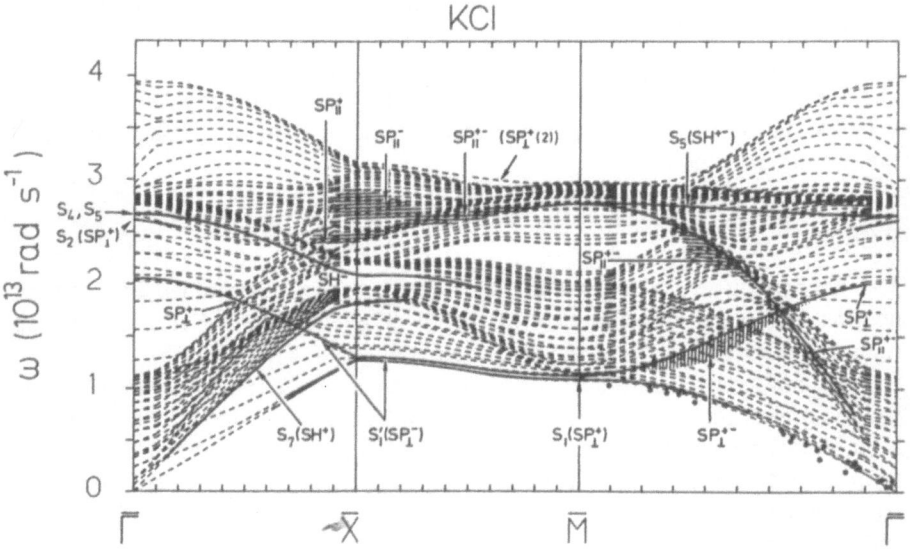

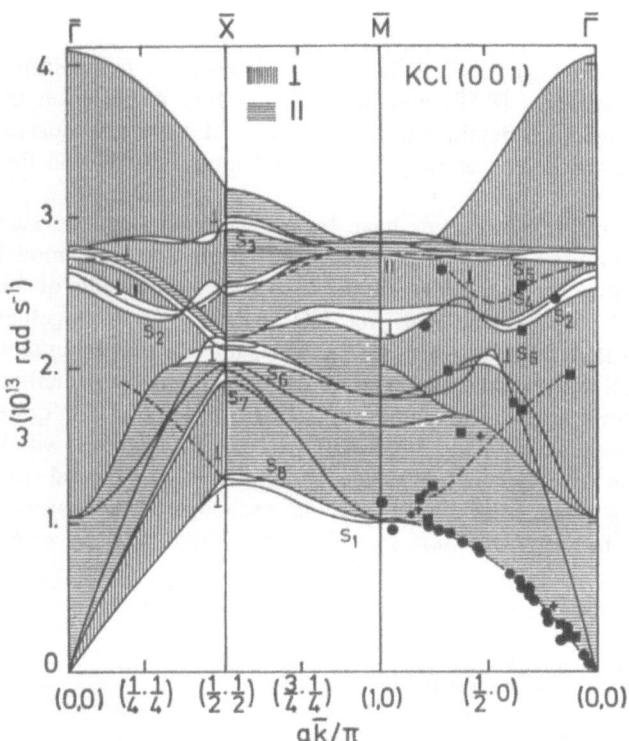

Fig. 8.16. KCl(001): Surface phonon dispersion curves (after [8.68]). Calculations [8.68]: Green's function method (unrelaxed surface, breathing shell model based on macroscopic bulk data); *solid lines*: surface localized modes with shear horizontal (∥) and sagittal (⊥) polarization; *dashed lines*: resonant modes; S_i: labels surface localized modes and resonances (notation as in [8.4]); *horizontally hatched*: surface projected bulk modes with shear horizontal polarization; *vertically hatched*: surface projected bulk modes with sagittal polarization. Experimental data [8.21]: surface modes as determined by He-atom scattering measurements (*filled circles*). See also [8.23]

The Green's function method has been used to calculate the surface phonon dispersion curves of the unrelaxed (001) surface [8.68]. The computations are based on a breathing shell model. The parameters of the model have been determined from the macroscopic data of the bulk material. The calculated surface phonon dispersion curves are shown in Fig. 8.16 together with the results of inelastic He-atom scattering measurements [8.23, 68].

Fig. 8.15. KCl(001): Surface phonon dispersion curves (after [8.19]). Calculations [8.19]: slab method (15 relaxed layers, shell model based on bulk data); *dashed lines*: bulk-like modes; *solid lines*: surface localized modes; S_i: labels surface localized methods; *hatching*: surface resonances (notation as in [8.3]); SP$_\perp$ and SP$_\parallel$: sagittal-plane vibrations perpendicular and parallel to the surface, respectively; SH: shear horizontal modes; superscripts + and −: cation and anion displacements in the top layer, respectively

KBr(001)

The phonon dispersion curves in the bulk have been determined by inelastic neutron scattering measurements [8.58] and the collision free propagation of sound waves has been studied in detail [8.69]. The measured dispersion curves are well reproduced by calculations employing the shell model [8.59] and the breathing shell model [8.60].

The surface phonon dispersion curves have been calculated with the slab method [8.19] and the Green's function method [8.50] using the shell model and the breathing shell model, respectively. Figure 8.17 shows the results of the calculation for a relaxed 15-layer slab with free (001) surfaces [8.19] together with recent inelastic He-atom scattering data [8.70]. The same He-scattering data are compared in Fig. 8.18 to the results obtained by the Green's function method [8.50]. Comparison of both figures shows that the slab calculations are in quite good agreement with the measured data, in particular for the $SP_{\perp}^{-}$-modes, which are totally missing in the Green's function results [8.50]. The experimental data [8.70] clearly support the anticrossing mechanism with exchange of the character of TA_1 and TO_1 modes in the Σ direction (see NaF,NaCl). It should, however,

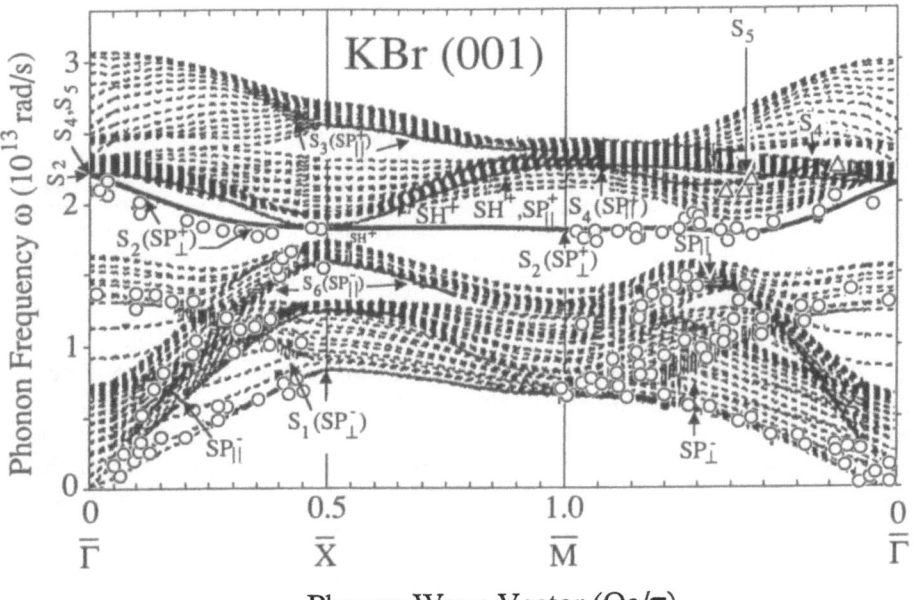

Phonon Wave Vector (Qa/π)

Fig. 8.17. KBr(001): Surface phonon dispersion curves (after [8.70]). Calculations [8.19]: slab method (15 relaxed layers, shell model based on bulk data); *dashed lines*: bulk-like modes; *solid lines*: surface localized modes and resonances; S_i: labels surface localized modes; *hatching*: surface resonances (notation as in [8.3]); $SP_\perp$ and $SP_\parallel$: sagittal-plane vibrations perpendicular and parallel to the surface, respecitvely.; SH: shear horizontal modes; superscripts + and −: cation and anion displacements in the top layer, respectively. Experimental data [8.70]: surface modes as determined by He-atom scattering measurements (*filled circles*: strong scattering intensities; *triangles*: weak scattering intensities)

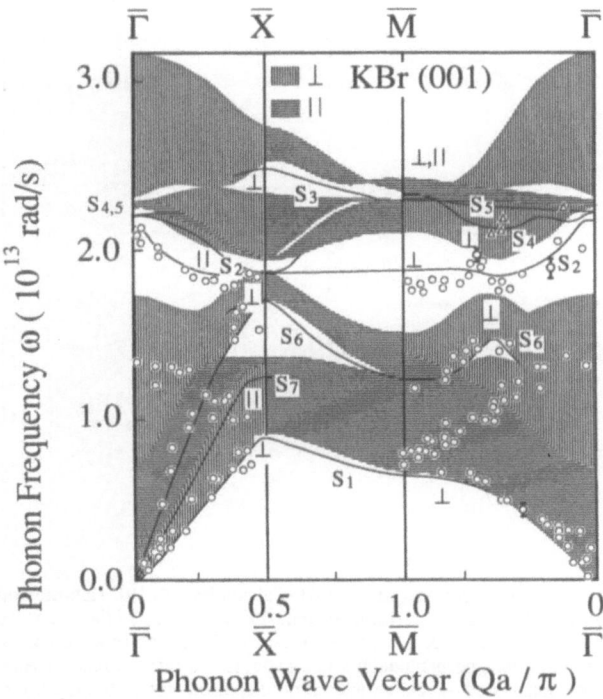

Fig. 8.18. KBr(001): Surface phonon dispersion curves (after [8.70]). Calculations [8.50]: Greens function method (unrelaxed surface, breathing shell model based on macroscopic bulk data); *solid lines*: surface localized modes with shear horizontal (∥) and sagittal (⊥) polarization; *dashed lines*: resonant modes; S_i: labels surface localized modes and resonances (notation as in [8.4]); *horizontally hatched*: surface projected bulk modes with shear horizontal polarization; *vertically hatched*: surface projected bulk modes with sagittal polarization. Experimental data [8.70]: surface modes as determined by He-atom scattering measurements (*filled circles*: strong scattering intensities; *triangles*: weak scattering intensities)

be mentioned that the peak width of the experimental data for the SP$_\bot^-$ branch is smaller than that estimated from the slab calculations. A calculation of a 15-layer-slab is, on the other hand, not sufficient for a realistic estimation of the width of the SP$_\bot^-$ peaks. It might be possible to reduce the remaining small differences between measured and calculated data by a refinement of the underlying shell model.

KI(001)

The phonon dispersion curves in the bulk have been determined by inelastic neutron scattering measurements [8.71]. Shell-model calculations [8.71] reproduce the experimental data satisfactorily.

Based on model III of [8.71], calculations for a relaxed 15-layer slab with free (001) surfaces have been carried out [8.19]. Figure 8.19 shows the calculated surface phonon dispersion curves. The frequencies of the S_2 modes are shifted

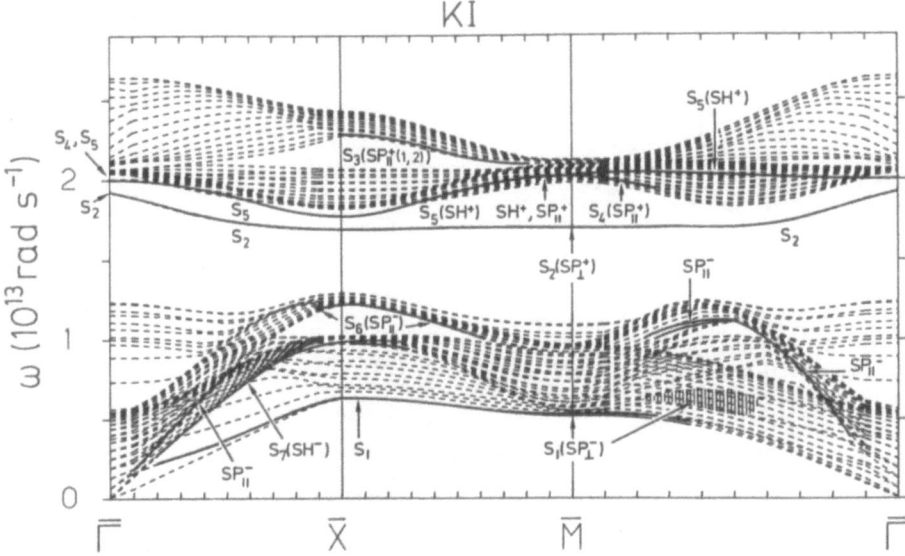

Fig. 8.19. KI(001): Surface phonon dispersion curves (after[8.19]). Calculations [8.19]: slab method (15 relaxed layers, shell model based on bulk data); *dashed lines*: bulk-like modes; *solid lines*: surface localized modes; S_i: labels surface localized modes; *hatching*: surface resonances (notation as in [8.3]); $SP_\perp$ and $SP_\parallel$: sagittal-plane vibrations perpendicular and parallel to the surface, respectively; SH: shear horizontal modes; superscripts + and −: cation and anion displacements in the top layer, respectively

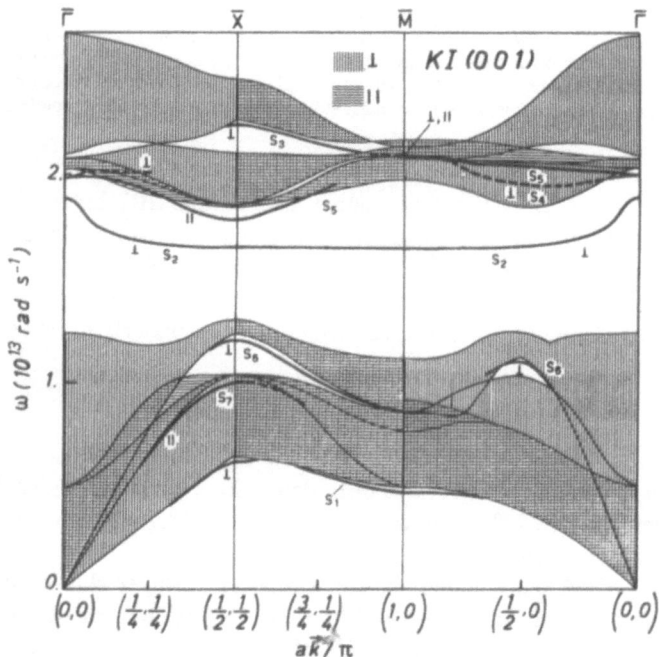

upwards by 5 to 8% [8.19] when the ions of the slab are allowed to relax into their equilibrium positions [8.2]. The shift for the other modes are small ($\leq$ 1%).

The results obtained with the Green's function method [8.50] for an unrelaxed 001 surface are shown in Fig. 8.20. These calculations use a breathing shell model. The parameters of this model have been obtained from macroscopic bulk data.

RbF(001)

The phonon dispersion curves in the bulk have been determined from inelastic neutron scattering measurements [8.72]. The experimental results are well reproduced by calculations using various types of shell models [8.41, 73, 74].

Based on shell model 2 of [8.41], the surface phonon dispersion curves of a relaxed 15-layer slab have been calculated. The results are shown in Fig. 8.21. Most of the surface modes of this model are slightly shifted upwards in frequency ($\leq$ 2%) [8.19] when the surface is allowed to relax into its true equilibrium positions [8.2]. The frequencies of the $S_4(\overline{X})$ and $S_4(\overline{M})$ modes increase, however, by 10 and 15 %, respectively, on relaxation.

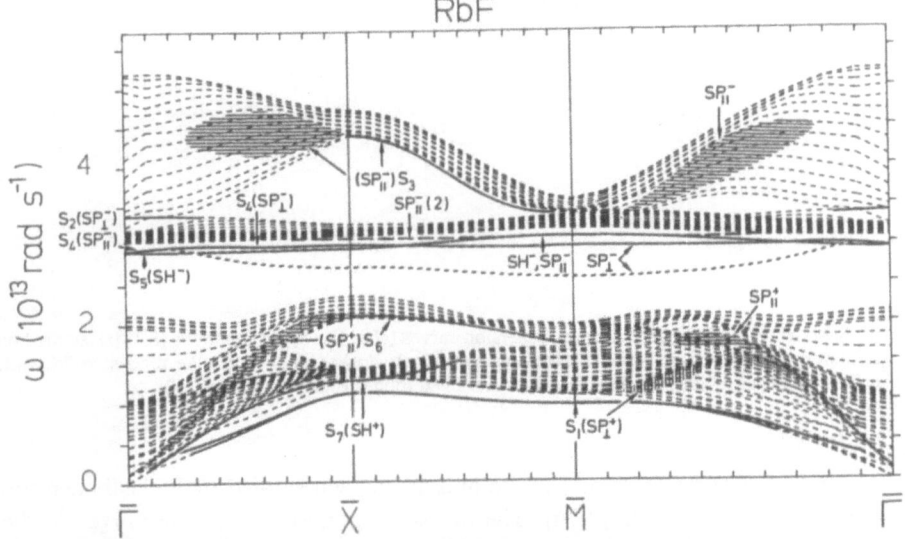

Fig. 8.21. RbF(001): Surface phonon dispersion curves (after [8.19]). Calculations [8.19]: slab method (15 relaxed layers, shell model based on bulk data); *dashed lines*: bulk-like modes; *solid lines*: surface localized modes; S_i: labels surface localized modes; *hatching*: surface resonances (notation as in [8.3]); SP$_\perp$ and SP$_\parallel$: sagittal-plane vibrations perpendicular and parallel to the surface, respectively; SH: shear horizontal modes; superscripts + and $-$: cation and anion displacements in the top layer, respectively

Fig. 8.20. KI(001): Surface phonon dispersion curves (after [8.50]). Calculations [8.50]: Green's function method (unrelaxed surface, breathing shell model based on macroscopic bulk data); *solid lines*: surface localized modes with shear horizontal ($\parallel$) and sagittal ($\perp$) polarization; *dashed lines*: resonant modes; S_i: labels surface localized modes and resonances (notation as in [8.4]); *horizontally hatched*: surface projected bulk modes with shear horizontal polarization; *vertically hatched*: surface projected bulk modes with sagittal polarization

RbCl(001)

The phonon dispersion curves in the bulk have been investigated by inelastic neutron scattering measurements [8.72]. The experimental results are well reproduced by calculations using various types of shell models [8.41, 73, 74].

The surface phonon dispersion curves of an unrelaxed 15-layer slab have been calculated using the shell model 2 of [8.41]. The results are shown in Fig. 8.22. All modes of this model show a frequency shift towards higher frequencies [8.19] when the surface is allowed to relax into its true equilibrium positions [8.2]. The shifts are +8% for S_4 $(\overline{M})$, +4% for S_2 $(\overline{M})$ and S_2 $(\overline{X})$, and less or equal to 2% for the other modes at the high symmetry points, $\overline{\Gamma}, \overline{X}$, and $\overline{M}$ of the surface Brillouin zone.

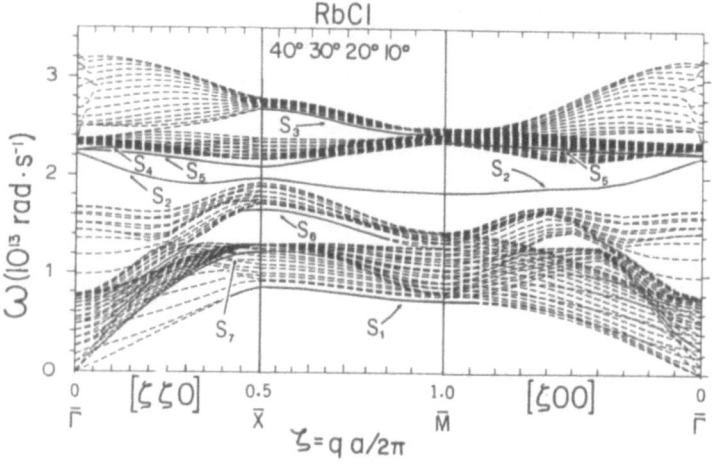

Fig. 8.22. RbCl(001): Surface phonon dispersion curves (after [8.18]. Calculations [8.18]: slab method (15 unrelaxed layers, shell model based on bulk data); *dashed lines*: bulk-like modes; *solid lines*: surface localized modes; S_i: labels surface localized modes (notation as in [8.3])

RbBr(001)

The phonon dispersion curves in the bulk have been studied by inelastic neutron scattering measurements [8.76]. The measured phonon dispersion curves in the main symmetry directions of the Brillouin zone are rather well reproduced by calculations employing various types of shell models [8.73, 74, 76].

The surface phonon dispersion curves of an unrelaxed and a relaxed [8.2] 15-layer slab with free (001) surfaces have been calculated [8.19, 52] on the basis of shell model IV of [8.76]. The results of both the calculation for the unrelaxed and the relaxed slab are compared in Fig. 8.23 and Fig. 8.24, respectively, with recent results of inelastic He-atom scattering experiments [8.77] which covered the whole frequency range up to the highest optical modes.

The main effect of the relaxation is the occurrence of a surface localized $SP_\perp$ mode, which is peeled off the top of the optic bulk band. This mode, in which

RbBr unrelaxed

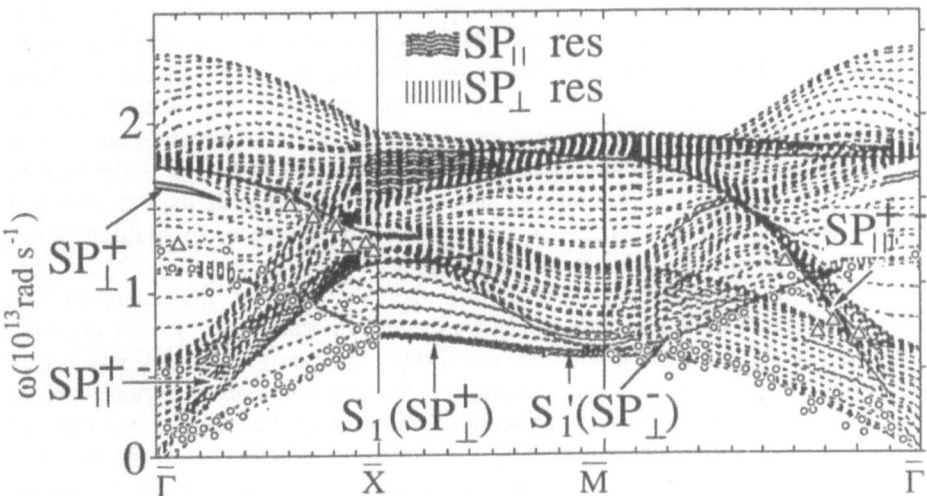

Fig. 8.23. RbBr(001): Surface phonon dispersion curves (after [8.77]). Calculations [8.19]: Slab method (15 *unrelaxed* layers, shell model based on bulk data); *dashed lines*: bulk-like modes; *solid lines*: surface localized modes; S_i: labels surface localized modes; *hatching*: surface resonances (notation as in [8.3]; $SP_{||}$ and $SP_{\perp}$: sagittal-plane vibrations perpendicular and parallel to the surface, respectively; SH: shear horizontagl modes; superscripts + and −: cation and anion displacements in the top layer, respectively. Experimental data [8.77]: Surface modes as determined by He-atom scattering measurements (*filled circles*: strong scattering intensities; *triangles*: weak scattering intensities)

RbBr relaxed

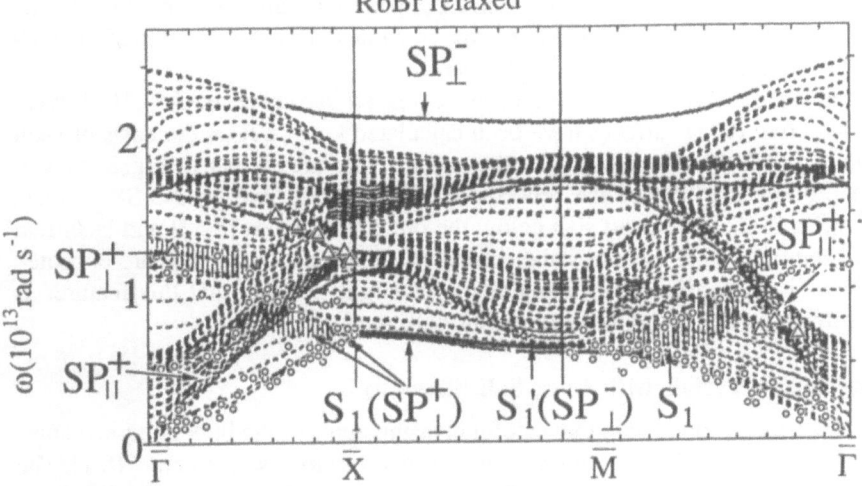

Fig. 8.24. RbBr(001): Surface phonon dispersion curves (after [8.19]). Calculations [8.19]: slab method (15 *relaxed* layers, shell model based on bulk data); *dashed lines*: bulk-like modes; *solid lines*: surface localized modes; S_i: labels surface localized modes; *hatching*: surface resonances (notation as in [8.3]; $SP_{\perp}$ and $SP_{||}$: sagittal-plane vibrations perpendicular and parallel to the surface, repectively; SH: shear horizontal modes; superscripts + and −: cation and anion displacements in the top layer, respectively. Experimental data [8.77]: surface modes as determined by He-atom scattering measurements (*filled circles*: strong scattering intensities; *triangles*: weak scattering intensities)

predominantly the anions vibrate perpendicular to the (001) surface, should be clearly visible in He-atom scattering experiments (cf. KBr). This mode has not, however, been detected in inelastic He-atom scattering measurements [8.77] in spite of a careful search. The calculations show that this mode is quite sensitive to details of the model and in particular to the polarizabilities. Calculated and measured linewidths of the sagittal resonance in the $\overline{\Gamma}$–$\overline{M}$ direction are also in disagreement, at least for the relaxed slab. The relaxation broadens the narrow sagittal resonance of the unrelaxed slab considerably whereas the experimental lines are much sharper. A few more minor differences occur in addition to the more serious discrepancies mentioned here.

The experimental results indicate that the calculation for the unrelaxed slab describes the observed surface modes better than the calculation for the properly relaxed slab. This leads to the conclusion that the underlying model, which was fitted in the early days of lattice dynamics to the measured bulk dispersion curves, does not describe the physics correctly. A new fit of the model parameters to the bulk and surface data may help to clarify the situation. It might turn out that the somewhat uncommon values of the parameters related to the electronic polarizabilities and the way the total polarizability is split into contributions of the individual ions is at the origin of the discrepancies described above. This might also apply to RbI, which shows a similar behavior. Inelastic He-scattering data for RbI, however, are not available at present.

RbI(001)

The bulk dispersion curves of the bulk have been measured by inelastic neutron scattering experiments [8.78]. The experimental data are well reproduced by various shell models [8.14, 73, 74, 78, 79].

The surface dispersion curves of an unrelaxed and an relaxed [8.2] 15-layer slab with free (001) surfaces have been calculated [8.19, 80] on the basis of shell model 2 of [8.78]. The results for the relaxed slab are shown in Fig. 8.25. The most prominent effect of the relaxation is an upward shift of the $SP_\perp^+$ mode, which lies in the unrelaxed slab below the optic bulk bands [8.52] and is shifted above the optic bulk bands by the relaxation. Since RbBr and RbI are expected to show similar surface dispersion curves it might turn out that this feature, as in RbBr, is an artifact of the underlying bulk model.

Other Ionic Crystals with Rock Salt Structure

It should be mentioned here that careful measurements of the bulk phonon dispersion curves are available for many ionic crystals with rock salt structure [8.1]. The measured dispersion curves are well reproduced by various shell models, which may serve as a basis for reliable calculations of the surface phonon dispersion curves. Such calculations could provide a valuable guide for further inelastic He-atom scattering investigations. It should be iterated here that measurements of the surface vibrations give in general more insight into the interparticle interactions and the on-site electronic deformabilities than measurements of the dynamical

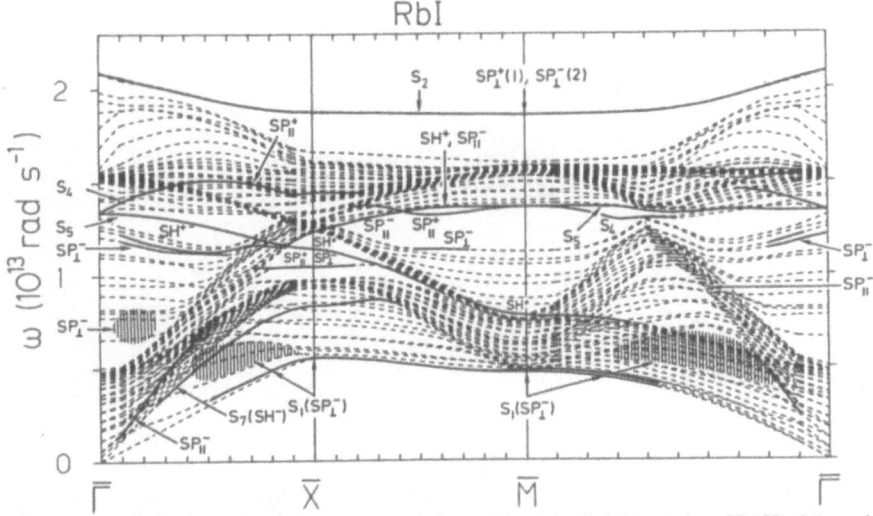

Fig. 8.25. RbI(001): Surface phonon dispersion curves (after [8.19]). Calculations [8.19]: slab method (15 relaxed layers, shell model based on bulk data); *dashed lines*: bulk-like modes; *solid lines*: surface localized modes; S_i: labels surface localized modes; *hatching*: surface resonances (notation as in [8.3]); $SP_\perp$ and $SP_\parallel$: sagittal-plane vibrations perpendicular and parallel to the surface, respectively; SH: shear horizontal modes; superscripts + and −: cation and anion displacements in the top layer, respectively

properties of the bulk, since the lower symmetry at the surface unmasks properties which are hidden by cancellation effects due to the high symmetry in the bulk.

8.1.2 Metal Oxides

Many of the numerous metal oxides with rock salt structure have been investigated by inelastic neutron scattering [8.1]. In general, the measured bulk phonon dispersion curves are well reproduced by various shell models. The overlap shell model [8.1], in particular, is very well suited to describe the lattice dynamics of these systems.

It should be mentioned here that the O^{--} ion is not stable in a free state, whereas the O^{--} electronic configuration is stabilized in the bulk by the Coulomb field of the surrounding ions. The surface might thus have a stronger influence on the properties of the oxygen ions (ionic charge, polarizability and even the short range potential) than on halogen ions. The experimental study of the surface vibrations and the surface relaxation would, therefore, be of interest as a first step in the direction towards metals where the situation is still more complicated. Up to now only the surface modes of MgO have been studied in detail.

MgO(001)

The phonon dispersion curves in the bulk of MgO have been measured by inelastic neutron scattering [8.81]. The measured data are well reproduced by various shell-model calculations [8.31, 81–83].

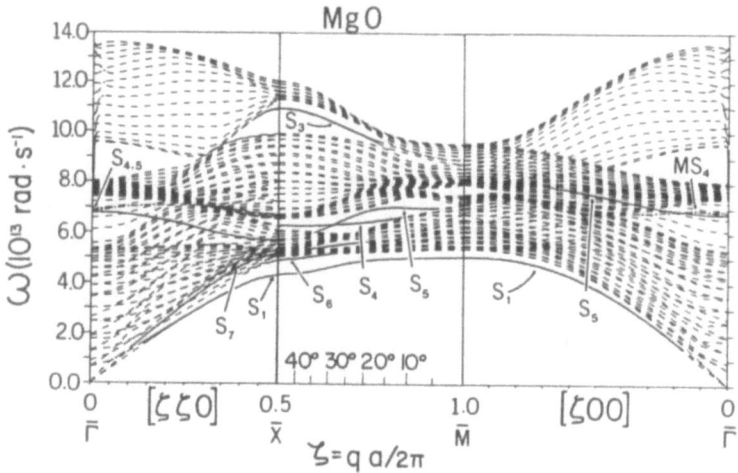

Fig. 8.26. MgO(001): Surface phonon dispersion curves (after [8.18]). Calculations [8.18]: slab method (15 unrelaxed layers, shell model based on bulk data); *dashed lines*: bulk-like modes; *solid lines*: surface localized modes; S_i and MS_i: label surface localized modes and resonances respectively (notation as in [8.3])

Both the shell model [8.18] and the breathing shell model [8.84] have served, respectively, as a basis for calculations of the surface phonon dispersion curves of a (001) surface by the slab method. Figure 8.26 shows the surface vibrations of the unrelaxed 15-layer slab with free (001) surfaces [8.18]. The relaxation pattern has been calculated in [8.2] on the assumption that the interaction potentials and the polarization parameters are the same throughout the whole slab [8.2]. The resulting frequency shifts [8.19] are small ($\leq 2\%$) everywhere.

A different situation arises, however, when changes of the bulk parameters at the surface are considered in the relaxation process [8.85]. Figure 8.27 shows the surface phonon dispersion curves of a relaxed 15-layer slab with free (001) surfaces [8.86]. A small charge transfer at the surface induces large shifts in the surface modes (lower part of Fig. 8.27). The frequencies of the Lucas modes S_4 and S_5 are lowered by 25% and the frequencies of the Rayleigh waves in the $\overline{\Gamma}$–$\overline{M}$ direction are lowered by 27%.

Up to now only a few He-scattering data have been published [8.87]. Figure 8.28 shows the experimental results for surface waves propagating in the (100) direction. Dashed and dotted lines correspond to the sound velocities of longitudinal and transverse bulk waves. The solid line corresponds to the sound velocity of Rayleigh waves.

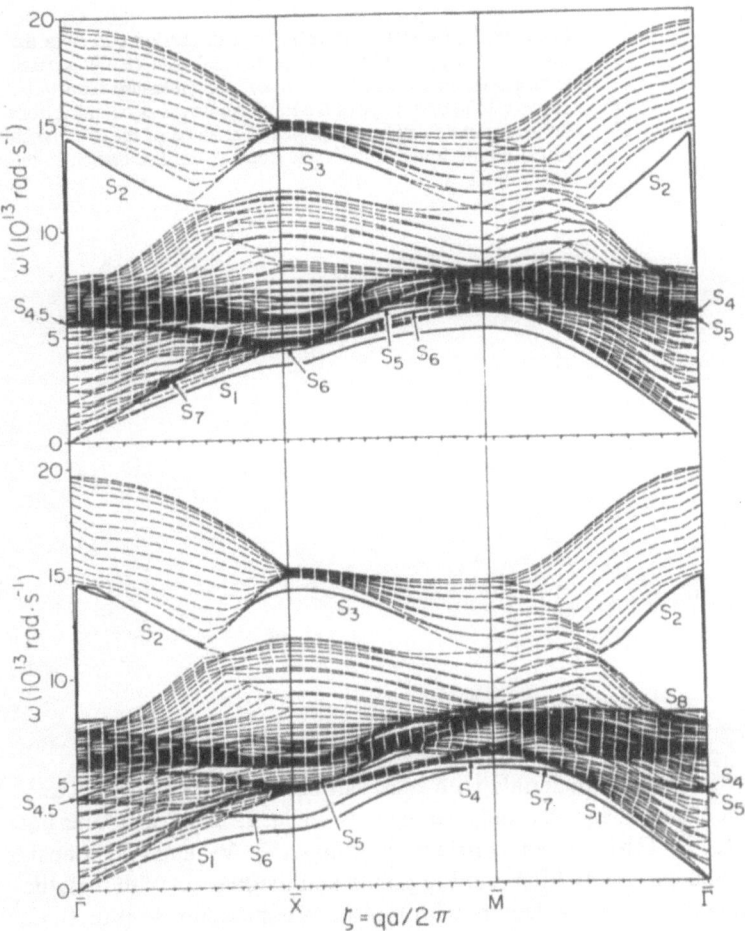

Fig. 8.27. MgO(001): Surface phonon dispersion curves (after [8.86]). Calculations [8.86]: slab method (15 relaxed layers); *upper panel:* calculations based on rigid ion model 1 of [8.85] (static relaxation only) *lower panel:* calculations based on rigid ion model 4 of [8.85] (static relaxation and charge transfer at the surface); *dashed lines:* bulk-like modes; *solid lines:* surface localized modes; S_i: labels surface localized modes (notation as in [8.3])

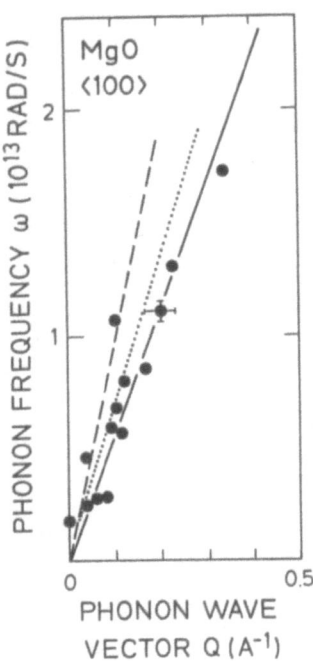

Fig. 8.28. MgO(001): Surface phonon dispersion curves in the elastic range (after [8.87]). Experimental data [8.87]: surface modes as determined by He-atom scattering measurements (*filled circles*); *solid line*: slope of the Rayleigh waves as obtained from the bulk elastic constants; *dashed line*: slope of the TA branch in the elastic limit; *dotted line*: slope of the LA branch in the elastic limit

8.1.3 Refractory Compounds

Another class of interesting materials with rock salt structure are the refractory compounds, which have been carefully studied by inelastic neutron scattering experiments [8.1, 88]. Carbides and nitrides of group IV_b V_b and VI_b transition metals combine extremely high melting points and hardnesses with metallic conductivity. Some are superconductors with transition temperatures up to 18 K. The superconducting refractory compounds show pronounced anomalies in the phonon dispersion curves. These anomalies are already present in the metallic phase and do not change when the transition into the superconducting phase takes place. Phonon anomalies and superconductivity correlate strongly with the number of valence electrons per unit cell. Compounds with 8 valence electrons per unit cell show neither anomalies nor superconductivity, while compounds with 9 valence electrons show anomalies and become superconducting at low temperatures.

The origin of the phonon anomalies is a strong electron–phonon coupling, which leads, together with the phonon softening, to transition temperatures which are much higher than those of simple metals. The dependence of the electron–phonon coupling on the valence electron density can be explained in a rigid band picture by the shift of the Fermi energy from a low density of metal d and nonmetal p states into a high density of p and d states when the number of valence electrons is increased from 8 to 9. This picture is essentially confirmed by first principles calculations [8.89–92].

The measured phonon dispersion curves of the bulk are well reproduced by shell models, double shell models [8.93], and cluster deformation models [8.94], which all include the Lindhard screening of a dilute free electron gas. One of the most interesting questions, still under discussion, is whether or not the phonon anomalies in the longitudinal acoustic bulk branch show up also in the surface phonon dispersion curves. Refractory compounds are very well suited for electron energy loss spectroscopy (EELS) since they are metallic and have much steeper acoustic branches than elementary metals and ionic crystals, with zone-boundary frequencies of acoustic modes in the range 5–10 THz and optic bands the range 12–20 THz.

TiC(001)

TiC belongs to the class of refractory compounds with 8 valence electrons per unit cell. It does not show phonon anomalies and does not become superconducting. The phonon dispersion curves in the bulk have been measured by inelastic neutron scattering [8.95]. The measured data are well reproduced by calculations employing a 12 parameter shell model with free electron screening [8.95].

The surface modes of the (001) surface have been investigated by EELS [8.96]. Figures 8.29 and 8.30 show the measured surface modes in the $\overline{\Gamma}$–$\overline{M}$ and

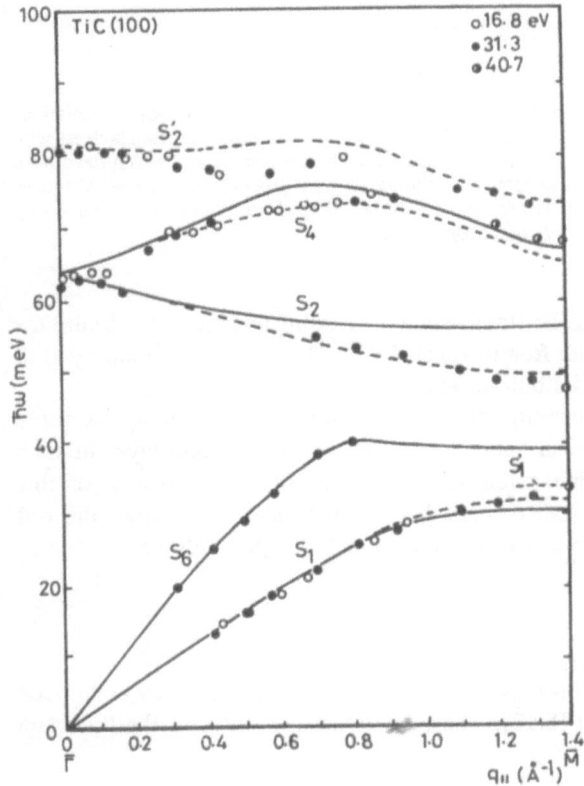

Fig. 8.29. TiC(001): Surface phonon dispersion curves in the $\overline{\Gamma}$–$\overline{M}$ direction (after [8.96]). Calculations [8.96]: slab method (16 unrelaxed layers, screened shell model); *solid lines*: results obtained with unmodified bulk parameters; *dashed lines*: results obtained with bulk parameters which have been modified at the surface to account for surface relaxation effects. S_i: labels surface localized modes and resonances (notation as in [8.3]). Experimental data [8.96]: surface modes as determined EELS measurements (*open circles, half-filled circles, filled circles*)

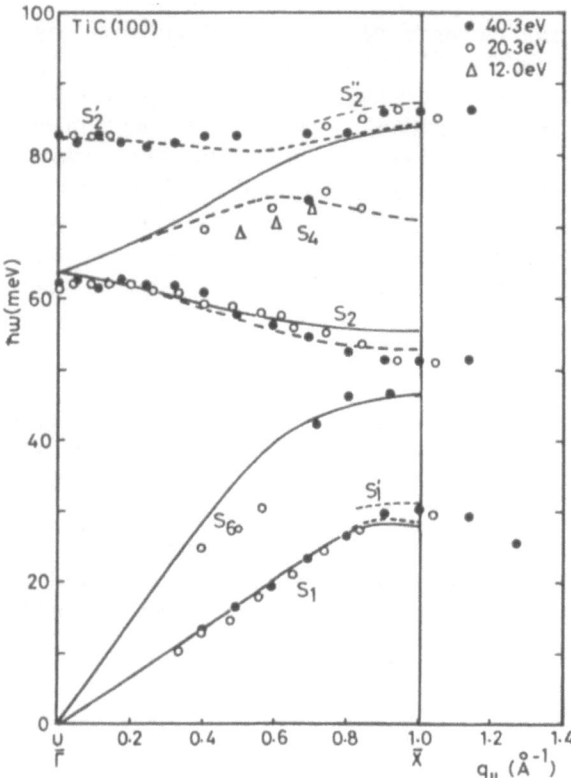

Fig. 8.30. TiC(001): Surface phonon dispersion curves in the $\overline{\Gamma}$-$\overline{X}$ direction (after [8.96]). Calculations [8.96]: slab method (16 unrelaxed layers, screened shell model); *solid lines*: results obtained with unmodified bulk parameters; *dashed lines*: results obtained with bulk parameters which have been modified at the surface to account for surface relaxation effects. S_i: labels surface localized modes and resonances (notation as in [8.3]). Experimental data [8.96]: surface modes as determined by EELS measurements (*open circles*, *filled circles*, *open triangles*)

$\overline{\Gamma}$-$\overline{X}$ directions of the surface Brillouin zone. The solid lines are the results for an unrelaxed 16-layer slab with free (001) surfaces. The parameters employed in this calculation are those of the bulk model.

The effect of a surface relaxation have been simulated by changing the force constants in the first layer and in between the first and the second layer in such a way as to reproduce the measured surface modes best. The results of this calculation are indicated by dashed lines. Figures 8.31 and 8.32 shows the full surface phonon dispersion curves for unchanged and changed bulk parameters at the surface, respectively.

TiC(110)

The surface vibrations of an unrelaxed slab with free (110) surfaces have been calculated with a shell model [8.97]. The parameters are those of the bulk and

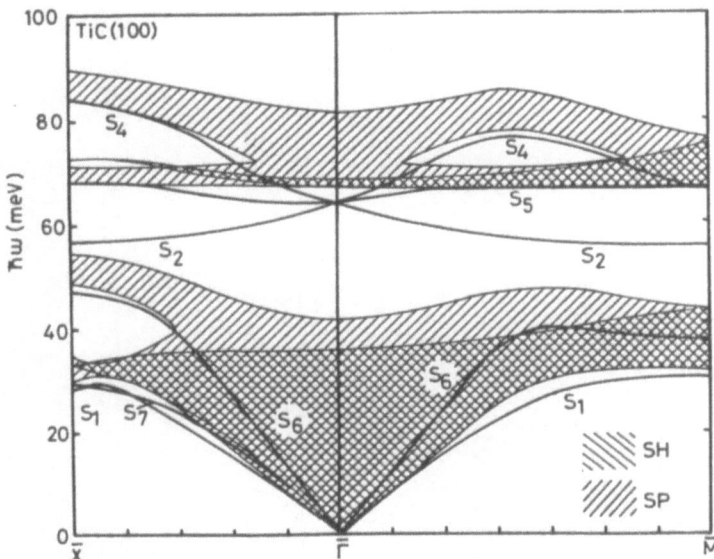

Fig. 8.31. TiC(001): Surface phonon dispersion curves (after [8.96]). Calculations [8.96]: slab method (16 unrelaxed layers, screened shell model based on unmodified bulk parameters); *solid lines*: surface localized modes; S_i: labels surface localized modes and resonances (notation as in [8.3]); *hatched*: surface projected bulk modes with shear horizontal (SH) and sagittal plane (SP) polarization

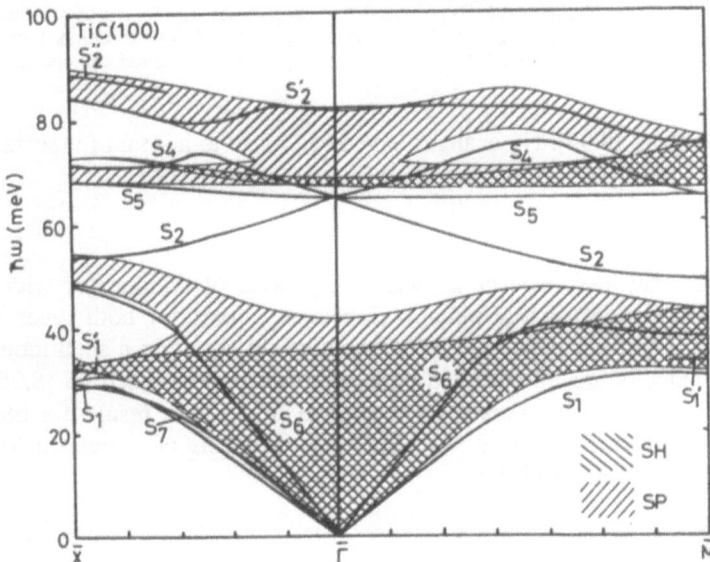

Fig. 8.32. TiC(001): Surface phonon dispersion curves (after [8.96]). Calculations [8.96]: slab method (16 unrelaxed layers, screened shell model based on bulk parameters which have been modified at the surface to account for relaxation effects); *solid lines*: surface localized modes; S_i: labels surface localized modes and resonances (notation as in [8.3]); *hatched*: surface projected bulk modes with shear horizontal (SH) and sagittal plane (SP) polarization

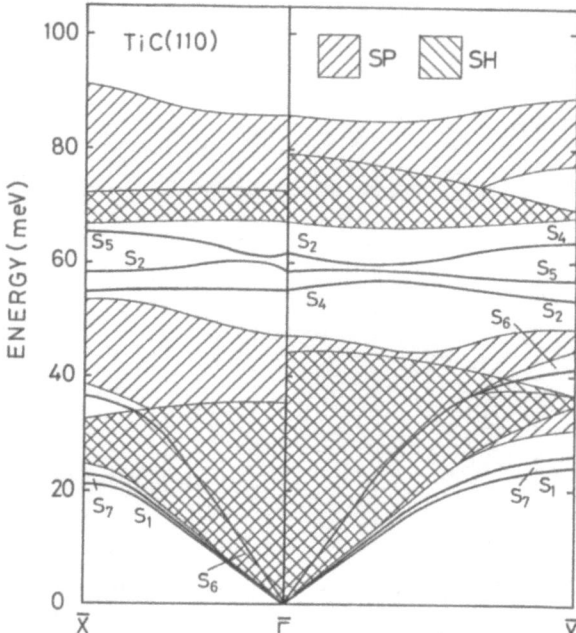

Fig. 8.33. TiC(110): Surface phonon dispersion curves (after [8.97]). Calculations [8.97]: slab method (25 unrelaxed layers, screened shell model based on unmodified bulk parameters); *bold lines*: surface localized modes. S_1, S_2, S_4, and S_6, have sagittal plane (SP) polarization along $\overline{\Gamma}$-$\overline{Y}$, while S_5 and S_7 have in this direction shear horizontal (SH) polarization. Along $\overline{\Gamma}$ -$\overline{X}$, S_1, S_2, S_5, and S_6, have sagittal plane (SP) polarization, while S_4 and S_7 have shear horizontal (SH) polarization. Surface projected bulk modes with shear horizontal (SH) and sagittal plane (SP) polarization are represented by hatched areas

have been taken from [8.95]. No changes have been made to account for surface effects. The results are shown in Fig. 8.33.

TiC(111)

A TiC(111) slab may terminate either with a top layer of titanium or with a top layer of carbon. The surface phonon dispersion curves for both types of (111) surface have been calculated. The calculations are based on a shell model [8.97]. The parameters are those of the bulk and have been taken from [8.95]. No changes have been made to account for surface effects. The results for both (111) surfaces are shown in Fig. 8.34. EELS measurements of a stepped TiC (310) surface are reported in [8.98].

TiN(001)

TiN has 9 valence electrons per unit cell. As a consequence, it shows pronounced anomalies in the phonon dispersion curves of the bulk and becomes superconducting at about 5.5 K. The phonon dispersion curves in the bulk have been determined by inelastic neutron scattering measurements [8.99]. The experimen-

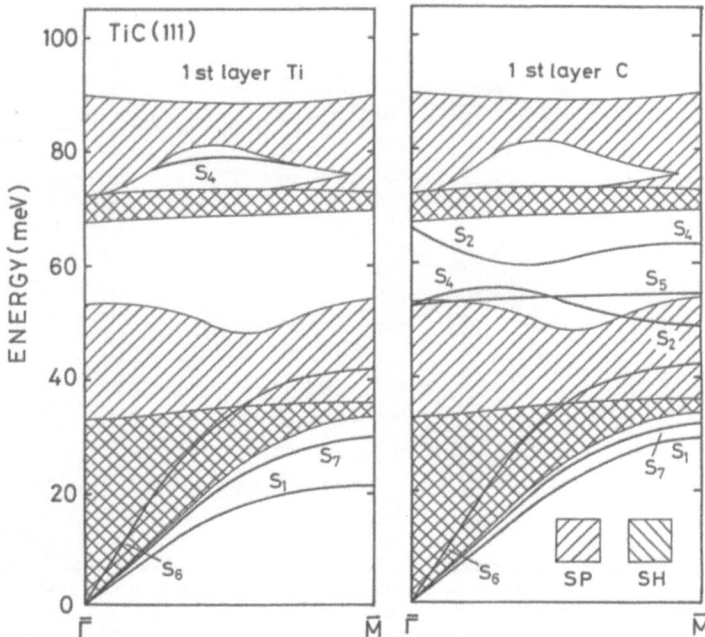

Fig. 8.34. TiC(111): surface phonon dispersion curves (after [8.97]. Calculations [8.97]: slab method (28 unrelaxed layers, screened shell model based on unmodified bulk parameters); *bold lines*: surface localized modes, *hatched*: surface projected bulk modes with shear horizontal(SH) and sagittal plane (SP) polarization. Note that under normal experimental conditions the slab terminates with a Ti-layer

tal data are equally well reproduced by the double shell model [8.99] and the cluster deformation model [8.94]. Both models take into account the free electron screening.

The cluster deformation model has been used to calculate the surface dispersion curves of an unrelaxed (001) surface in the framework of the Green's function method [8.100, 101]. The results are shown in Fig. 8.35. The Rayleigh wave S_1, the quasi-longitudinal resonance S_6, and the optical modes S_2 and S_4 are sagittal modes, while the mode S_5 has shear horizontal polarization.

The Rayleigh mode shows a very weak anomaly. The sagittal resonance S_6, however, shows a strong anomaly. It should be mentioned here that the double shell model does not lead to any anomaly in the Rayleigh modes of TaC, which also has 9 valence electrons and is in many respects very similar to TiN.

In order to fit the surface excess phonon density of TiN measured by incoherent neutron scattering from powder samples with different grain sizes [8.102] and to account for surface relaxation, the dipolar and quadrupolar polarizabilities at the surface have to be increased slightly. Figure 8.36 shows a comparison between the neutron scattering data and the surface excess density calculated with unchanged (full line) and with increased dipolar and quadrupolar deformabilities (dashed line) at the surface. Figure 8.37 shows the surface dispersion curves for the unchanged (full line) and changed (dashed line) polarizabilities.

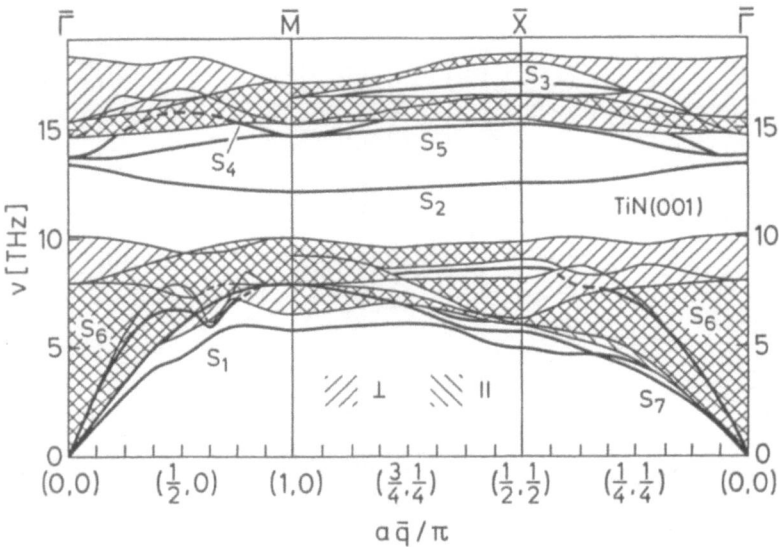

Fig. 8.35. TiN(001): Surface phonon dispersion curves (after [8.101]). Calculations [8.101]: Green's function method (unrelaxed surface, cluster deformation model based on bulk parameters); *bold lines*: surface localized modes and resonances: S_i: labels surface localized modes and resonances (notation as in [8.4]); *hatched*: surface projected bulk modes with shear horizontal (||) and sagittal (⊥) polarization

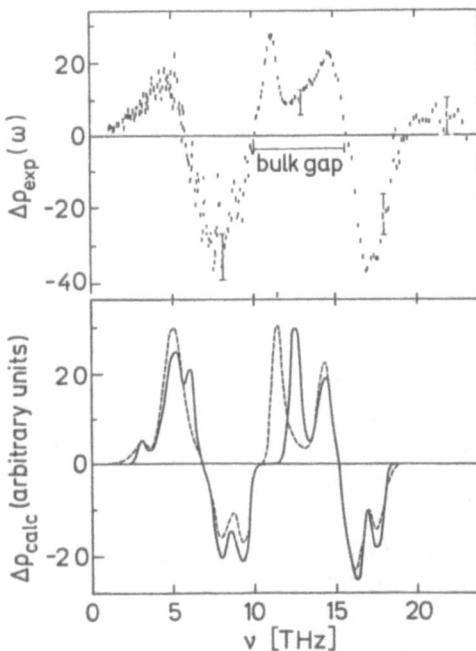

Fig. 8.36. TiN(001): Surface-excess phonon density of states (after [8.101]). *Top panel*: experimental data as obtained from incoherent neutron scattering measurements on powder samples [8.102]. Calculations [[8.101]: Green's function method (unrelaxed surface, cluster deformation model); *full lines*: results obtained with unchanged bulk parameters; *dashed lines*: results obtained with changed dipolar and quadrupolar deformabilities of the nitrogen ions at the surface

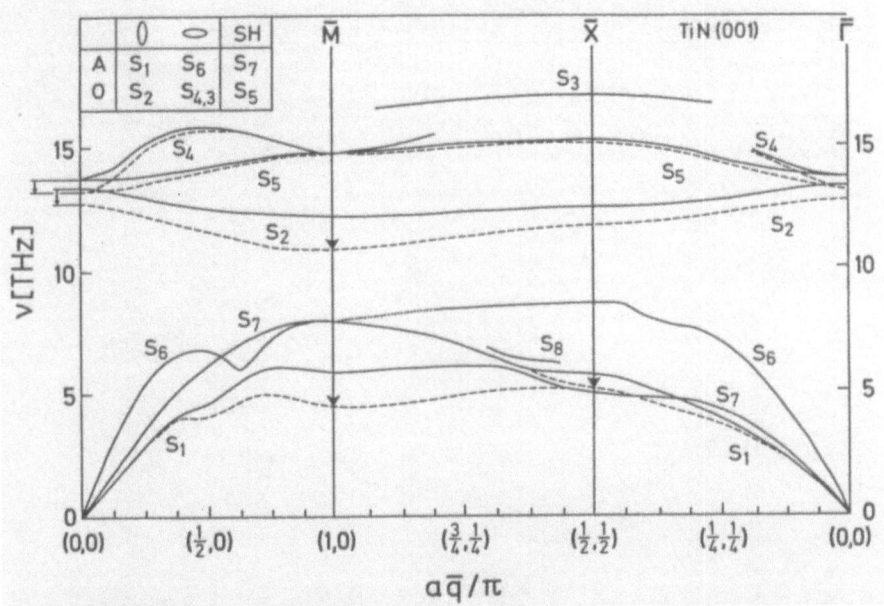

Fig. 8.37. TiN(001): Surface phonon dispersion curves (after [8.101]). Calculations [8.101]: Green's function method (unrelaxed surface, cluster deformation model); *full lines*: results obtained with unchanged bulk parameters; *dashed lines*: results obtained with changed dipolar and quadrupolar deformabilities of the nitrogen ions at the surface

HfC(001)

HfC has 8 valence electrons per unit cell. It shows no phonon anomalies in the bulk and does not become superconducting at low temperatures. The dispersion curves in the bulk have been determined by inelastic neutron scattering experiments [8.63, 104]. The results of these experiments are well reproduced by model calculations employing a shell model with free electron screening [8.93].

Using this model the surface phonon dispersion curves for an unrelaxed 16-layer slab with free (001) surfaces have been calculated [8.103]. The values of the parameters used in this calculations were those determined from the phonon dispersion curves in the bulk. The results are shown in Fig. 8.39.

In order to reproduce the measured surface phonon dispersion curves [8.103], shown in Fig. 8.38, the force constants between the first layer Hf and the second layer C had to be increased by 20%, the force constants between the first layer C and the second layer Hf had to be decreased by 40% and the force constants between Hf and C in the first layer had to be decreased by 40%. These force constant changes are consistent with the assumption of a surface rumpling in which the carbon atoms in the first layer relax outwards and the hafnium atoms in the first layer relax inwards. The results of these calculations are indicated by solid and dashed lines in Fig. 8.39. A proper calculation of the surface relaxation has not yet been carried out since the interaction potentials and their changes at and close to the surface are not known at present.

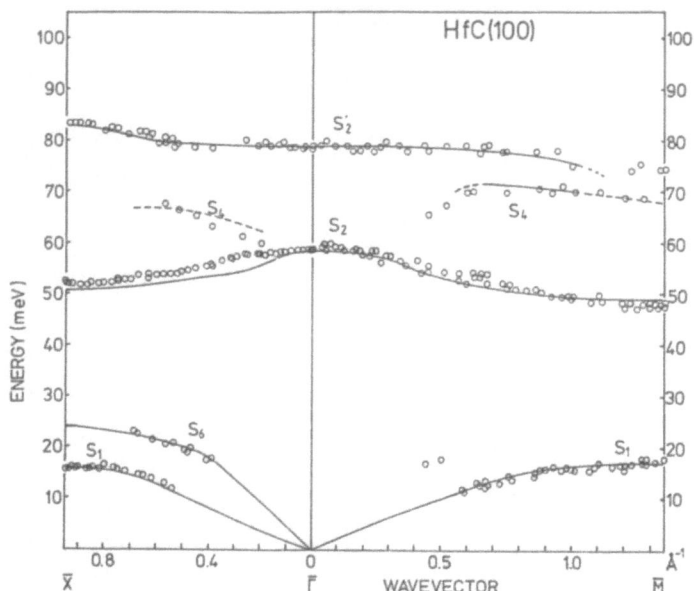

Fig. 8.38. HfC(001): Surface phonon dispersion curves (after [8.103]). Calculations [8.103]: slab method (16 unrelaxed layers, screened shell model); *lines*: surface localized modes as obtained with bulk parameters, which have been modified at the surface to account for surface relaxation effects; S_i: labels surface localized modes and resonances (notation as in [8.3]). Experimental data [8.103]: surface modes as determined by EELS measurements (*open circles*)

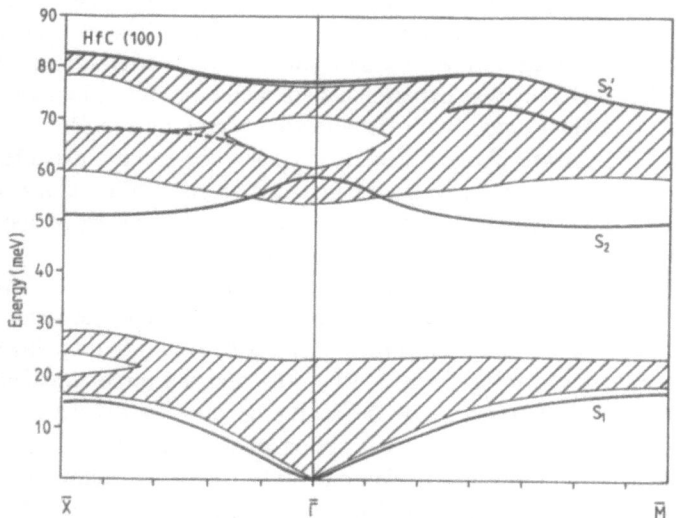

Fig. 8.39. HfC(001): Surface phonon dispersion curves (after [8.103]). Calculations [8.103]: slab method (16 unrelaxed layers, screened shell model); *bold lines*: surface localized modes as obtained with unmodified bulk parameters; S_i; labels surface localized modes and resonances (notation as in [8.3]); *hatched*: surfce projected bulk modes

NbC(001)

NbC has 9 valence electrons per unit cell, exhibits pronounced phonon anomalies in the phonon dispersion curves of the bulk material, and becomes superconducting at about 11.1 K. The phonon dispersion curves in the bulk have been determined by inelastic neutron scattering measurements [8.63, 104, 105] The experimental data are well reproduced by the double shell model [8.93].

The surface phonons of the (001) surface in the $\overline{\Gamma}$–$\overline{M}$ direction have been investigated by EELS [8.106] measurements. The results are shown in Fig. 8.40. The surface phonon dispersion curves of an unrelaxed 24 layer slab with free (001) surfaces have been calculated [8.107] in the framework of a double shell model. The results are shown in Fig. 8.41. Also shown are the EELS data of [8.106, 108].

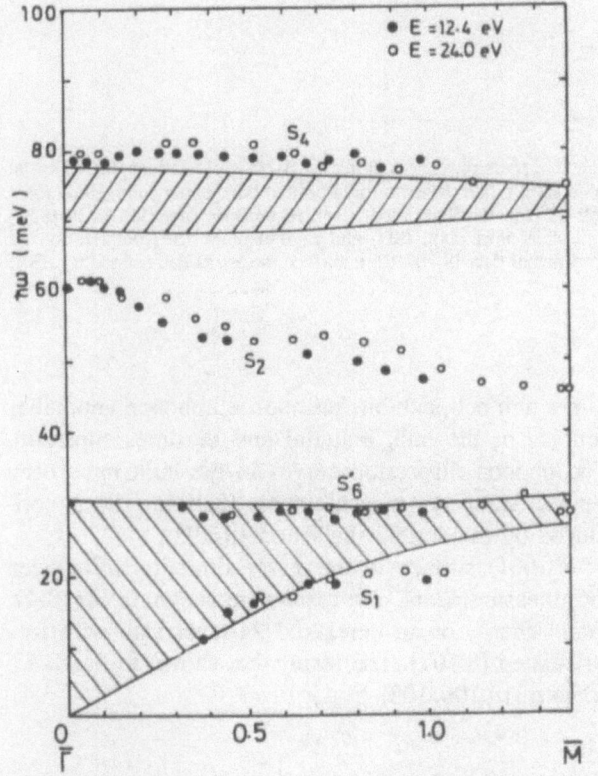

Fig. 8.40. NbC(001): Surface phonon dispersion curves (after [8.106]). ($\overline{\Gamma}$-$\overline{M}$ direction). Experimental data [8.106, 108]: surface modes as determind by EELS measurements (*open circles, filled circles*); S_i: labels the measured surface modes (notation as in [8.3]). The hatched areas represent surface projected bulk bands

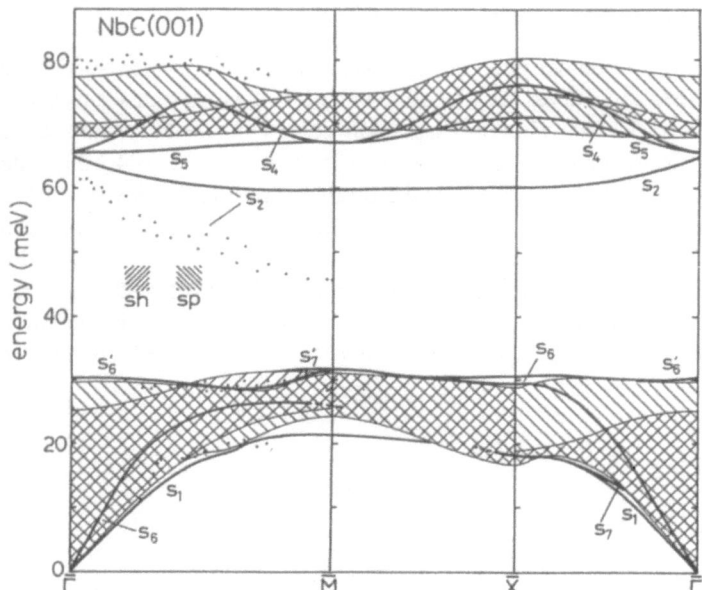

Fig. 8.41. NbC(001): Surface phonon dispersion curves (after [8.107]). Calculations [8.107]: slab method (24 unrelaxed layers, double shell model based on unmodified bulk parameters); *bold lines*: surface localized modes; S_i: labels surface localized modes and resonances (notation as in [8.3]). Surface projected bulk modes with shear horizontal (SH) and sagittal plane (SP) polarization are represented by hatched areas. Experimental data [8.106,108]: surface modes as determined by EELS measurements (*points*)

TaC(001)

TaC has 9 valence electrons per unit cell, exhibits pronounced phonon anomalies in the phonon dispersion curves of the bulk material and becomes superconducting at about 10.4 K. The phonon dispersion curves in the bulk have been determined by inelastic neutron scattering measurements [8.104]. The experimental data are well reproduced by the double shell model [8.93].

The surface modes of the (001) surface in the $\overline{\Gamma}$–$\overline{M}$ direction have been investigated by EELS [8.106] measurements. The results are shown in Fig. 8.42. The surface phonon dispersion curves of an unrelaxed 24-layer slab with free (001) surfaces have been calculated [8.107]. The results are shown in Fig. 8.43. Also shown are the EELS data of [8.106, 108].

Fig. 8.43. TaC(001): Surface phonon dispersion curves (after [8.107]). Calculations [8.107]: slab method (24 unrelaxed layers, double shell model based on unmodified bulk parameters); *bold lines*: surface localized modes; S_i: labels surface localized modes and resonances (notation as in [8.3]). Surface projected bulk modes with shear horizontal (SH) and sagittal plane (SP) polarization are represented by hatched areas. Experimental data [8.106,108]: surface modes as determined EELS measurements (*points*)

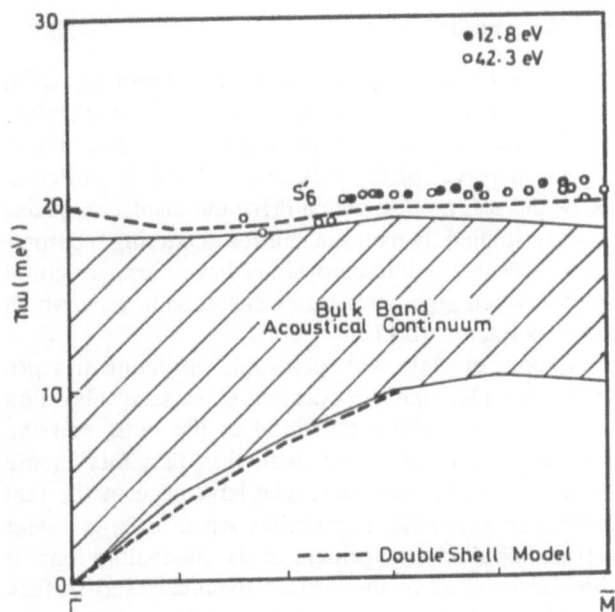

Fig. 8.42. TaC(001): Surface phonon dispersion curves in the $\overline{\Gamma}$-$\overline{M}$ direction (after [8.106]). Experimental data [8.106]: surface modes as determined by EELS measurements (*open circles, filled circles*); S_i; labels the measured surface modes (notation as in [8.3]). The hatched areas represent surface projected bulk bands

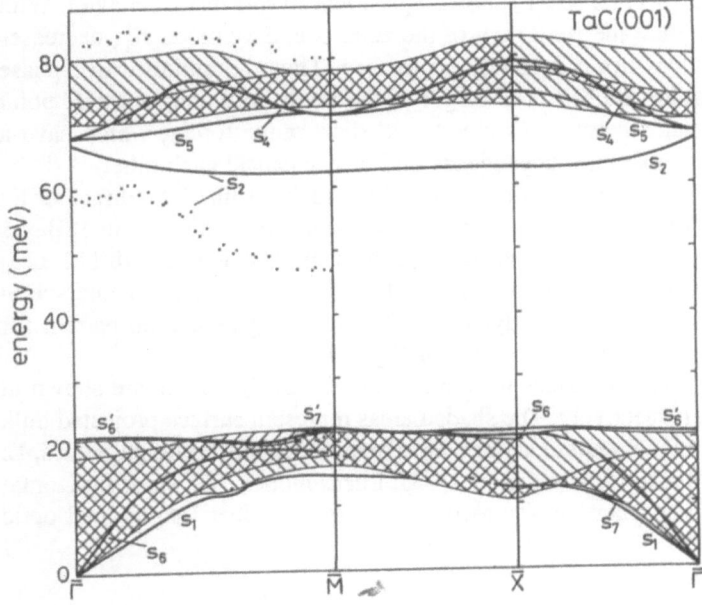

8.1.4 Perovskite Structure Compounds

The bulk dynamics of perovskites has attracted considerable interest in recent years. Some of the perovskites become ferroelectric or are at least incipient ferroelectric, and most crystals with perovskite structure undergo various phase transitions. The phonon dispersion curves of the bulk material and in particular the temperature dependence of the soft modes, which drive the displacive phase transitions, have been intensively studied by inelastic neutron scattering measurements [8.1]. It is interesting to note that oxidic perovskites have close structural relationships to the recently discovered high-temperature superconductors, which are currently attracting enormous interest [8.110].

The surface dynamics of ionic crystals with perovskite structure has not received much attention until now, although the question of surface relaxation and surface reconstruction close to the phase transition in the bulk, and the relation between soft surface modes and surface reconstruction, may merit some attention. Theoretical investigations may, moreover, take advantage of the fact that the short range interactions in nonoxidic perovskites are to a large extent independent of the particular geometrical arrangement of the interacting ions. It is thus believed that reliable calculations of the surface dynamics and surface relaxations can be carried out in the framework of shell models based on long range Coulomb and short range Born–Mayer potentials.

$KZnF_3(001)$

The bulk dynamics of $KZnF_3$ has been investigated by inelastic neutron scattering [8.111]. The phonon dispersion curves in the bulk are well reproduced by shell-model calculations [8.111]. $KZnF_3$ shows a pronounced soft-mode behavior. With decreasing temperature the frequency of the zone-boundary mode R'_{15} decreases considerably but does not completely go to zero. Thus the antidistortive phase transition is not quite reached. The origin of the mode softening at the R point is the delicate balance between Coulomb and short range forces, which have a slightly different temperature dependence and nearly cancel each other.

A $KZnF_3(001)$ slab may terminate either by a rocksalt-like KF surface or by a ZnF_2 surface. The relaxation pattern and the surface dynamics of an 18-layer slab with free (001) surfaces of either type have been calculated [8.112, 113] in the framework of a shell model. The calculations are based on interaction potentials derived from the bulk dynamics. The resulting relaxation pattern for both surfaces is schematically exhibited in Fig. 8.44.

The calculated surface modes for the KF and the ZnF_2 surface are shown in Figs. 8.45 and 46, respectively. The shaded areas represent surface projected bulk modes while the full curves are surface localized modes (R: Rayleigh modes, L: Love modes, FK: Fuchs–Kliewer modes, LU: Lucas modes, SH: shear horizontal modes, SP_{LO}: sagittal plane modes associated with the upper longitudinal optic (LO) bulk band).

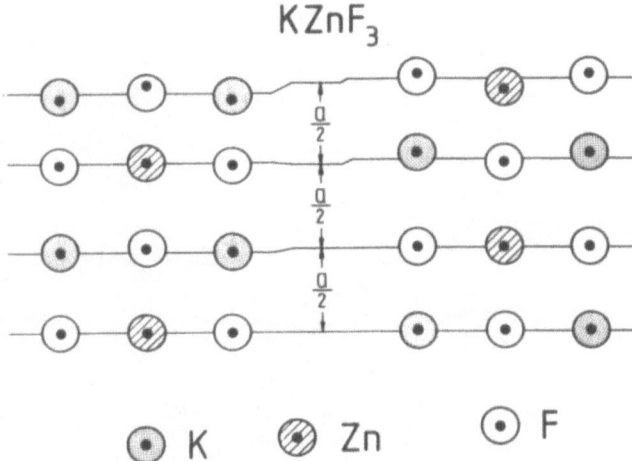

Fig. 8.44. KZnF$_3$(001): Surface relaxation (after [8.113]). Calculations [8.113]: slab method (18 layers, shell model based on interaction potentials). *Left part*: slab terminating with a KF layer; *right part*: slab terminating with a ZnF$_2$ layer; *dots*: cores; *circles*: shells; *lines*: (001)-planes. The displacements in the z-direction are enlarged by a factor of 3.

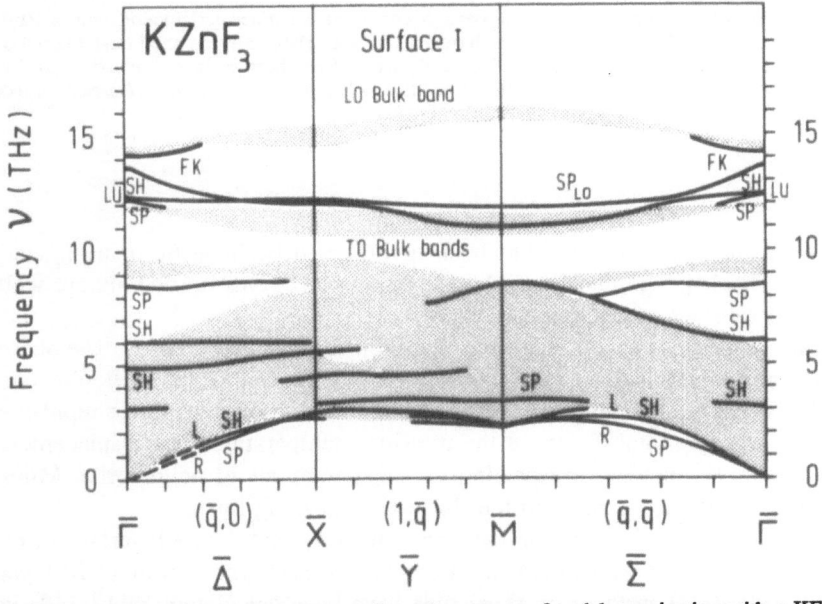

Fig. 8.45. KZnF$_3$(001): Surface phonon dispersion curves of a slab terminating with a KF surface (after [8.113]). Calculations [8.113]: slab method (relaxed slab, 25-layers, shell model based on interaction potentials); *solid lines*: surface localized modes (SH: shear horizontal polarization; SP: sagittal plane polarization; FK: Fuchs–Kliewer modes; LU: Lucas modes); *shaded areas*: surface projected bulk bands

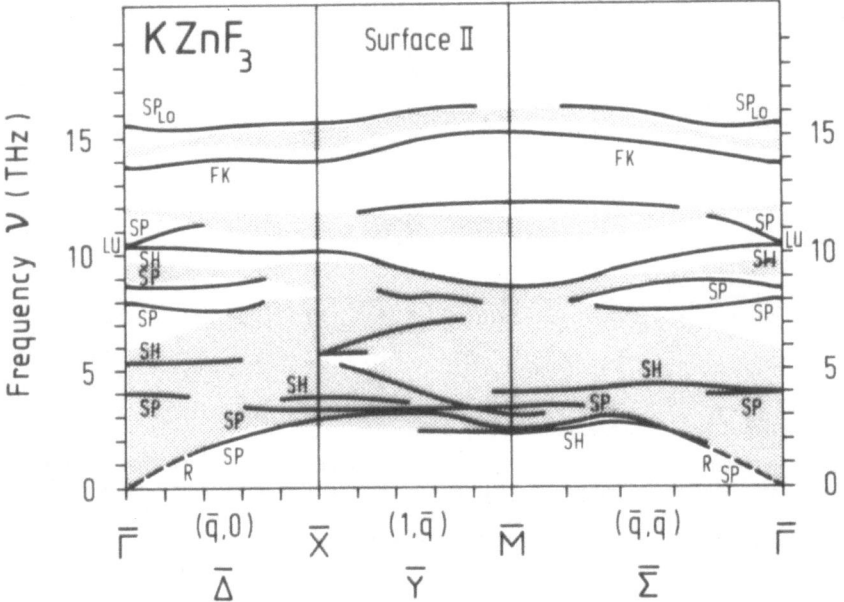

Fig. 8.46. KZnF$_3$(001): Surface phonon dispersion curves of a surface terminating with a ZnF$_2$ layer (after [8.113]). Calculations [8.113]: slab method (relaxed slab, 25-layer, shell model based on interaction potentials); *solid lines*: surface localized modes (SH: shear horizontal polarization; SP: sagittal plane polarization; FK: Fuchs–Kliewer modes; LU: Lucas modes); *shaded areas*: surface projected bulk bands.

KMnF$_3$(001)

The bulk dynamics of KMnF$_3$ has been investigated by inelastic neutron scattering [8.111, 114]. The measured phonon dispersion curves in the bulk are well reproduced by shell-model calculations [8.111].

KMnF$_3$ undergoes an antidistortive phase transition at about 186 K. The phase transition of the bulk is driven by a soft mode at the R point of the Brillouin zone (R'_{15}). This soft mode is due to a complete cancellation of short range repulsive and long range Coulomb forces at the transition temperature. The displacement pattern of the R'_{15} mode corresponds to counter-rotations of neighboring MnF$_6$ octahedra around axes parallel to the F-Mn-F connecting lines.

A KMnF$_3$(001) slab may terminate either by a rocksalt-like KF surface or by an MnF$_2$ surface. The relaxation pattern and the surface dynamics of an 18-layer slab with free (001) surfaces of either type have been calculated [8.112, 113] in the framework of a shell model. The interaction potentials are derived from the bulk dynamics above the antidistortive phase transition.

The temperature dependence of the soft bulk mode R'_{15} [8.111, 114] has been used to determine the temperature dependence of the quasiharmonic radial force constant A_1 for the K–F coupling [8.115]. It is sufficient to take into account only the temperature dependence of this single force constant A_1 in order to reproduce

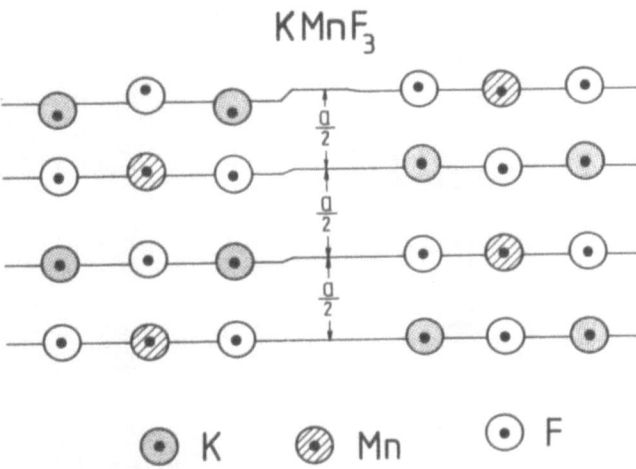

Fig. 8.47. KMnF$_3$(001): Surface relaxation (after 8.113]). Calculations [8.113]: slab method (18 layers, shell model based on interaction potentials). *Left part*: slab terminating with a KF layer; *right part*: slab terminating with a MnF$_2$ layer; *dots*: cores; *circles*: shells; *lines*: (001)-planes. The displacements in the z-direction are enlarged by a factor of 3

the phonon dispersion curves in the bulk in the temperature range from room temperature down to the transition temperature. All other model parameters are assumed to be temperature independent. Applying this model to a slab, it has been found [8.115] that the (001) surface which terminates with a KF layer undergoes a $\sqrt{2} \times \sqrt{2}$ surface reconstruction at a temperature slightly above the bulk transition temperature. The relaxation pattern for both the KF and the MnF$_2$ surface is shown schematically in Fig. 8.47.

The calculated surface modes [8.113] for both the KF and the MnF$_3$ surface are shown in Figs. 8.48 and 8.49, respectively. The shaded areas represent surface projected bulk modes while the solid lines indicate surface localized modes (R: Rayleigh modes, L: Love modes, FK Fuchs–Kliewer modes, LU: Lucas modes, SH: shear horizontal modes, SP$_{LO}$: sagittal plane modes associated with the upper longitudinal optic (LO) bulk band). The surface reconstruction of the KF surface is driven by a soft surface mode at the $\bar{M}$ point of the surface Brillouin zone. Surface relaxation tends to stabilize this mode. Thus the phase transition for a properly relaxed slab is reached at a lower temperature than that obtained for an unrelaxed slab. Nonetheless the surface reconstruction occurs prior to the bulk transition.

SrTiO$_3$

SrTiO$_3$ is an incipient ferroelectric which undergoes an antidistortive structural phase transition driven by a soft R'_{15} mode. The phonon dispersion curves in the bulk material have been determined by inelastic neutron scattering measurements [8.116–123]. The experimental results are well reproduced by calculations using various shell models [8.120, 124–126].

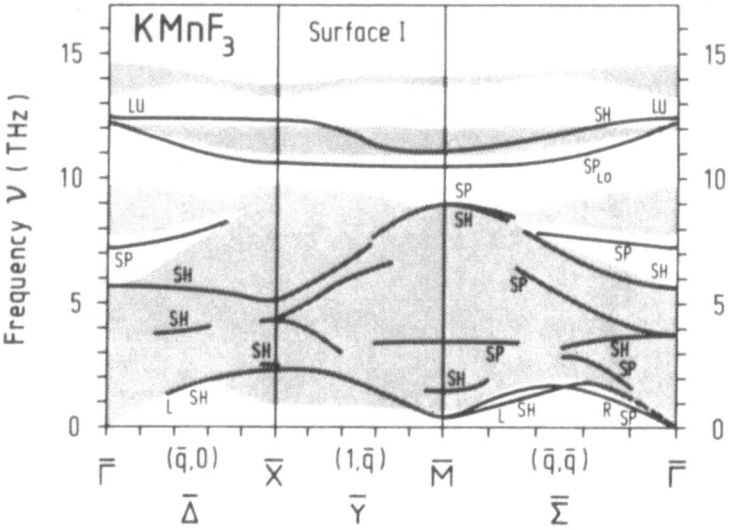

Fig. 8.48. KMnF$_3$(001): Surface phonon dispersion curves of a slab terminating with a KF surface (after [8.113]). Calculations [8.113]: slab method (relaxed slab, 25-layers, shell model based on interaction potentials); *solid lines*: surface localized modes (SH: shear horizontal polarization; SP: sagittal plane polarization; FK: Fuchs–Kliewer modes; LU: Lucas modes); *shaded areas*: surface projected bulk bands

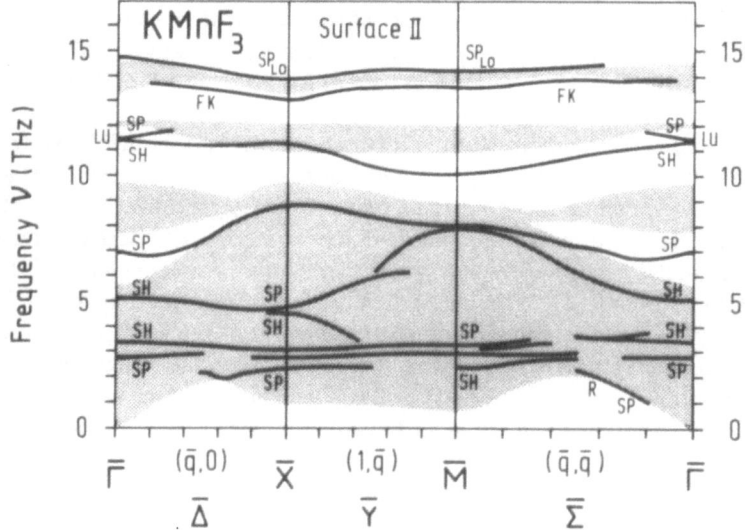

Fig. 8.49. KMnF$_3$(001): Surface phonon dispersion curves of a slab terminating with a MnF$_2$ surface (after [8.113]). Calculations [8.113]: slab method (relaxed slab, 25-layers, shell model based on interaction potentials); *solid lines*: surface localized modes (SH: shear horizontal polarization; SP: sagittal plane polarization; FK: Fuchs–Kliewer modes; LU: Lucas modes); *shaded areas*: surface projected bulk bands

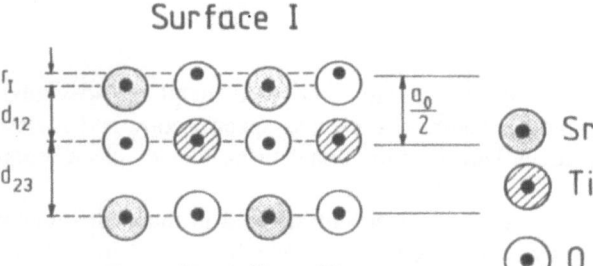

Fig. 8.50. SrTiO$_3$(001): Surface relaxation of a slab terminating with a SrO surface (after [8.128]). Calculations [8.128]: slab method (18 layers, shell model based on interaction potentials). *Dots*: cores; *circles*: shells; *lines*: (001)-planes. The displacements in the z-direction are enlarged by a factor of 2

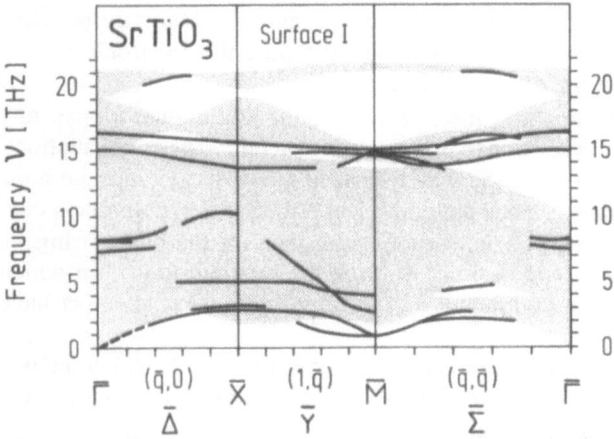

Fig. 8.51. SrTiO$_3$(001): Surface phonon dispersion curves of a slab terminating with a SrO surface (after [8.128]). Calculations [8.128]: slab method (relaxed slab, 25-layers, shell model based on interaction potenials); *solid lines*: surface localized modes(SH: shear horizontal polarization, SP: sagittal plane polarization; FK: Fuchs–Kliewer modes; LU: Lucas modes); *shaded areas*: surface projected bulk bands

It has been confirmed experimentally [8.129] that the (001) surface of SrTiO$_3$ may terminate either with a SrO or a TiO$_2$ layer. The surface relaxation of a (001) surface terminating with a SrO layer has been calculated [8.128] with the slab method in the framework of a shell model based on a set of parameters which has been determined from the measured bulk phonon dispersion curves. The results are represented schematically in Fig. 8.50. The calculated changes in interlayer distances and the surface rumpling [8.128] are in good agreement with the experimental values [8.129].

Figure 8.51 shows the calculated surface phonon dispersion curves of a slab which terminates with a SrO surface [8.128]. The calculations have been based on a shell model. The parameters are those employed for the calculation of the relaxation.

8.2 Metals

The dynamical properties of clean and adsorbate covered metal surfaces have been intensively studied by both electron energy loss spectroscopy and inelastic He-atom scattering. We focus here on vibrational properties of clean metal surfaces.

The surface vibrations of metals have been investigated in many theoretical studies. Most of these studies are based on models which are able to describe the measured phonon dispersion curves in the bulk within experimental error. However, this does not necessarily imply that reliable predictions of the surface phonons can be made on the basis of these models. It has to be expected that the force constants and even the interaction potentials, on which these models are based, undergo quite drastic changes at and close to the surface. These changes are due to the "relaxation" of the electronic ground state and can only be taken into account properly by calculations which also consider the electronic system and its changes due to the presence of a surface.

First principles calculations are, therefore, of major importance for the understanding of the surface vibrational properties of metals. These calculations suffer, however, from the fact that the way a surface mode decays into the bulk is not determined by symmetry. Thus the numerical efforts in frozen-phonon calculations for a slab are an order of magnitude larger than for the bulk. Stringent approximations have to be made in order to bring the problem to a form which can be handled with modern computers. Up to now only a few attempts have been made in this direction.

The next best way to handle the problem is to develop models which include electronic relaxation in a more phenomenological way. Not many attempts have been made so far in this direction either.

The most commonly used method is to base the surface phonon calculations on bulk force constants and then to allow for changes of these force constants at and close to the surface in such a way as to reproduce the experimentally determined surface modes. Since the changes required to reproduce the experimental data may be as much as 50%, the predictive power of surface calculations based on bulk data alone is rather limited.

We concentrate in this chapter on those surfaces for which experimental data are available. It should be mentioned here that many metals show spontaneous (e.g. Au(111), Au(110), Au(001), Ir(110), Pt(110)) or temperature induced (e.g. W(001) and Mo(001)) surface reconstructions. The tendency towards reconstruction increases, in general, with decreasing packing density at the surface. The relaxation follows the same pattern; it is small for surfaces with high packing densities and increases with decreasing packing density.

8.2.1 Body Centered Cubic Metals

The bulk phonon dispersion curves of Li, Na, K, Rb, Co, Ba, V, Nb, Ta, Cr, Mo, W, and α-Fe have been determined by inelastic neutron scattering measurements [8.130]. The measured phonon dispersion curves in the main symmetry directions of the bulk Brillouin zone are well described by various force-constant models and in most cases also by calculations based on empirical and occasionally even on first principles pseudopotentials.

The surface vibrations of most of the bcc metals have so far only been investigated theoretically for both (110) and (001) surfaces. The results of slab calculations based on force constants of the bulk are reported in [8.127, 131]. Some EELS measurements have been reported for Fe(111) [8.132]. By far the best investigated, however, is the (001) surface of bcc-W.

W(001)

The phonon dispersion in the bulk has been determined by inelastic neutron scattering experiments [8.133, 134]. The measured data are well reproduced by a force-constant model which includes interactions up to third nearest neighbors.

W has an ideal (001) surface at temperatures above about 280K. Below, the (001) surface exhibits a $\sqrt{2} \times \sqrt{2}\,R45°$ reconstruction [8.135]. The surface vibrations above and below the surface reconstruction temperature have been investigated by inelastic He-atom scattering measurements [8.136, 137]. The results of these investigations are shown in Figs. 8.52 and 8.53.

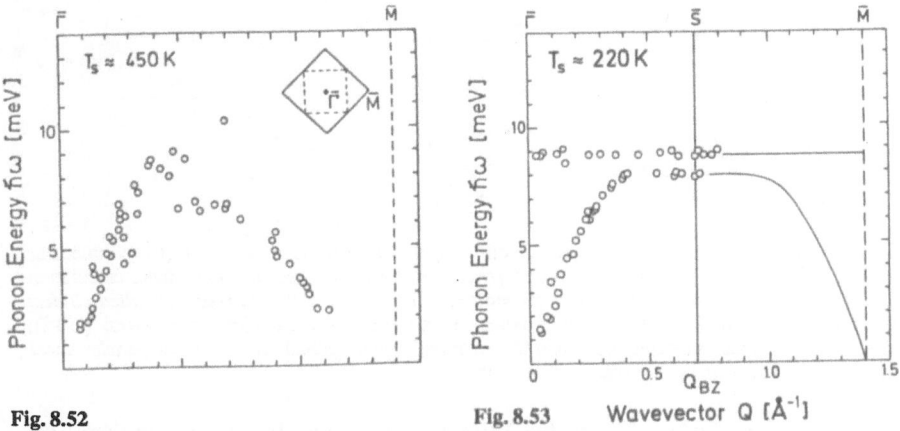

Fig. 8.52   **Fig. 8.53**   Wavevector Q [Å⁻¹]

Fig. 8.52. W(001): Surface phonon dispersion curves above the $\sqrt{2} \times \sqrt{2}\,R45°$ reconstruction (after [8.137]). Experimental data [8.136,137]: surface modes as determined by He-atom scattering measurements at a surface temperature of $T_s = 450\,K$ (*open circles*)

Fig. 8.53. W(001): Surface phonon dispersion curves below the $\sqrt{2} \times \sqrt{2}\,R45°$ reconstruction (after [8.137]). Experimental data [8.136,137]: surface modes as determined by He-atom scattering measurements at a surface temperature of $T_s = 220\,K$ (*open circles*). Q_{BZ} indicates the zone boundary of the surface Brillouin zone for the reconstructed surface. The lines show the continuation of the surface phonon dispersion curves into the next zone

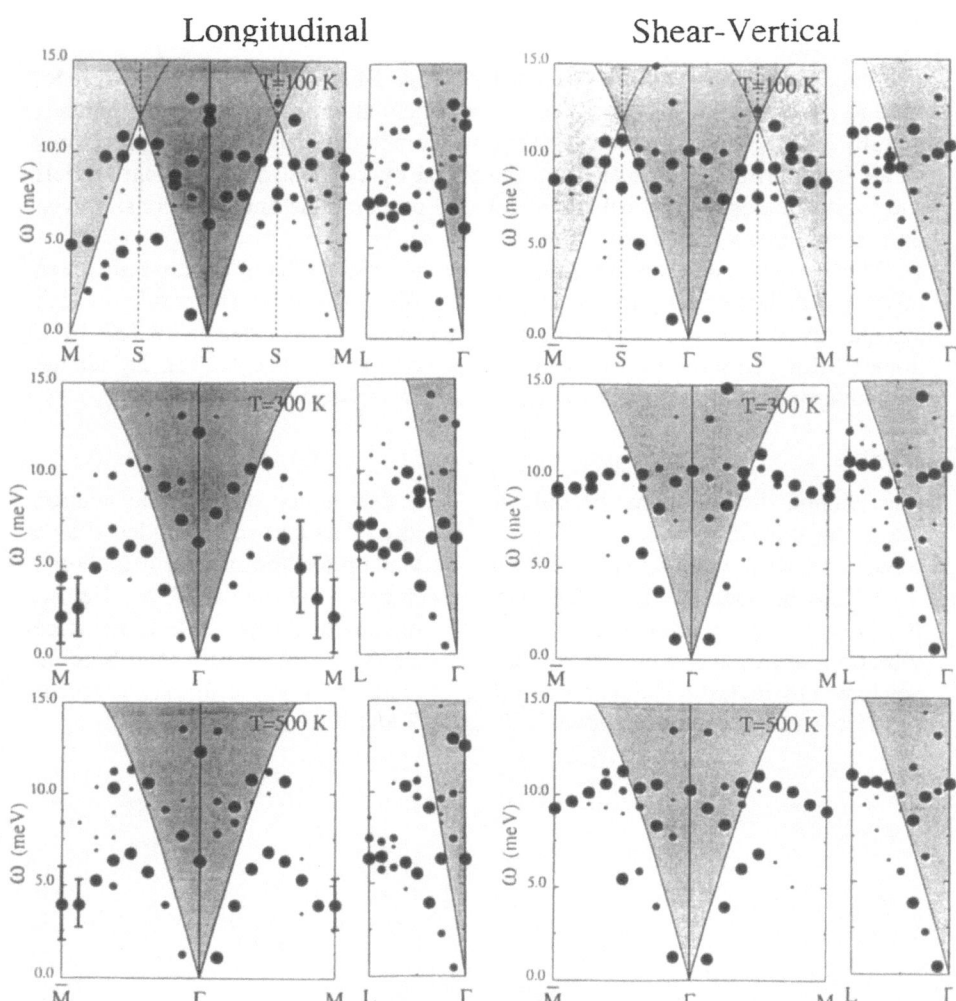

Fig. 8.54. W(001): Surface phonon dispersion curves (after [8.139]). Calculations [8.139]: molecular dynamics method (semi-empirical interaction potentials in the bulk and at the surface, calculation of the surface phonon spectral functions for various temperatures and polarizations); *filled circles*: peaks in the spectral function (large circles indicate strong peaks, small circles indicate weak peaks); *shaded areas*: surface projected bulk bands; *left panels*: longitudinal modes; *right panels*: shear vertical modes; *opposite page*: shear horizontal modes

The dynamical properties of the W(001) surface can be nicely simulated by molecular dynamics calculations which are based on empirical interaction potentials [8.138, 139]. Bulk and surface properties are adequately taken into account in these calculations. The surface modes extracted from the velocity–velocity correlation function are shown in Fig. 8.54. Note the softening and broadening of the longitudinal and shear horizontal modes close to the surface reconstruction, which takes place at about 280K. Further theoretical studies of the surface reconstruction have been reported in [8.140, 141].

Fig. 8.54. (cont.)

Shear-Horizontal

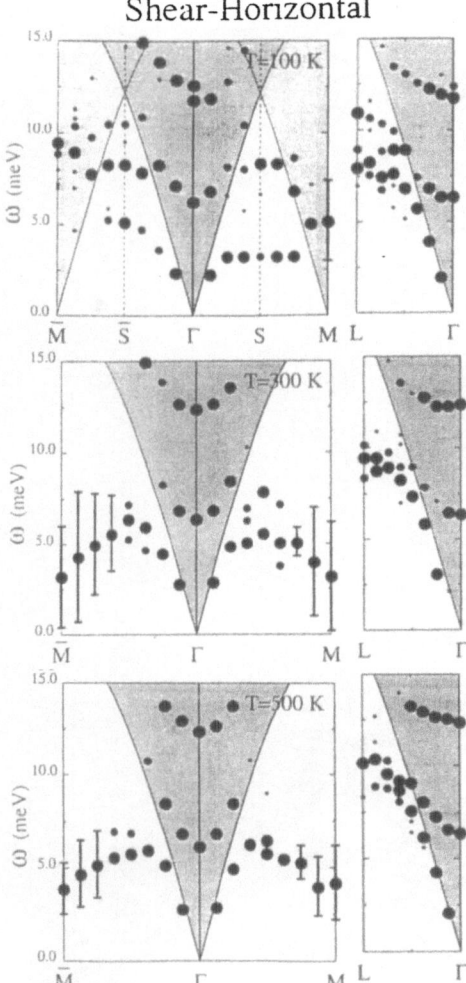

8.2.2 Face Centered Cubic Metals

The phonon dispersion curves in the bulk of Ca, Sr, La, Ce, Yb, Al, Pb, Ni, Pd, Pt, Cu, Ag, and Au have been determined by inelastic neutron scattering measurements [8.130]. Measurements for rhodium and iridium are still not available. The measured phonon dispersion curves in the main symmetry directions of the bulk Brillouin zone are well reproduced by various force-constant models and in most cases also by calculations based on empirical and occasionally even on first principles pseudopotentials [8.130].

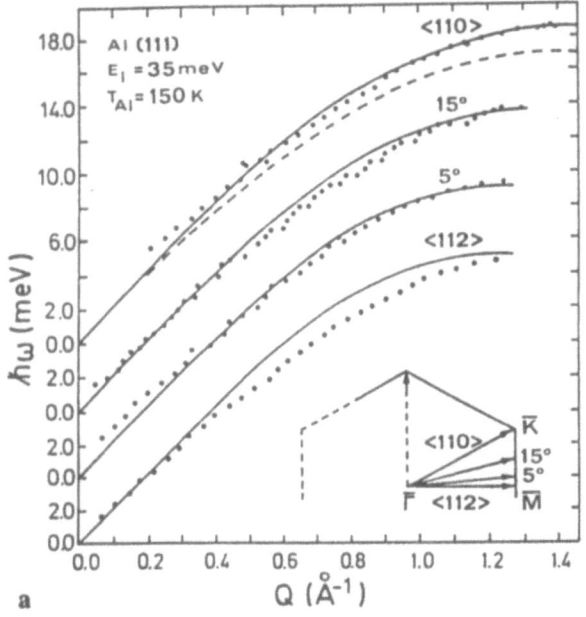

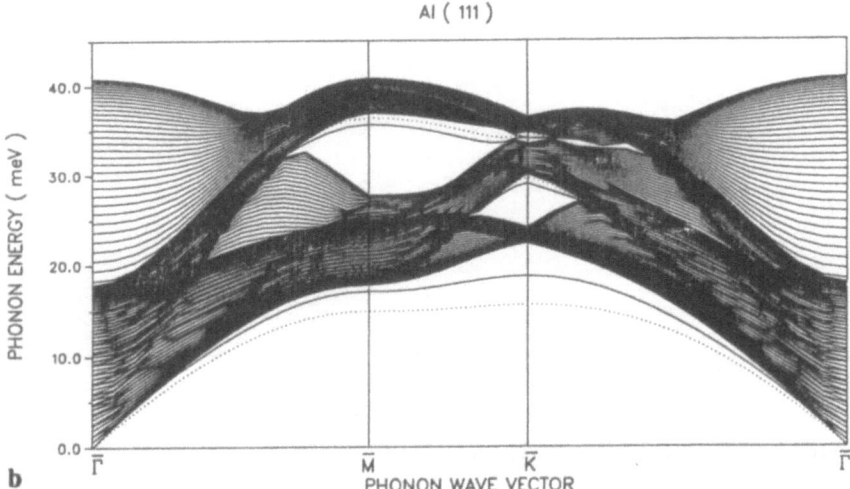

Al (111)

Fig. 8.55. Al(111): (a) Dispersion of Rayleigh waves (after [8.145]). Calculations [8.145]: slab method (unrelaxed 90-layer slab, force-constant model, parameters fitted to the measured phonon dispersion curves of the bulk material, nearest neighbor force constants at the surface changed in order to reproduce the experimental data and to account for relaxation effects); *solid lines*: surface localized modes (interactions to tenth nearest neighbors); *dashed lines*: surface localized modes (interactions to second nearest neighbors). Experimental data [8.145]: surface modes along various directions of the surface Brillouin zone (see inset) as determined by inelastic He-atom scattering measurements (*filled circles*) (b) Surface phonon dispersion curves (after [8.165b]). Calculations [8.165b]: slab method. *Solid lines*: 51 relaxed layers, first principles calculations based on a model pseudopotential of the Heine–Abarenkov type, dielectric response formalism; *dotted lines*: surface localized modes as obtained from the force constants in the bulk

Al(111)

The phonon dispersion curves in the bulk material have been determined by inelastic neutron scattering [8.142]. The measured phonon dispersion curves are not only well reproduced by various calculations based on force constants or on empirical pseudopotentials [8.130], but also by first principles calculations [8.143]. These calculations employ the frozen-phonon method. They are based on first principles pseudopotentials [8.144] and carried out in the framework of the local-density-functional formalism.

The surface vibrations of the Al(111) surface have been measured by inelastic He-atom scattering [8.145]. The results are shown in Fig. 8.55a. The dots represent the experimental results for the Rayleigh waves. Solid lines and the dashed lines represent the results obtained from force-constant models which take into account interactions up to 10th and 2nd nearest neighbors, respectively. The radial force constants at the surface have been adjusted to the experimental data.

Al(110)

The surface relaxation of Al(110) has been calculated [8.144] in the framework of the local-density-functional formalism from first principles pseudopotentials [8.144]. The frozen-phonon method has then been used to calculate – in the same framework – the surface phonon dispersion curves of the relaxed 15-layer slab with free (110) surfaces [8.146, 147]. These calculations have been carried out prior to the experimental determination of the surface phonon dispersion curves by inelastic He-atom scattering measurements [8.148]. Figure 8.56 shows the

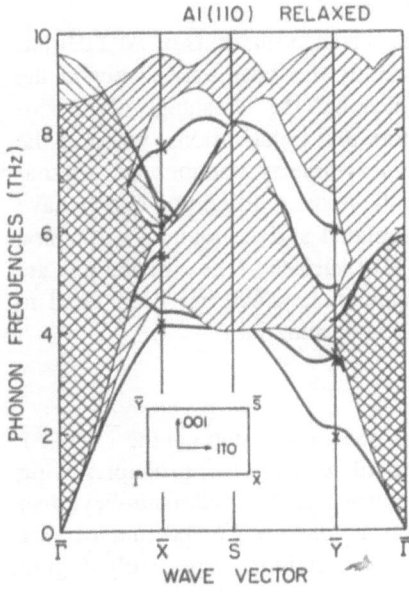

Fig. 8.56. Al(110): Surface phonon dispersion curves (after [8.146]). Calculations [8.146]: slab method (15 relaxed layers, first principles calculations based on a norm conserving first principles pseudopotential, local density functional formalism); *bold lines*: surface localized modes; *crosses*: "frozen phonon" results for surface modes at $\overline{X}$ and $\overline{Y}$, *hatched*: surface projected bulk modes (different symmetries are represented by different hatching types). The calculations have been performed prior to the inelastic He-atom scattering measurements shown in Fig. 8.57

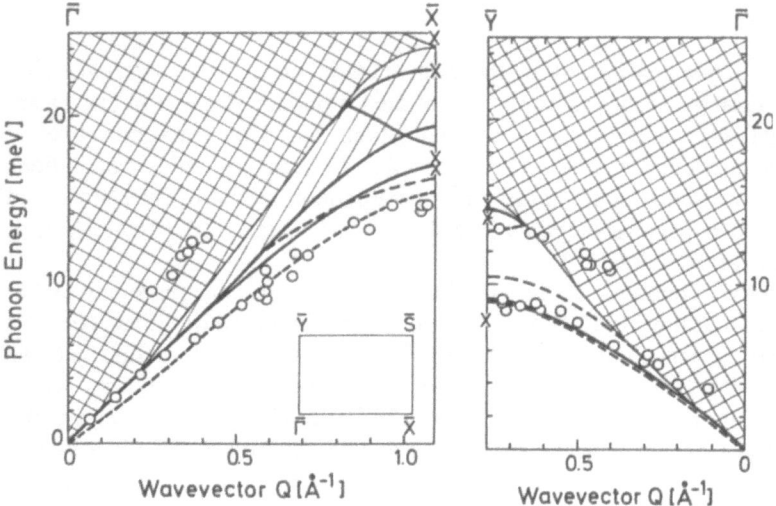

Fig. 8.57. Al(110): Surface phonon dispersion curves (after [8.148]), Calculations [8.146]: slab method (15 relaxed layers, first principles calculations based on a norm conserving first principles pseudopotential, local density functional formalism); *bold lines*: surface localized modes; *crosses*: "frozen phonon" results for surface modes at $\overline{X}$ and $\overline{Y}$; *hatched*: surface projected bulk modes (different symmetries are represented by different hatching types). Experimental data [8.148]: Surface modes as determined by inelastic He-atom scattering measurements (*filled circles*)

results of the calculations. The inelastic He scattering data [8.148] are compared in Fig. 8.57 with the calculated surface phonon dispersion curves. There are small discrepancies but the agreement is still rather gratifying if one takes into account the first principles nature of the calculations and the approximations, which had to be made in order to make these calculations feasible.

A second first principles calculation of the surface vibrations of Al(110) has been reported in the literature [8.149]. This study has been carried out in the framework of dielectric response theory, which has the advantage of providing lattice vibrations for arbitrary wave vectors and polarizations, whereas the frozen-phonon approach yields primarily high symmetry phonons or, alternatively, interplanar force constants from which the phonon dispersion curves have to be calculated. Figure 8.58a shows the calculated surface phonon dispersion curves of a relaxed 27-layer slab with free (110) surfaces. The calculations are based on a Heine–Abarenkov model pseudopotential, which has been fitted to the measured phonon dispersion curves of the bulk material.

Al(001)

Surface phonon frequencies and surface phonon densities of states at the $\overline{\Gamma}$ and $\overline{M}$ point of a relaxed Al(001) surface have been calculated from first principles using the local-density-functional total-energy formalism and the Hellmann–Feynman theorem [8.150]. The reported frequencies at the $\overline{X}$ point are 3.6, 3.0, and 8.1 THz for modes polarized in the [110], [110], and [001] directions, respectively. Figure

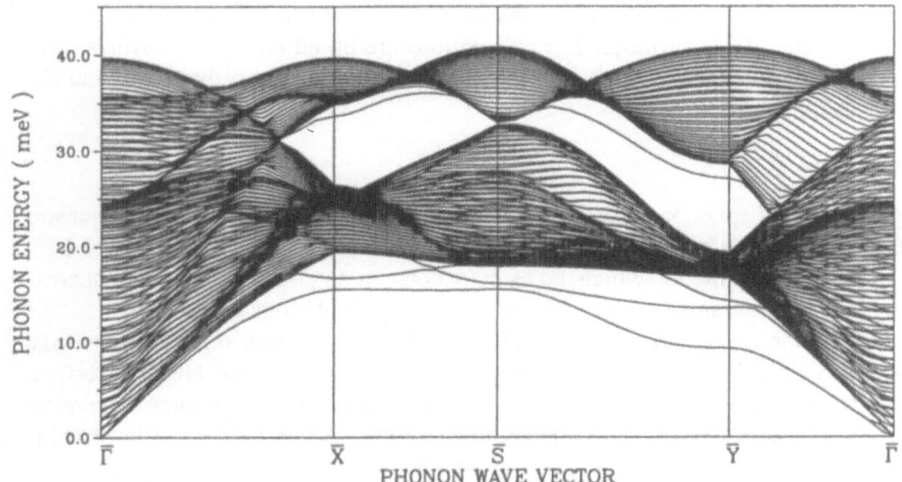

Fig. 8.58a. Al(110): Surface phonon dispersion curves (after [8.165b]. Calculations [8.149,165b]: slab method (51 relaxed layers, first principles calculations based on a model pseudopotential of the Heine–Abarenkov type, dielectric response formalism)

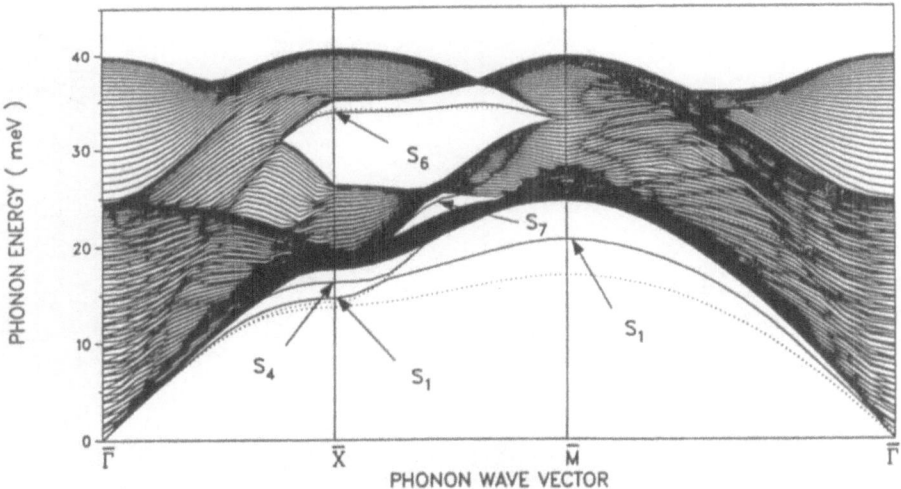

Fig. 8.58b. Al(001): Surface phonon dispersion curves (after [8.165c]). Calculations [8.165c]: slab method. *Solid lines* : 51 relaxed layers, first principles calculations based on a model pseudopotential of the Heine–Abarenkov type, dielectric response formalism); *dotted lines*: surface localized modes as obtained from the force constants in the bulk

261

8.58b shows the calculated surface phonon dispersion curves of a relaxed 51 layer slab with free (001) surfaces. The calculations are based on a Heine–Abarenkov model pseudopotential, which has been fitted to the measured dispersion curves of the bulk material.

Ni(001)

The phonon dispersion curves in the bulk have been determined by inelastic neutron scattering experiments [8.151, 152]. The results are well reproduced by a force-constant model which takes into account interactions up to 5 nearest neighbors (15 parameters) and by various other calculations [8.130].

Clean and adsorbate covered Ni(100) surfaces have been studied intensively by EELS experiments. The experimental results for the clean Ni(100) surface [8.153–157] are shown in Fig. 8.59. The calculated surface phonon dispersion curves of an unrelaxed Ni(001) slab with free (001) surfaces are given in [8.158].

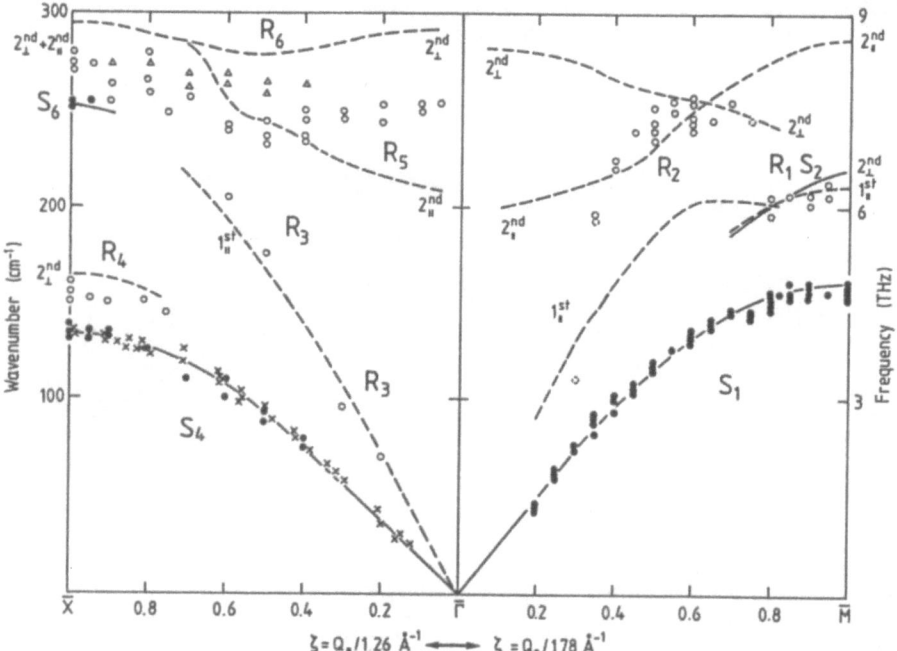

Fig. 8.59. Ni(001): Surface phonon dispersion curves (after [8.157]. Calculations [8.157]: Green's function method (unrelaxed surface, force-constant model, nearest neighbors interactions only, force constants determined from bulk data); *solid lines*: surface localized modes and resonances; *dashed lines*: resonances with maxima in the spectral density in the first (1st) and second (2nd) layer. Experimental data [8.157]: surface modes as determined by EELS measurements (*filled circles*: surface localized modes, *open circles* and *open triangles*: resonances, *crosses*: EELS data of [8.153]

Pd(110)

The phonon dispersion curves in the bulk of palladium have been determined by inelastic neutron scattering measurements [8.159–161]. The TA_1 branch in the (110) direction exhibits small, but rather pronounced, Kohn anomalies. The overall dispersion of the measured data is very well reproduced by calculations based on simple force-constant models [8.160], on modified empirical pseudopotentials [8.162], and on empirical ion–ion interaction potentials of the Morse type [8.163].

Pd exhibits a strongly relaxed (110) surface. The spacing between the first and second layer is about 95% of the bulk value [8.164]. The vibrations of clean (110) surfaces have been studied by inelastic He-atom scattering [8.165]. Figure 8.60 shows the measured data together with the results of calculations for an unrelaxed 61-layer slab with free (110) surfaces. The calculations are based on a force-constant model, which takes into account radial and transverse force constants up to second nearest neighbors and in addition an angular bending force constant for the angle formed by two different nearest neighbors of the apex particle.

The outmost left and right parts of Fig. 8.60 show the experimental results (open circles) together with the calculated surface phonon dispersion curves [8.165a]. The solid lines are obtained with drastically changed force constants at the surface. The first nearest neighbor radial force constant in the surface is softened to 70% of its bulk value. The corresponding transverse force constant assumes −0.60 times its bulk value and the second nearest neighbor radial and

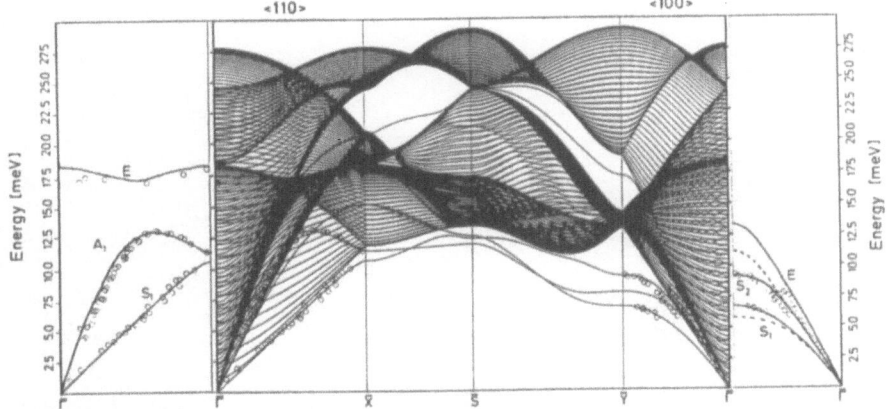

Fig. 8.60. Pd(110): Surface phonon dispersion curves (after [8.165]). Calculations [8.165]: slab method (61 unrelaxed layers, force-constant model). *Left* and *right panel*: surface localized modes as obtained with bulk force constats, which have been modified at the surface to account for surface relaxation effects (*solid lines*) and surface localized modes obtained with unmodified bulk force constats (*dashed lines*). S_i: labels surface localized modes and resonances. Experimental data [8.165]: surface modes as determined by inelastic He-atom scattering measurements (*open circles*). *Inner panel*: Calculated surface phonon dispersion curves (*solid lines*) for an unrelaxed 61-layer slab with force constants which have been changed at the surface in order to reproduce the He scattering data (*circles*) and to account for surface relaxation effects

transverse force constants in the surface layer assume −0.30 and +0.35 times their bulk values, respectively. The bulk force constants are those which have been determined from the measured phonon dispersion curves in the bulk (model III of [8.166]). The dashed lines indicate the results which have been obtained without changing the bulk force constants at the surface. The middle panel of Fig. 8.60 shows the complete results of the calculations, which include the surface projected bulk modes in addition to the surface phonon dispersion curves shown in the right and left panel.

Pt(111)

The phonon dispersion curves in the bulk of material have been determined by inelastic neutron scattering measurements [8.167]. The measured data are well reproduced by calculations based on a force-constant model which takes into account coupling constants up to sixth nearest neighbors [8.167].

The normal modes of clean (111) surfaces have been studied by inelastic He-atom scattering [8.168–170]. Figure 8.61 shows the measured Rayleigh dispersion curve in the $\overline{\Gamma}$-$\overline{K}$ direction. The low temperature measurements show weak structures, which exhibit at first glance strong parallels with similar structures in the dispersion of the TA_1 branch in the (110) direction of the bulk material. The weak structures of the bulk phonon dispersion curves have been attributed

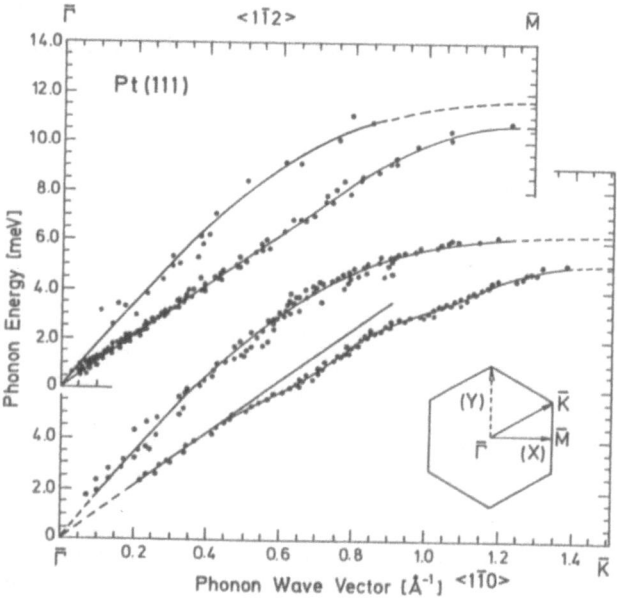

Fig. 8.61. Pt(111): Dispersion of the acoustic surface waves (after [8.168]). Experimental data [8.145]: surface modes along the $\overline{\Gamma}$-$\overline{M}$ (two upper curves, sample temperature $T_s = 400$ K) and along the $\overline{\Gamma}$-$\overline{K}$ (two lower curves, sample temperature $T_s = 160$ K) as determined by inelastic He-atom scattering measurements (*filled circles*). Calculations [8.168] (*solid lines*): least squares fit of the first nine coefficients of a Fourier expansion to the measured data

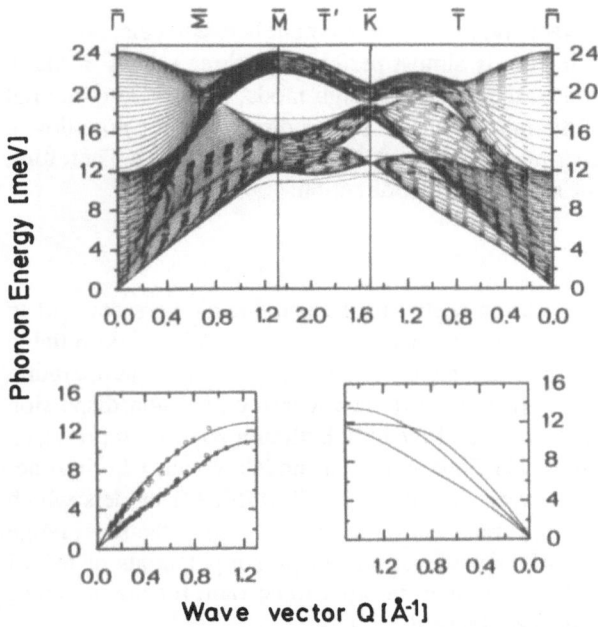

Fig. 8.62. Pt(111): Surface phonon dispersion curves (after [8.171]). *Upper panel*: calculations [8.171]. Slab method (183 unrelaxed layers, force-constant model, force constants in the bulk fitted to the neutron scattering data [8.167], force constants at the surface changed in order to reproduce the He-atom scattering data [8.171] and to account for surface relaxation effects); dashed lines in the $\overline{\Sigma}$-direction: shear horizontal modes; dashed lines in the $\overline{T}$-direction: modes odd with respect to the central layer; solid lines in the $\overline{\Sigma}$-direction: sagittal modes: solid lines in the $\overline{\Sigma}$-direction: modes even with respect to the central layer. *Lower panels*: calculated surface localized modes (*full lines*). Experimental data [8.171]: Surface modes as determined by He-atom scattering measurements (*open circles*)

to Kohn anomalies [8.167]. Therefore, it has been speculated that the structures in the surface phonon dispersion curves could also be interpreted as weak Kohn anomalies. It is shown in [8.171] that the Rayleigh modes in the $\overline{\Gamma}$–$\overline{K}$ direction are separated from the band edge of the acoustic bulk bands by a very small frequency gap only. Thus interactions with the bulk bands and not surface Kohn effects lead to the observed modulation of the Rayleigh dispersion curve. It should be pointed out that, for the first time, a slight, but well resolved temperature shift of the Rayleigh modes has been observed in these measurements. As in the bulk, this shift is caused by anharmonic terms in the interaction potentials.

Figure 8.62 shows the calculated surface vibrations of an unrelaxed 180-layer slab with free (111) surfaces. The calculations are based on force constants obtained from a fit to the measured phonon dispersion curves in the bulk. In order to reproduce the measured surface data and in particular the longitudinal resonance, both radial and transverse force constants had to be modified at the surface. The radial nearest neighbor force constant at the surface assumes only 39% of its bulk value. Drastic changes are also assumed for the transverse and for the three-body force constants at the surface. The bottom of Fig. 8.62 shows the localized and resonant surface modes. The agreement between calculated and measured curves in the $\overline{\Sigma}$ direction is almost perfect. The three modes in the $\overline{T}$ direction are the Rayleigh mode, the quasi-Rayleigh mode, and the longitudinal resonance. The resolution of the measurements in this direction does not allow a proper assignment of the experimental data points to the three modes. Therefore, the experimental data are not shown for this direction.

Cu(111)

Many of the pioneering early neutron scattering measurements were devoted to the study of the bulk vibrations in copper. The most recent studies yield reliable measurements of the bulk phonon dispersion curves for the whole temperature range between 49 K and 1336 K [8.172–175]. The measured phonon dispersion curves in the main symmetry directions of the bulk Brillouin zone are reproduced well by three classes of models: (1) Force-constant models which take into account interactions up to sixth nearest neighbors [8.172–175], (2) models which take into account the response of a free electron gas in addition to the short range force constants, and (3) calculations based on model pseudopotentials or model pseudopotentials which have been modified in order to account for the influence of d electrons and s-d hybridization [8.130].

Inelastic neutron scattering studies have revealed weak Kohn anomalies in the phonon dispersion curves of the bulk material [8.176, 177]. However, the observed anomalies of the TA$_1$ branch in the (110) direction of the bulk Brillouin zone are not Kohn anomalies but rather have to be attributed to the transition from the collision dominated first sound regime to the collision free second sound regime [8.178].

The dispersion curves of the Rayleigh waves have been determined by inelastic He-atom scattering measurements [8.179]. The experimental data exhibit clearly a second surface phonon branch above the Rayleigh modes. Slab calculations based on force-constant models, which reproduce the measured bulk phonon dispersion curves rather nicely, show in the range of the experimental data only surface projected transverse acoustic bulk bands, which cannot lead to the observed resonances [8.180]. In order to reproduce the experimental data, the nearest neighbor radial force constant in the surface plane has to be reduced by about 30%.

A quite different approach is adopted by the pseudoparticle model [8.181]. This model incorporates the relevant electronic properties in a phenomenological way. The four-parameter model reproduces the measured bulk phonon dispersion curves within experimental error [8.181]. It also describes well the measured surface phonon dispersion curves, if the relaxation of the electronic charge density at the surface is taken into account by a change of the pseudoparticle–ion coupling constant. The change of this coupling constant is calculated from the measured work function using a phenomenological relationship [8.182]. Figure 8.63 shows the calculated surface phonon dispersion curves [8.182] together with the experimental data [8.179]. The agreement is convincing.

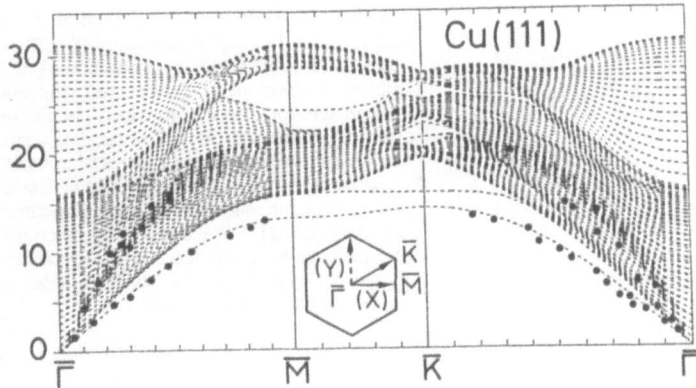

Fig. 8.63. Cu(111): Surface phonon dispersion curves (after [8.182]). Calculations [8.182]: slab method (35 unrelaxed layers, pseudoparticle model, parameters fitted to the measured phonon dispersion curves of the bulk material, relaxation of the electronic charge density taken into account in a phenomenological way). Experimental data [8.179]: surface modes as determined by He-atom scattering measurements (*filled circles*)

Cu(110)

The surface phonon dispersion curves of the Cu(110) have been studied by inelastic He-atom scattering measurements [8.183]. Figure 8.64 shows the experimental data together with the results of calculations for an unrelaxed 151-layer slab with free (110) surfaces. The calculations are based on a model which takes into account coupling constants up to sixth nearest neighbors and reproduces the bulk phonon dispersion curves within experimental error. The transverse force constants for nearest neighbor interactions between particles in the surface plane and in the first and second inner plane are reduced to 80% and 20% of the bulk values, respectively. In this way the surface stress, which arises from the redistribution of the electronic charge density at and close to the surface, is taken into consideration. (Note that the transverse force constants of a central potential are proportional to the first derivative of the interaction potential with respect to the distance between the interacting particles.) It should be mentioned that EELS [8.184] measurements yield a resonant $\overline{\Gamma}$-point mode at 20 meV, which does not show up in Fig. 8.64 because of its weak localization.

Cu(001)

The surface phonon dispersion curves of Cu(100) have been studied by EELS measurements [8.185, 186]. The results in the $\overline{\Gamma}$–$\overline{X}$ and $\overline{\Gamma}$–$\overline{M}$ direction are shown in Fig. 8.65. Inelastic He-atom scattering data have been reported for the $\overline{\Gamma}$–$\overline{M}$ direction [8.187]. The experimental data are well reproduced by a central force model with nearest neighbor interactions only. The force constant between the particles in the first and second layer, however, has to be increased to 20% above the bulk value (solid lines). The dashed lines are obtained with unchanged force constants at the surface.

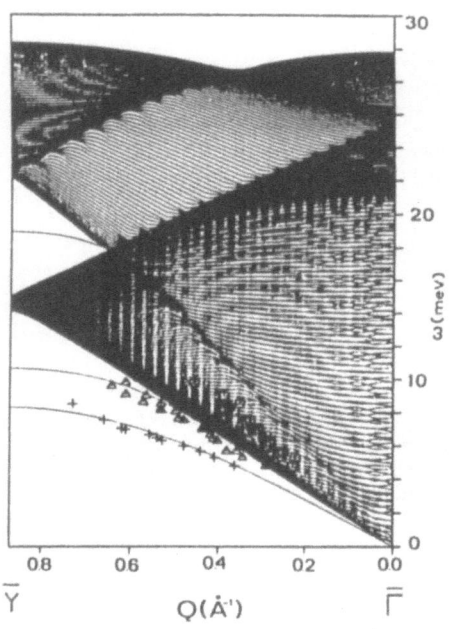

Fig. 8.64. Cu(110): Surface phonon dispersion curves (after [8.258]). Calculations [8.180]: slab method (unrelaxed 151-layer slab, force-constant model, interactions to sixth nearest neighbors, parameters fitted to the measured phonon dispersion curves of the bulk material, nearest neighbor force constants at the surface changed in order to reproduce the experimental data and to account for relaxation effects). Experimental data [8.183]: surface modes as determined by inelastic He-atom scattering measurements (*crosses*: Rayleigh modes, *open triangles* and *open circles*: sagittal modes)

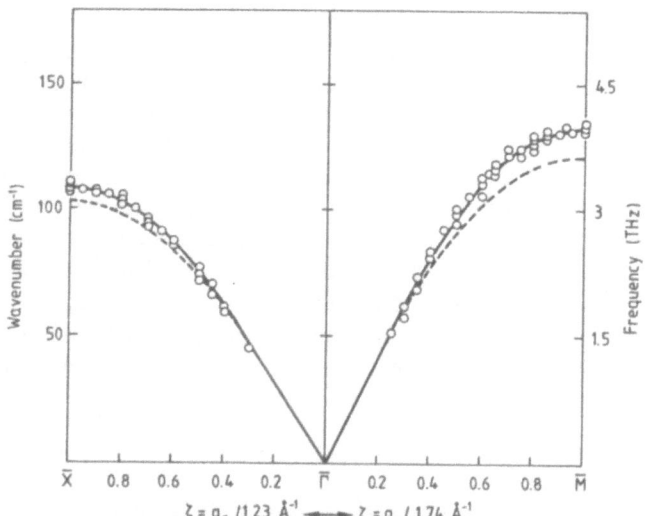

Fig. 8.65. Cu(001): Surface phonon dispersion curves (after [8.186]). Calculations [8.186]: slab method (unrelaxed 65-layer slab, force-constant model, nearest neighbors interactions only, parameters fitted to the measured phonon dispersion curves of the bulk material, nearest neighbor force constants at the surface changed in order to reproduce the experimental data and to account for relaxation effects); *solid lines*: surface localized modes as obtained with changed force constants at the surface; *dashed lines* surface localized modes as obtained with unchanged force constants. Experimental data [8.186]: surface modes as determined by EELS measurements (*open circles*)

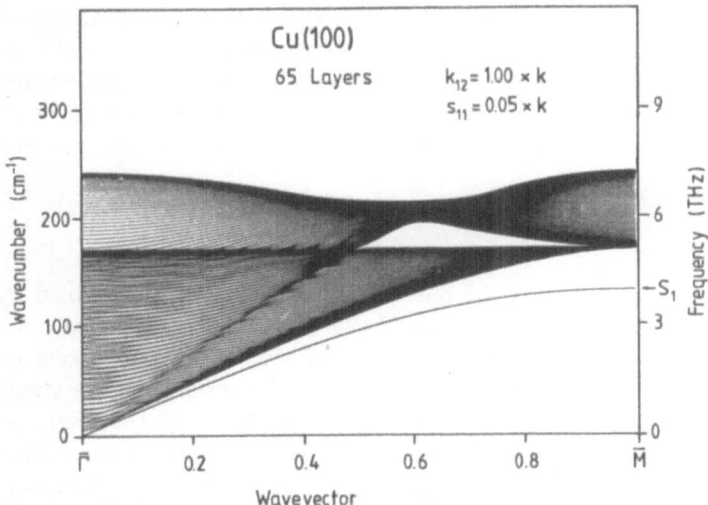

Fig. 8.66. Cu(001): Surface phonon dispersion curves in the $\overline{\Gamma}\text{-}\overline{M}$ direction (after [8.186]). Calculations [8.186]: slab method (unrelaxed 65-layer slab, force-constant model, nearest neighbors interactions only, parameters fitted to the measured phonon dispersion curves of the bulk material)

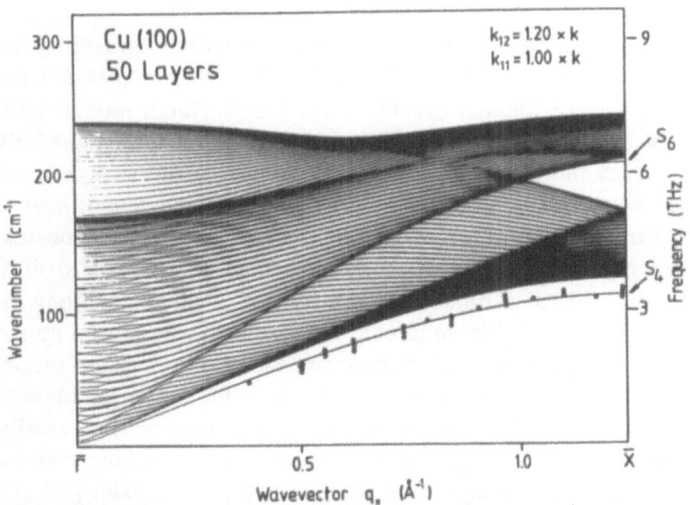

Fig. 8.67. Cu(001): Surface phonon dispersion curves in the $\overline{\Gamma}\text{-}\overline{X}$ direction (after [8.185]). Calculations [8.185]: slab method (unrelaxed 50-layer slab, force-constant model, nearest neighbors interactions only, parameters fitted to the measured phonon dispersion curves of the bulk material, nearest neighbor force constants at the surface changed in order to reproduce the experimental data and to account for relaxation effects)

Figure 8.66 shows the sagittal plane (even) modes in the $\overline{\Gamma}-\overline{M}$ direction of a 65-layer slab. The calculations are based on the nearest neighbor model with unchanged force constants at the surface. Figure 8.67 shows the sagittal plane modes in the $\overline{\Gamma}-\overline{X}$ direction of a 50-layer slab. The calculations are based on the nearest neighbor model with an increased force constant between the first and second layer.

Ag(111)

The phonon dispersion curves in the bulk of silver have been determined by inelastic neutron scattering measurements [8.188–190]. The measured phonon dispersion curves in the main symmetry directions of the bulk Brillouin zone are reproduced well by three classes of models: (1) Force-constant models which take into account interactions up to forth nearest neighbors [8.188, 189], (2) models which take into account the response of a free electron gas in addition to the short range force constants [8.130], and (3) calculations based on model pseudopotentials or model pseudopotentials which have been modified in order to account for the influence of d electrons and s-d hybridization [8.130].

The TA$_1$ branch in the (110) direction exhibits [8.190] anomalies similar to those found in Cu. These anomalies have to be attributed to the transition from the collision dominated first sound regime to the collision free second sound regime [8.178].

The dispersion curves of the Rayleigh waves have been determined by inelastic He-atom scattering measurements [8.179, 192]. The experimental data exhibit clearly a second surface phonon branch above the Rayleigh modes. Slab calculations, based on force-constant models that reproduce the measured bulk phonon dispersion curves rather nicely show in the range of the experimental data only surface projected transverse acoustic bulk bands, which cannot lead to the observed resonances. In order to reproduce the experimental data, the nearest neighbor radial force constant in the surface plane had to be reduced to 48% of the bulk value [8.193]. Figure 8.68 shows the results of such calculations together with the experimental data. The measured data are well reproduced by the calculations. The ad-hoc change of some parameters of a force-constant model which employs a large number of parameters to reproduce the measured phonon dispersion curves in the bulk and at the surface, however, does not necessarily provide a deeper understanding of the physical origin of the observed resonances.

Quite a different approach is adopted by the pseudoparticle model [8.181]. This model incorporates the relevant electronic properties in a phenomenological way. It takes into account the lowest order multipole expansion terms of the displacement-induced deformations of the electronic charge density in addition to the nearest neighbor forces. The four-parameter model reproduces the measured bulk phonon dispersion curves within experimental error [8.181]. It also describes well the measured surface phonon dispersion curves, if the relaxation of the electronic charge density at the surface is taken into account by a change of the pseudoparticle–ion coupling constant. The change of this coupling con-

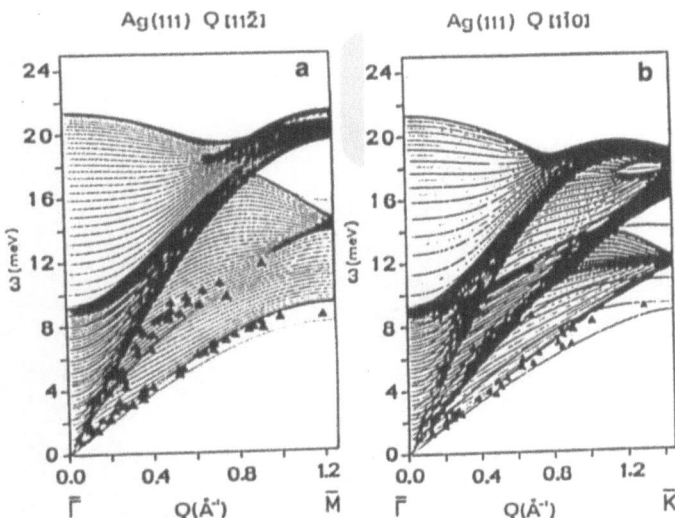

Fig. 8.68. Ag(111): Surface phonon dispersion curves (after [8.180]). Calculations [8.180]: slab method (unrelaxed 45-layer slab, force-constant model, interactions to second nearest neighbors, parameters fitted to the measured phonon dispersion curves of the bulk material, nearest neighbor force constants at the surface changed in order to reproduce the experimental data and to account for relaxation effects). Experimental data [8.192]: Surface modes as determined by inelastic He-atom scattering measurements (*filled triangles*)

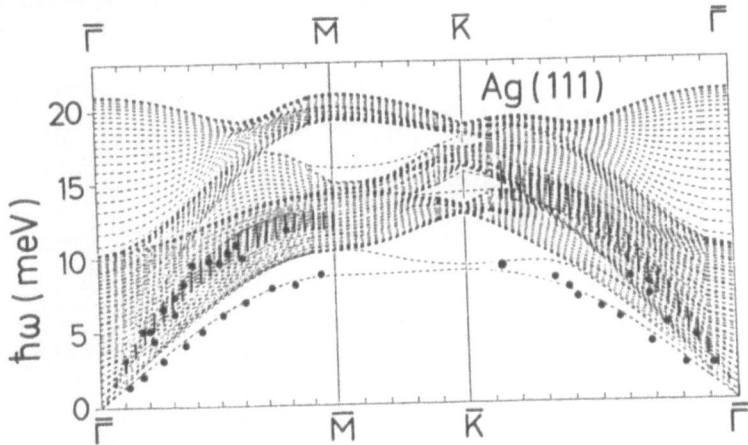

Fig. 8.69. Ag(111): Surface phonon dispersion curves (after [8.182]). Calculations [8.182]: slab method (35 unrelaxed layers, pseudoparticle model, parameters fitted to the measured phonon dispersion curves of the bulk material, relaxation of the electronic charge density taken into account in a phenomenological way). Experimental data [8.179]: surface modes as determined by He-atom scattering measurements (*filled circles*)

stant is calculated from the measured work function using a phenomenological relationship [8.182]. Figure 8.69 shows the calculated surface phonon dispersion curves together with the experimental data. The agreement is rather convincing.

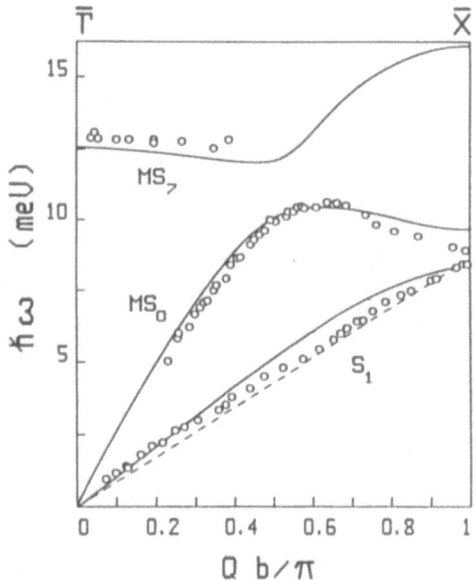

Fig. 8.70a. Ag(110): Surface phonon dispersion curves (after [8.194a]). Calculations [8.194a]: slab method (unrelaxed slab, force-constant model, nearest neighbor interactions only, parameters fitted to the phonon dispersion curves of the bulk material): *solid lines*: surface localized modes as obtained from the force-constant model; *dashed lines*: dispersion of the Rayleigh waves as obtained from elastic continuum theory. Experimental data [8.194a]: surface modes as determined by inelastic He-scattering measurements (*open circles*)

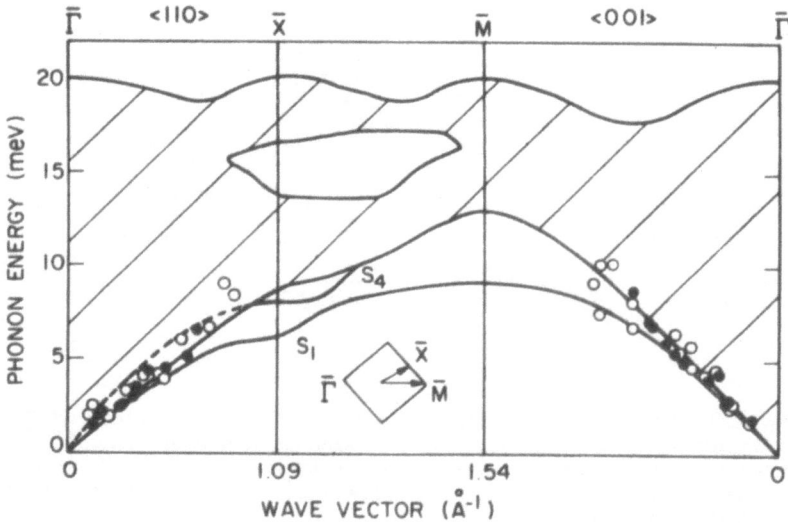

Fig. 8.70b. Ag(001): Surface phonon dispersion curves (after [8.194b]. Experimental data [8.194b]: surface modes as determined by inelastic He-scattering measurements. The open and filled circles refer to energy gain and energy loss measurements, respectively. The dashed line represents an estimate for the location of the S_4 vertical surface mode within the shear horizontal polarized region of the surface projected bulk modes. The hatched areas represent the surface projected bulk modes as obtained from slab calculations [8.195, 196]

Ag(110)

The surface phonon dispersion curves of the Ag(110) have been studied by inelastic He-atom scattering measurements [8.194a]. The results are shown in Fig. 8.70a. The experimental data are astonishingly well reproduced by first nearest neighbor model which takes only into account a single radial force constant [8.195] (solid lines). The dashed line represents the dispersion curve of the Rayleigh mode as obtained from continuum theory. The agreement between measured and calculated curves is good except for the MS_0 branch close to the $\overline{X}$ point where interaction of longer range might become important. The missing longer range interactions also show up in deviations between the measured and calculated Rayleigh modes.

Ag(001)

The surface phonon dispersion curves of Ag(001) have been studied by inelastic He-atom scattering measurements [8.194b]. The results are shown in Fig. 8.70b.

Au(111)

The phonon dispersion curves in the bulk of gold have been determined by inelastic neutron scattering measurements [8.191]. The measured phonon dispersion curves in the main symmetry directions of the bulk Brillouin zone are reproduced rather well by three classes of models: (1) Force-constant models which take into account interactions up to fifth nearest neighbors [8.191], (2) models which take into account the response of a free electron gas in addition to the short range force constants [8.130], and (3) calculations based on model pseudopotentials or model pseudopotentials which have been modified in order to account for the influence of d electrons and s-d hybridization [8.130].

The electronic ground state energy of gold is so strongly perturbed by the formation of a surface that the (111) surface undergoes an immediate (23 × 1) reconstruction [8.197].

The dispersion curves of the Rayleigh waves have been determined by inelastic He-atom scattering measurements [8.179]. The experimental data exhibit sharp structures in the scattering intensity above the Rayleigh modes [8.179, 198]. Slab calculations based on force-constant models which reproduce the measured bulk phonon dispersion curves rather nicely, show in the range of these experimental structures only surface projected transverse acoustic bulk bands, which cannot lead to the observed spectra. In order to reproduce the experimental data the nearest neighbor radial force constant and the angular force constant in the surface plane have to be reduced by 70% and 50% respectively [8.193, 198]. Figure 8.71 shows the results of such calculations together with the experimental data. The measured data are well reproduced by the calculations. The ad-hoc change of some parameters of a model that employs a large number of force constants to reproduce the measured phonon dispersion curves in the bulk and at the surface does, however, not necessarily provide a deeper understanding of the physical origin of the observed spectra.

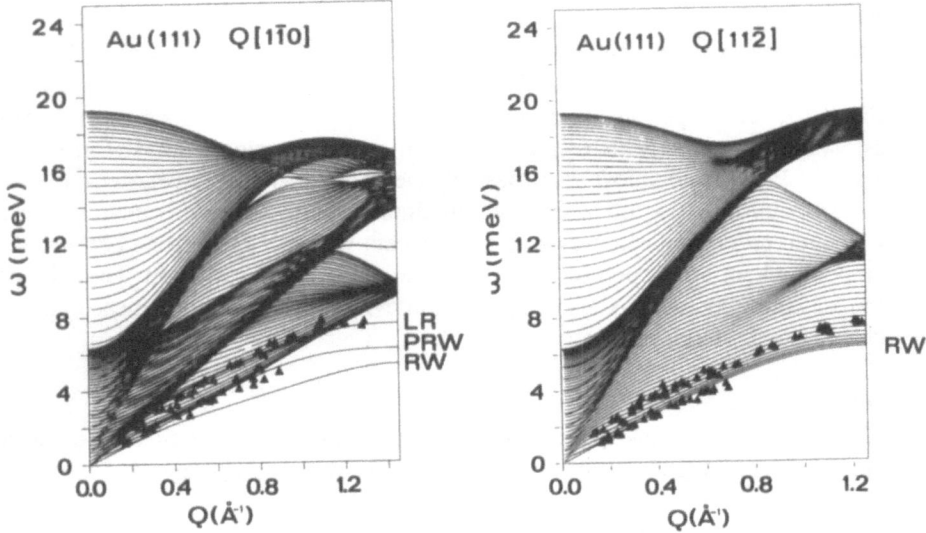

Fig. 8.71. Au(111): Surface phonon dispersion curves (after [8.193]). Calculations [8.193]: slab method (unrelaxed 45-layer slab, force-constant model, interactions to second nearest neighbors, parameters fitted to the measured phonon dispersion curves of the bulk material, nearest neighbor force constants at the surface changed in order to reproduce the experimental data and to account for relaxation effects). Experimental data [8.193]: surface modes as determined by inelastic He-atom scattering measurements (*filled triangles*)

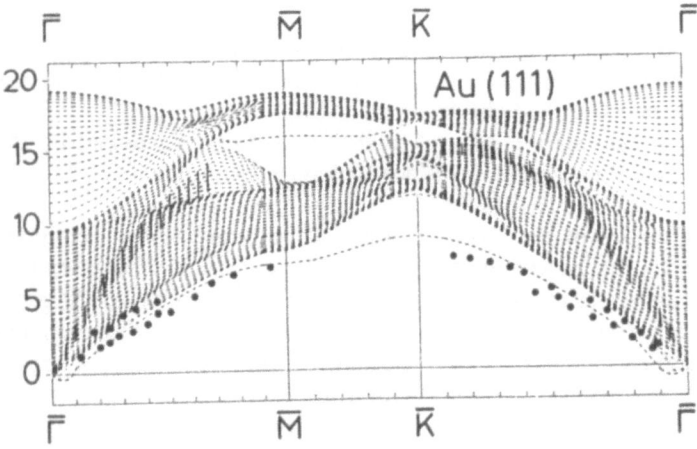

Fig. 8.72. Au(111): Surface phonon dispersion curves (after [8.182]). Calculations [8.182]: slab method (35 unrelaxed layers, pseudoparticle model, parameters fitted to the measured phonon dispersion curves of the bulk material, relaxation of the electronic charge density taken into account in a phenomenological way). Experimental data [8.179]: surface modes as determined by He-atom scattering measurements (*filled circles*)

Quite a different approach is adopted by the pseudoparticle model [8.181]. This model incorporates the relevant electronic properties in a phenomenological way. It takes into account the lowest order multipole expansion terms of the displacement induced deformations of the electronic charge density in addition to the nearest neighbor forces. The six parameter model reproduces the measured bulk phonon dispersion curves within experimental error [8.181]. It also describes well the measured surface phonon dispersion curves, if the relaxation of the electronic charge density at the surface is taken into account by a change of the pseudoparticle–ion coupling constant. The change of this coupling constant is calculated from the measured work function using a phenomenological relationship [8.182]. Figure 8.72 shows the calculated surface phonon dispersion curves together with the experimental data. The calculations have been carried out for an unrelaxed 35-layer slab. The agreement between measured and calculated data is quite satisfactory. The calculations even show, for small wave vectors $\bar{q}$, a lattice instability. The wave vector at which the frequency becomes zero is compatible with the observed (23×1) reconstruction. The reconstruction might also lead to a broader distribution of scattering intensity in the surface Brillouin zone of the unreconstructed surface.

Au(110)

The Au(110) surface exhibits a spontaneous missing row reconstruction together with a contraction of about 18% between the first and second layer [8.166, 199–204]. The missing row reconstruction and the surface contraction have been predicted by calculations [8.205] based on first principles pseudopotentials [8.144], which have been carried out in the framework of the local-density-functional formalism. The same formalism has been used to determine the surface phonon dispersion curves of a reconstructed and properly relaxed 153-layer slab with free (110) surfaces [8.206, 207]. The employed method differs from the frozen-phonon procedure, since the coupling constants for the first three layers have been directly calculated from first principles rather than the restoring force constants for some high symmetry modes (frozen phonons). The eigenmodes of the slab have then been calculated on the basis of the first-principles force constants for the first three layers and the bulk force constants for the remaining layers of the slab. The bulk force constants have been taken from a fit to the measured phonon dispersion curves in the bulk.

Inelastic He-atom scattering [8.206] and EELS [8.207] measurements have been employed in the experimental study of the surface phonon dispersion curves of Au(110). Figure 8.73 shows the results of both techniques which are to a large extent complementary and agree nicely with each other in the range in which the data points overlap. The EELS data are represented by filled circles while the He scattering data are represented by open circles. Also shown are the results of the first principles calculations (crosses) at the $\bar{\Gamma}$ and $\bar{M}$ points. The mutual agreement is gratifying. In addition, results obtained from molecular dynamics calculations [8.208] are shown by dashed lines. It is fair to mention that these

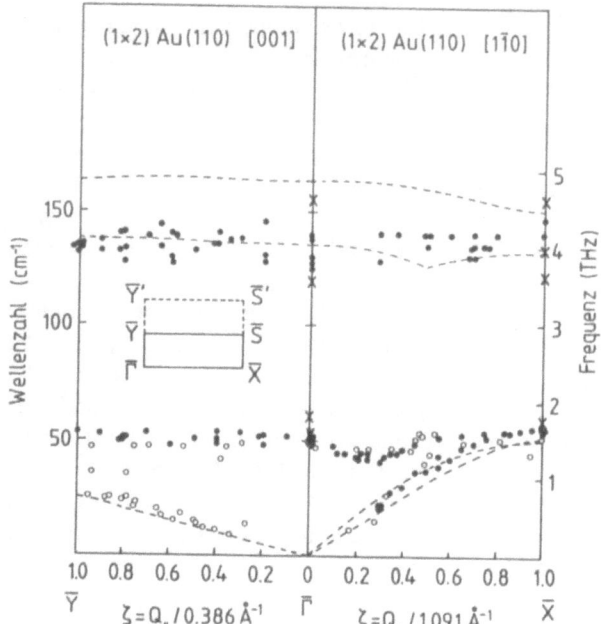

Fig. 8.73. Au(110): Surface phonon dispersion curves (after [8.207]. Calculations [8.206,207]: slab method (153 layers, reconstructed and relaxed, first principles calculations based on a norm conserving first principles pseudopotential, local density functional formalism, surface force constants for 3 surface layers from first principles, bulk force constants from fit to the phonon dispersion curves of the bulk material); *crosses*: surface localized modes as obtained from first principles; *dashed lines* results of Monte Carlo calculation based on semi-empirical interaction potentials. Experimental data: surface modes as determined from EELS measurements [8.207] (*filled circles*) and as determined by inelastic He-atom scattering measurements [8.206] (*open circles*)

calculations have been carried out prior to the measurements. These calculations are based on semi-empirical interaction potentials ("glue model"). The model reproduces the Rayleigh modes, but fails to account for the flat branch at about 50 cm^{-1}, which is due to reconstruction-induced folding. The character of the highest measured modes is also in disagreement with the experimental studies, which do not yield dipole active modes in that range, whereas the calculated modes are dipole-active.

Au(110) is an example which nicely shows, that models which do not treat correctly the influence of the surface on the electronic system are not able to predict the correct surface modes. Nonetheless these models are flexible and can reproduce the surface phonon dispersion curves once they are measured. This implies, however, that some of the model parameters have to be fitted to the measured surface data.

8.3 Miscellaneous

8.3.1 Diamond Structure Crystals

Si(111)

Lattice vibrations of silicon has attracted continued interest during the last 25 years. The first inelastic neutron scattering measurements of the phonon dispersion curves of the bulk material [8.209] were published in 1963. In 1972 more data [8.210] became available. The experimental data were reproduced by many calculations [8.209–222], which hallmark the progress in our understanding of lattice dynamics during the past few decades.

The surface dynamics of Si(111) is complicated by the fact that Si(111) undergoes a spontaneous (2×1) reconstruction on cleaving. The Si(111) (2×1) reconstructed surface is metastable and transforms upon annealing to T$\approx$500 K irreversibly into a stable (7×7) structure. Various models have been proposed for the arrangement of Si atoms at the reconstructed (2×1) surface. The model which is in best agreement with most experimental data is the π-bonded chain model [8.223].

The surface phonon dispersion curves have been investigated by inelastic He-atom scattering [8.224] and by EELS measurements [8.225]. The EELS measurements exhibit a well-defined peak at 55 meV. Various attempts have been made to deduce the origin of this peak from calculations of the surface vibrations [8.226–228]. Good agreement between the measured and calculated mode is found but the calculations yield a full azimuthal anisotropy of the EELS intensity, whereas recent refined EELS studies reveal a large isotropic component of the observed structure at 56 meV [8.229, 230].

The He scattering measurements [8.224] exhibit in addition to the Rayleigh modes a flat branch at about 10 meV. The experimental results are nicely reproduced by calculations of the surface vibrations of a reconstructed 24 layer slab with free (111) surfaces. These calculations are based on a bond charge model. The bulk parameters are taken from a fit [8.217] to the inelastic neutron scattering data [8.209, 210]. The bond charge of the dangling bonds at the surface is removed and added to the bond charges of the π-bonded chains. The radial force constants at the surface are kept at their bulk values, since the variation of the bond length throughout the whole slab is small and has been neglected. The transverse force constants change drastically, however, since they are proportional to the first derivatives of the interaction potentials and thus have to fulfill the equilibrium conditions for the tilted-chain configuration at the surface. The parameters for the stiffness of the bond angles are kept at their bulk values. Their contribution to the dynamical matrix, however, is different at the surface, since the geometrical configuration of the atoms involved in the formation of the bond angles changes at the surface. In this way the parameters of the model have been fixed without any fit to the measured surface phonon dispersion curves.

Figure 8.74 shows the calculated phonon dispersion curves together with the inelastic He-atom scattering data. The agreement is gratifying. Not only are the

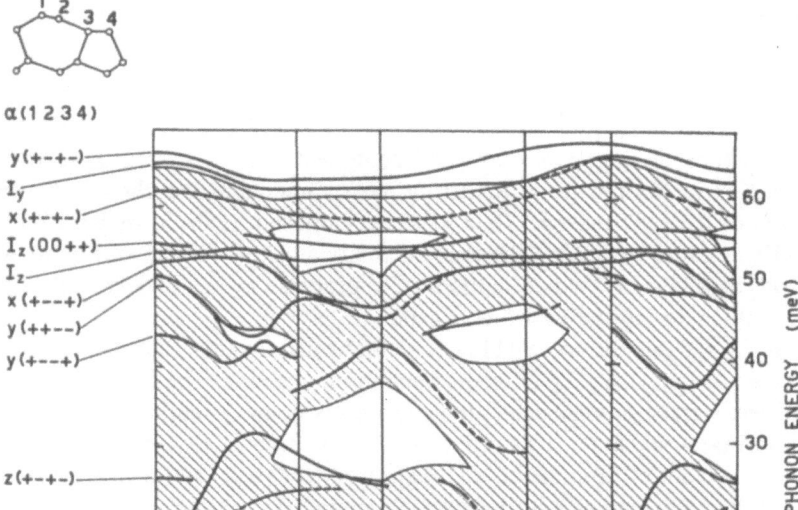

Fig. 8.74. Si(111): Surface phonon dispersion curves (after [8.231]). Calculations [8.231]: slab method (24 layers, reconstructed surface as determined by LEED measurements [8.259], bond charge model based on bulk parameters, parameters at the surface changed according to the equilibrium conditions); *solid lines*: surface localized modes; *dashed lines* weak resonances; *hatched*: surface projected bulk bands. Labels at the left side indicate the polarization vectors of the surface modes. Experimental data [8.224]: surface modes as determined by inelastic He-scattering measurements (*filled circles*; cross hatched regions indicate bands of high scattering intensity)

Rayleigh modes reproduced but also the nearly dispersionless branch at 10 meV is explained as an intrinsic property of the Si(111) surface. It is found that this branch originates from Rayleigh modes of the unreconstructed (111) surface, which are folded back towards the origin around the boundary of the Brillouin zone of the reconstructed surface. The 56 meV structure in the EELS spectra is due to the Lucas mode (I_z) at about 55 meV. The displacement pattern of the calculated surface modes are indicated on the left hand side of Fig. 8.74. The labeling refers to the surface structure shown on top.

Si(111): H(1 × 1)

The complications linked to the surface reconstruction, which result in drastic changes of the charge distribution and of the interactions at and close to the surface are largely avoided if the dangling bonds at the surface are saturated by a surface layer of hydrogen. The vibrational properties of the topmost

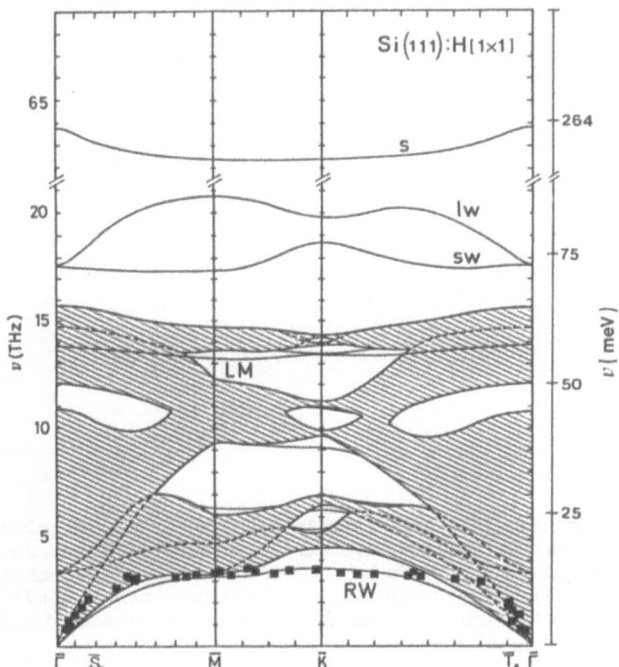

Fig. 8.75. Si(111):H(1 × 1): Surface phonon dispersion curves (after [8.260]). Calculations [8.260]: slab method (22 unrelaxed layers, bond charge model based on bulk parameters, parameters at the surface changed in order to account for mass changes, charge transfers and changes in force constants due to the H absorption); *solid lines*: surface localized modes (LM: Lucas mode, RW: Rayleigh waves); *hatched*: surface projected bulk bands; hydrogen modes: s (stretching modes), lw (longitudinal wagging modes), and sw (shear horizontal wagging modes). Experimental data [8.260]: surface modes as determined by inelastic He-scattering measurements (*filled squares*)

Si-layer of this Si(111):H(1 × 1) structure is very close to an ideal, unreconstructed Si(111) surface, since the large mass difference drives the localized hydrogen vibrations above the spectrum of the silicon vibrations. Si(111):H(1 × 1) is thus a realization of an ideal, geometrical Si(111) surface and calculations based on unchanged bulk parameters at an unrelaxed (111) surface should yield reliable results. For this reason, the rule that only results for clean surfaces would be included is broken and the results for Si(111):H(1 × 1) are presented.

The surface phonon dispersion curves of Si(111):H(1 × 1) have been determined by inelastic He-atom scattering measurements [8.224]. Figure 8.75 shows the surface phonon dispersion curves. The filled squares represent inelastic He-scattering data while the solid lines are the calculated surface modes of an unrelaxed 22-layer slab with free (111) surfaces [8.231]. The surface projected bulk modes are indicated by hatched areas. The calculations are based on the bond charge model [8.217] for the bulk dynamics, in which the masses and coupling constants of the topmost bond charges have been changed in order to describe the hydrogen coverage. The hydrogen modes are labelled s (stretching vibration of

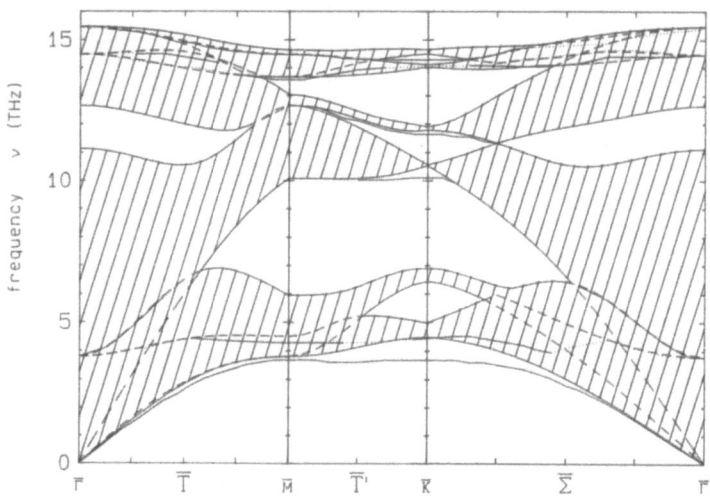

Fig. 8.76. Si(111):H(1 × 1): Surface phonon dispersion curves (after [8.232]). Calculations [8.232]: Green's function method (unrelaxed surface, force-constant model with additional dipole interactions, parameters fitted to the phonon dispersion curves of the bulk material); *solid lines*: surface localized modes; *hatched*: surface projected bulk bands. The calculations reproduce the Si modes of Si(111): H(1 × 1) rather nicely (see Fig. 8.75). Note that the adsorbed hydrogen layer is not considered in the model and that therefore the hydrogen modes are missing

the Si-H bond), lw (longitudinal wagging modes), and sh (shear horizontal wagging modes). The Lucas mode and the Rayleigh wave are labelled LM and RW, respectively. Results very similar to those presented here have been obtained for the Si modes by calculations based on a force-constant model which takes into account, in addition, interactions between displacement induced dipoles [8.232]. The parameters of this model have been obtained by a fit to the measured phonon dispersion curves. The results are shown in Fig. 8.76.

It should be pointed out that this model yields a soft mode at the $\overline{M}$ point which drives a (2 × 1) reconstruction when the dipole parameters in the first layer are increased [8.256]. A weakening of the short range force constants at the surface, on the other hand, yields a soft mode in the $\overline{\Sigma}$ direction which drives the 7 × 7 reconstruction [8.256].

8.3.2 Zinc-Blende Structure Crystals

GaAs(110)

GaAs and most of the other crystals with zinc-blende structure exhibit a complex mixture of covalent and ionic properties together with a large electronic polarizability. These properties determine the lattice vibrations, which have been studied along the main symmetry directions of the bulk Brillouin zone by coherent inelastic neutron scattering measurements [8.233]. The experimental results are well reproduced by calculations based on an overlap shell model with a valence-force-field parametrization of the short range force constants [8.234, 235].

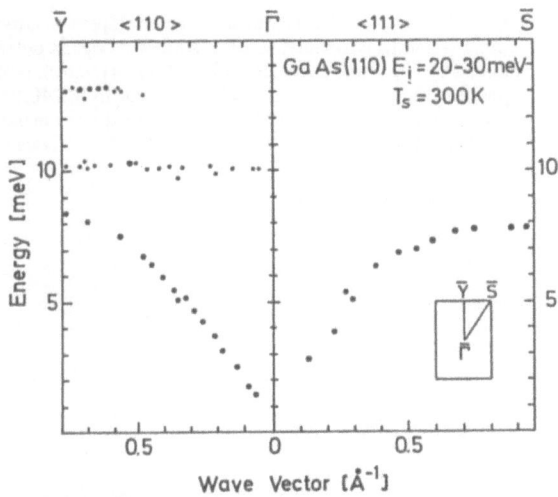

Fig. 8.77. GaAs(110): Surface phonon dispersion curves (after [8.236]). Experimental data [8.236]: surface modes as determined by He-atom scattering measurements at T_s =300 K (*filled circles*, diameter proportional to scattering intensities)

The surface phonon dispersion curves have been determined by inelastic He-atom scattering measurements [8.236]. The results are shown in Fig. 8.77. The lowest modes are the Rayleigh modes and lie at the zone boundary about 10% below the acoustic bulk bands. The assignment of the flat branches at 10 and 13 meV has to be left open until detailed surface phonon calculations become available. The Fuchs–Kliewer mode has been measured by EELS and is reported in [8.237].

8.3.3 Layered Structure Crystals

Graphite(0001)

Graphite is the textbook example for a layered structure. It is made out of covalent (0001) layers which are coupled to each other essentially by weak van der Waals interactions. The strongly anisotropic coupling leads to interesting dynamical properties. The frequencies of the intralayer modes extend up to about 50 THz while the inter-layer frequencies reach only about 4 THz. Vibrations in which adjacent layers move in phase and in antiphase are separated by an energy gap which is proportional to the interlayer coupling. The corresponding pair of modes is formed by the Γ-point and the zone-edge vibrations for modes propagating perpendicular to the layers. Thus the strong anisotropy leads only to a minor dispersion of the corresponding phonon branches. This means, on the other hand, that the bulk modes are localized in the layers. The missing of neighboring layers in the outward direction of a (0001) surface has only a very small influence on the layer vibrations. Thus surface and bulk sensitive inelastic scattering by lattice vibrations yields very much the same information.

281

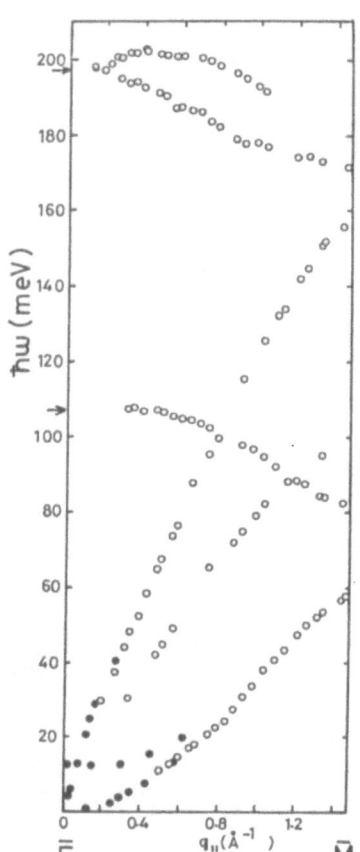

Fig. 8.78. Graphite(0001): Surface phonon dispersion curves (after [8.243]). Experimental data: surface modes as determined by EELS measurements [8.243] (*open circles*); bulk phonons as obtained from optical measurements [8.240, 241] (*arrows*); low energy bulk modes as obtained from inelastic neutron scatting measurements [8.238] (*filled circles*)

In the case of graphite, bulk studies of the phonon dispersion curves by inelastic neutron scattering measurements [8.238] were restricted by the thermal energy of incident neutrons to the low energy range up to about 50 meV. The inelastic neutron scattering measurements exhibit an upward curvature of the TA$_\perp$ branch in the Γ–M direction, which is just another consequence of the highly anisotropic coupling. The Γ-point vibrations in the high energy range have been explored by light scattering experiments [8.239–241]. The measured data are well reproduced by a force-constant model which takes into account axially symmetric forces up to sixth nearest neighbors [8.238]. A slightly more general force-constant model which also reproduces the measured neutron scattering data quite well has been used to calculate the surface vibrations of an unrelaxed 13-layer slab with free (0001) surfaces [8.242].

The surface phonon dispersion curves have been investigated in the high energy range by EELS measurements [8.243]. For lower energies inelastic He-atom scattering experiments have been carried out. Figure 8.78 shows the EELS data (open circles) [8.243] together with the neutron scattering data [8.238]. The optic

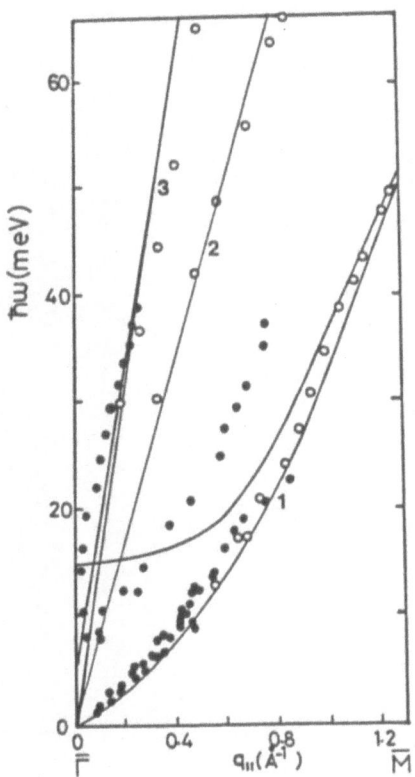

Fig. 8.79. Graphite(0001): Surface phonon dispersion curves (after [8.243]). Experimental data in the low energy range: surface modes as determined by EELS measurements [8.243] (*open circles*); surface modes as determined by inelastic He-atom scattering measurements [8.244] (*open circles*); phonon dispersion curves in the bulk as obtained from inelastic neutron scattering measurements [8.238] (represented by solid lines)

data [8.240, 241] are indicated by arrows. The mutual agreement between surface and bulk data proves that graphite is in fact an almost ideal layered compound with strongly anisotropic interaction potentials. Figure 8.79 shows the low energy range of the surface phonon dispersion curves in more detail. The open circles are again the EELS data [8.243]. The filled circles are the inelastic He-atom scattering data [8.244] and the solid lines are the phonon dispersion curves of the bulk as obtained by inelastic neutron scattering measurements [8.238]. The agreement between He-atom scattering data and EELS data is convincing except for the shear horizontal branch (2), for which large discrepancies occur. The origin of these discrepancies has not been yet established. It might be related to the different samples which have been used in the two experiments. It should also be pointed out that the parameters used in the calculations have been determined, prior to the EELS measurements, by a fit to the rather restricted data set of the inelastic neutron scattering measurements.

2H-TaSe₂(0001)

TaSe₂ is a strongly anisotropic material. The layered properties, however, are less pronounced than in graphite. As in graphite, the inelastic neutron scattering measurements of the bulk phonon dispersion curves [8.245] are restricted to the

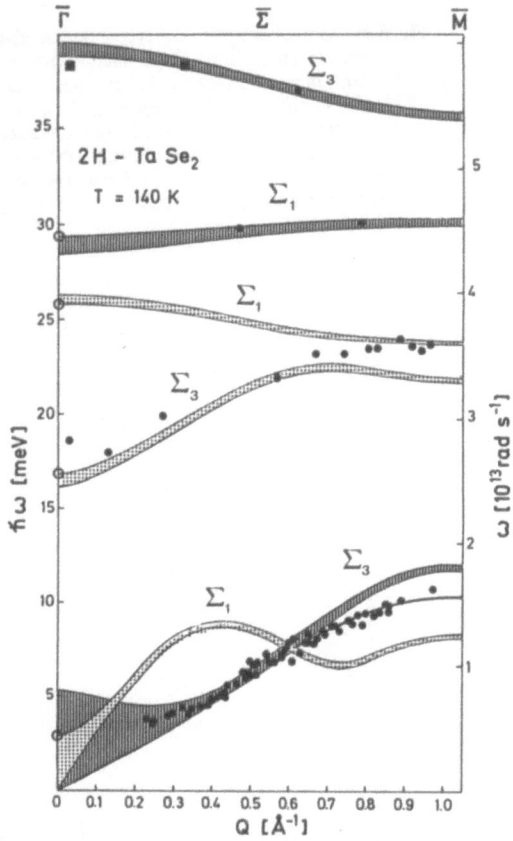

Fig. 8.80. 2H-TaSe₂(0001): Surface phonon dispersion curves (after [8.250]). Calculations [8.250]: dispersive linear chain model (parameters fitted to the bulk data; changed parameters at the surface to fit the experimental data and to allow for relaxation effects); *solid line:* Rayleigh modes; *shaded areas:* surface projected bulk bands. Experimental data: surface modes as determined by He-atom scattering measurements [8.250] (T_s =140 K: *filled circles*, T_s =60 K: *filled squares*); optical bulk phonons as determined by Raman scattering measurements [8.246, 247] (*open circles*) at Γ)

low frequency range up to about 10 meV (2.4 THz). In addition, some Γ-point phonons have been determined by Raman measurements [8.246, 247]. Most remarkable, however, is the observation of a temperature dependent softening of the longitudinal acoustic (Σ_1) phonons at about 2/3 of the direction (Γ–M) – a wave vector, which is just equal to 2 k_F. This phonon anomaly has a two-dimensional character, since it does not change when components perpendicular to the surface are admixed to the propagation vector in the Γ–M direction [8.245]. The soft-mode behavior is associated with the charge density wave transition in the bulk of 2H-TaSe₂, which takes place at T = 123 K. Neutron scattering measurements in 1T-TaSe₂ exhibit a central peak behavior near the charge density wave transition [8.248, 249].

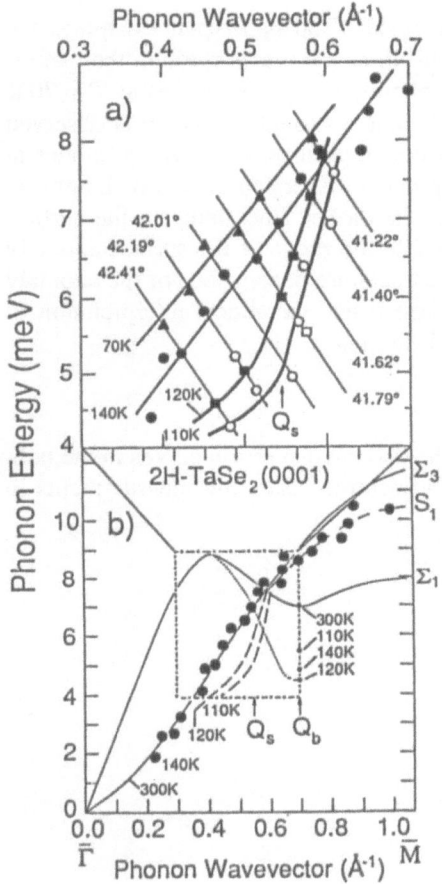

Fig. 8.81a,b. 2H-TaSe$_2$(0001): Anomalous temperature dependence of Rayleigh modes (after [8.251]). Experimental data [8.251]: surface modes as determined by inelastic He-atom scattering at various surface temperatures T_s. *Upper panel*: Collected data from various scans at five different temperatures (*filled circles*: T_s = 140 K; *filled squares*: T_s = 120 K, *open squares*: T_s = 115 K, *open circles*: T_s = 110 K, *filled triangles*: T_s = 70 K). Bold lines are drawn to guide the eye. *Lower panel*: Comparison between bulk and surface anomalies. Solid lines, small filled circles at Q_b and the dotted line indicate the phonon dispersion curves in the bulk and the temperature dependence of the Σ_1 mode. The large filled circles represent the data for the Rayleigh modes at T_s = 140 K. The dashed lines indicate the temperature dependence of the Rayleigh modes

The surface phonon dispersion curves of 2H-TaSe$_2$(0001) have been studied by inelastic He-atom scattering measurements [8.250 251]. Figure 8.80 shows the results at T = 140 K (filled circles) together with some high frequency data, taken at T = 60 K (filled rectangles). The Raman data for the bulk are shown by open circles. The shaded areas represent surface projected bulk bands. As expected, the surface phonon dispersion curves follow the bulk bands, in particular in the high frequency range. The Rayleigh mode at the zone boundary is, however, significantly pulled off the acoustic bulk bands. This indicates a strong localization at the surface. The measured data are reproduced within a dispersive-linear-chain model.

The temperature dependence of the surface phonon dispersion curves is in general very weak. An exception are the Rayleigh modes at about halfway between Γ and the zone boundary. These modes show a pronounced temperature dependence reflecting the phonon anomalies of the longitudinal acoustic bulk modes close to the charge density transition at T = 123 K. Figure 8.81 shows the temperature dependence of the Rayleigh modes. The upper panel displays the scattering data at various temperatures. Full lines are drawn through the experimental data points to guide the eye. The phonon dispersion curves at $T = 70\,\text{K}$ and $T = 140\,\text{K}$ show normal behavior, whereas a strongly anomaly is observed at $T = 110\,\text{K}$. The bottom panel shows a comparison of the Rayleigh modes at $T = 140\,\text{K}$ (filled circles) with the bulk phonon dispersion curves of longitudinal Σ_1 modes and transverse Σ_3 modes. The dashed lines indicate the surface phonon dispersion curves at 120 and 110 K in the range of the surface anomaly at Q_s while the points at Q_b indicate the temperature dependence of the anomaly in the bulk phonon dispersion curves. A conclusive theoretical interpretation of the observed phenomena is at present still missing.

GaSe(001)

GaSe is a strongly anisotropic material. The phonon dispersion curves in the bulk material have been determined by inelastic neutron scattering measurements in

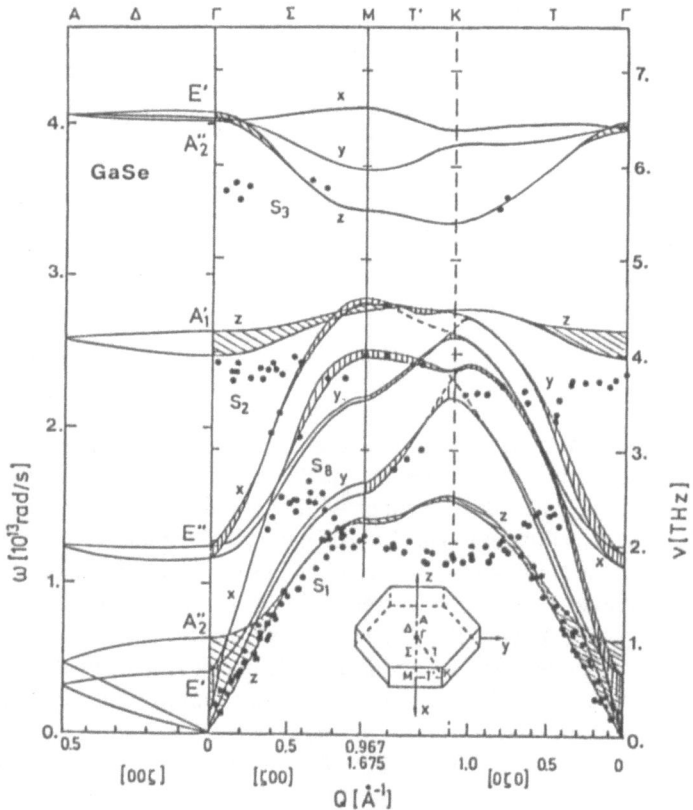

Fig. 8.82. GaSe(001): Surface phonon dispersion curves (after [8.253]. Experimental data [8.253]: surface modes as determined by He-atom scattering measurements (*filled circles*). Also shown are the bands of the surface projected bulk modes

the frequency range up to about 4 THz [8.252]. The measured phonon dispersion curves are well described by a force-constant model which takes into account 9 short-range parameters [8.252].

The surface phonon dispersion curves have been determined by inelastic He-atom scattering measurements [8.253]. The results are shown in Fig. 8.82 together with the surface projected bulk bands, which have been deduced from the force-constant model for the phonon dispersion curves in the bulk [8.252]. Most experimental data, except those of the Rayleigh waves at long wavelength, are in contradiction to the hand waving argument that the surface phonon dispersion curves should follow the band edges of the surface projected bulk bands and that they might, at most, be separated from these bands by a frequency gap in the order of the band width, which measures directly the interlayer coupling strength. Figure 8.82 shows that drastic changes have to be assumed in order to describe the measured data. It is shown in [8.253] that a dispersive-linear-chain model in which such drastic changes of force constants at the surface are assumed, is in fact able to reproduce the measured surface phonon dispersion curves. The authors argue that these changes are due to a rearrangement of electronic states at the surface which leads to major changes in the susceptibility.

References

8.1   B. Bilz, W. Kress: *Phonon Dispersion Relations in Insulators* (Springer, Berlin, Heidelberg 1979)
8.2   F.W. de Wette, W. Kress, U. Schröder: Phys. Rev. B32, 4143 (1985).
8.3   F.W. de Wette: this volume, p. 67
8.4   G. Benedek, L. Miglio: this volume, p. 37
8.5   J.P. Toennies: this volume, p. 111
8.6   D.L. Mills, S.Y. Tong, J.E. Black: this volume, p. 193
8.7   J.L. Verble, J.L. Warren, J.L. Yarnell: Phys. Rev. 168, 980 (1968)
8.8   R.K. Singh: J. Phys. C 7, 3473 (1974)
8.9   S.S. Jaswal, V.D. Dilly: Phys. Rev. B15, 2366 (1977)
8.10   D. Laplaze: J. Phys. C 10, 3499 (1977)
8.11   H. Wendel, R. Zeyher: Phys. Rev. B21, 5544 (1980)
8.12   G. Dolling, H.G. Smith, R.M. Nicklow, P.R. Vijayaraghavan, M.K. Wilkinson: Phys. Rev. 168, 970 (1968)
8.13   V. Nüsslein, U. Schröder: Phys. Stat. Sol. 21, 309 (1967)
8.14   K.V. Namjoshi, S.S. Mitra, J.F. Vetelino: Phys. Rev. B 3, 4398 (1971)
8.15   M.P. Verma, R.K. Singh: J. Phys. C 4, 2749 (1971)
8.16   A.N. Basu, S. Sengupta: J. Phys. C 5, 1158 (1972)
8.17   A. Rastogi, J.P. Hawranek, R.P. Lowndes: Phys. Rev. B 9, 1938 (1974)
8.18   T.S. Chen, F.W. de Wette, G.P. Alldredge: Phys. Rev. B 15, 1167 (1977)
8.19   W. Kress, F.W. de Wette, A.D. Kulkarni, U. Schröder: Phys. Rev. B 35, 5783 (1987)
8.20   G. Benedek, G.P. Brivio, L. Miglio, V.R. Velasco: Phys. Rev. B 26, 497 (1982)
8.21   R.B. Doak: Ph. D. thesis, MIT 1981
8.22   G. Benedek, J.P. Toennies, R.B. Doak: Phys. Rev. B 28, 7277 (1983)
8.23   G. Brusdeylins, R.B. Doak, J.P. Toennies: Phys. Rev. B 27, 3662 (1983)
8.24   G. Bracco, M. d'Avanzo, C. Salvo, R. Tatarek, S. Terreni, F. Tommasini: Surf. Sci. 189/190, 684 (1987)
8.25   A.D.B. Woods, B.N. Brockhouse, M. Sakamoto, R.N. Sinclair: In *Inelastic Scattering of Neutrons in Solids and Liquids*, (IAEA, Vienna 1961) p. 487

8.26   W.J.L. Buyers: Phys. Rev. **153**, 923 (1967)

8.27   L. Miglio, G. Benedek: In *Structure and Dynamics of Surfaces*, ed. by W. Schommers, P. von Blanckenhagen (Springer, Berlin, Heidelberg 1987), Vol. 2, p. 35

8.28   J.S. Melvin, J.D. Pirie, T. Smith: Phys. Rev. **175**, 1082 (1968)

8.29   A.M. Karo, J.R. Hardy: Phys. Rev. **181**, 1272 (1969)

8.30   R.K. Singh, M.P. Verma: Phys. Rev. B **2**, 4288 (1970)

8.31   K.V. Namjoshi, S.S. Mitra, J.F. Vetelino: Solid State Commun. **9**, 185 (1971)

8.32   A. Ghosh, A.N. Basu, S. Sengupta: Proc. Roy. Soc. (London) A**340**, 199 (1974)

8.33   Sneh, B. Dayal: Phys. Status Solidi: B **67**, 125 (1975)

8.34   R.K. Singh, K. Chandra: Phys. Rev. B **14**, 2625 (1976)

8.35   V.V.S. Nirwal, R.K. Singh: Phys. Rev. B **20**, 5379 (1979)

8.36   G. Benedek, L. Miglio, G. Brusdeylins, J.G. Skofronick, J.P. Toennies: Phys. Rev. B **35**, 6593 (1987)

8.37   G. Brusdeylins, R. Rechsteiner, J.G. Skofronick, J.P. Toennies, G. Benedek, L. Miglio: Phys. Rev. Lett. **54**, 466 (1985)

8.38   G. Benedek, F. Galimberti: Surf. Sci. **71**, 87 (1978)

8.39   F. W. de Wette, W. Kress, U. Schröder: Phys. Rev. B **33**, 2835 (1986)

8.40   G. Raunio, L. Almquist, R. Stedman: Phys. Rev. **178**, 1496 (1969)

8.41   G. Raunio, S. Rolandson: Phys. Rev. B **2**, 2098 (1979)

8.42   G. Bracco, E. Cavanna, A. Gussoni, C. Salvo, R. Tatarek, S. Terreni, F. Tommasini: Vuoto Sci. Tecn. **16** (1986)

8.43   R.K. Singh, M.P. Verma: Phys. Status Solidi **38**, 851 (1970)

8.44   R.E. Schmunk, D.R. Winder: J. Phys. Chem. Solids **31**, 131 (1970)

8.45   E.R. Cowley: J. Phys. C **5**, 1345 (1972)

8.46   O.N. Bolonin: Sov. Phys. Solid State **18**, 1415 (1976)

8.47   A.N. Basu, S. Sengupta: Phys. Rev. B **14**, 2635 (1976)

8.48   S.K. Sarkar, S. Sengupta: Phys. Status Solidi B **87**, 517 (1978)

8.49   F.W. de Wette, G.P. Alldredge: In *Methods of Computational Physics*, ed. by G. Gilat, B. Alder, S. Fernbach, M. Rotenberg (Academic, New York 1976) p.163

8.50   G. Benedek, L. Miglio: In *Ab Initio Calculation of Phonon Spectra*, ed. by J.T. Devreese, V.E. van Doren, P.E. van Camp (Plenum, New York 1983) p. 215.

8.51   G. Benedek: In *Collective Excitations in Solids*, ed. by B. di Bartolo, J. Danko (Plenum, New York 1983) p.523

8.52   F.W. de Wette, A.D. Kulkarni, U. Schröder, W. Kress: Phys. Rev. B **35**, 2476 (1987)

8.53   G. Raunio, S. Rolandson: Phys. Rev. B **6**, 2511 (1972)

8.54   G. Benedek, G. Brusdeylins, R.B. Doak, J.G. Skofronick, J.P. Toennies: Phys. Rev. B **28**, 2104 (1983)

8.55   J. S. Reid, T. Smith, W.J.L. Buyers: Phys. Rev. B **1**, 1833 (1970)

8.56   O.N. Bolin: Sov. Phys. Solid State **19**, 1088 (1977)

8.57   A.D.B. Woods, W. Cochran, B.N. Brockhouse: Phys. Rev. **119**, 980 (1960)

8.58   A.D.B. Woods, B.N. Brockhouse, R.A. Cowley, W. Cochran: Phys. Rev. **131**, 1025 (1963)

8.59   R.A. Cowley, W. Cochran, B.N. Brockhouse, A.D.B. Woods: Phys. Rev. **131**, 1030 (1963)

8.60   U. Schröder: Solid State Commun. **4**, 347 (1966)

8.61   A.N. Basu, S. Sengupta: Phys. Stat. Sol. **29**, 367 (1968)

8.62   T.I. Kucher, O.E. Tomasevich: Sov. Phys. Solid State **12**, 423 (1970)

8.63   H.G. Smith, W. Gläser: In *Phonons*, ed. by M.A. Nusimovici (Flammarion, Paris 1971) p. 145

8.64   W. Bührer: Phys. Status Solidi **41**, 789 (1970)

8.65   G. Raunio, L. Almquist: Phys. Status Solidi **33**, 209 (1969)

8.66   J.R.D. Copley, R.W. MacPherson, T. Timusk: Phys. Rev. **182**, 965 (1969)

8.67   H.N. Gupta, R.S. Upadhyaya: Phys. Status Solidi B **93**, 781 (1979)

8.68   G. Benedek, G. Brusdeylins, R.B. Doak, J.P. Toennies: J. Physique **42**, C6–793 (1981)

8.69   E.C. Svensson, W.L.J. Buyers: Phys. Rev. **165**, 1063 (1968)

8.70   G. Chern, J.G. Skofronick, W.P. Brug, S.A. Safran: Phys. Rev. B **39**, 12828 (1989)

8.71   G. Dolling, R.A. Cowley, C. Schittenhelm, J.M. Thorson: Phys. Rev. **147**, 577 (1966)

8.72   G. Raunio, S. Rolandson: J. Phys. C **3**, 1013 (1970)

8.73   H.H. Lal, M.P. Verma: J. Phys. C **5**, 543 (1972)

8.74   R.K. Singh, H.N. Gupta, M.K. Agrawal: Phys. Rev. B **17**, 894 (1978)

8.75   W. Kress, F.W. de Wette, A.D. Kulkarni, U. Schröder: Phys. Rev. B **35**, 2467 (1987)

8.76   S. Rolandson, G. Raunio: J. Phys. C **4**, 958 (1971)

8.77   G. Chern, J.G. Skofronick, W.P. Brug, S.A. Safran: Phys. Rev. B **39**, 12838 (1989)
8.78   G. Raunio, S. Rolandson: Phys. Stat. Sol. **40**, 749 (1970)
8.79   W. Kress: Phys. Status Solidi B **62**, 403 (1974)
8.80   N. Wakabayashi, H.G. Smith, K.C. Woo F.C. Brown: Solid State Commun. **28**, 923 (1978)
8.81   M.J.L. Sangster, G. Peckham, D.M. Saunderson: J. Phys. C **3**, 1026 (1970)
8.82   R.K. Singh, K.S. Upadhyaya: Phys. Rev. B **6**, 1589 (1972)
8.83   M.P. Verma, S.K. Agrawal: Phys. Rev. B **8**, 4880 (1973)
8.84   R.N. Barnett, R. Bass: Phys. Rev. B **19**, 4259 (1979)
8.85   A.J. Martin, H. Bilz: Phys. Rev. B **19**, 6593 (1979)
8.86   G. Lakshmi, F.W. de Wette: Phys. Rev. B **22**, 5009 (1980)
8.87   G. Brusdeylins, R.B. Doak, J.G. Skofronick, J.P. Toennies: Surf. Sci. **128**, 191 (1983)
8.88   W. Kress: In *Landolt-Börnstein, Numerical Data and Functional Relationships in Science and Technology*, New Series, Group III, Vol. 13, p.259 (Springer, Berlin, Heidelberg 1983)
8.89   W. Hanke, J. Hafner, H. Bilz: Phys. Rev. Lett. **37**, 1560 (1976)
8.90   C.M. Varma, W. Weber: Phys. Rev. Lett. **39**, 1094 (1977)
8.91   C.M. Varma, W. Weber: Phys. Rev. B **19**, 6142 (1979)
8.92   W. Weber: In *Physics of Transition metals*, 1980, ed. by P. Rhodes (The Institute of Physics, London 1981) p.**495**.
8.93   W. Weber: Phys. Rev. B **8**, 5082 (1973)
8.94   M. Miura, W. Kress, H. Bilz: Z. Phys. B **54**, 103 (1984)
8.95   L. Pintschovius, W. Reichardt, B. Scheerer: J. Phys. C **11**, 1557 (1978)
8.96   C. Oshima, T. Aizawa, M. Wuttig, R. Souda, S. Otani, Y. Ishizawa: Phys. Rev. B **36**, 7510 (1987)
8.97   H. Ishida, K. Terakura: Phys. Rev. B **36**, 4403 (1987)
8.98   M. Wuttig, C. Oshima, T. Aizawa, R. Souda, S. Otani, Y. Ishizawa: Surf. Sci. **193**, 180 (1988)
8.99   W. Kress, P. Roedhammer, H. Bilz, W.D. Teuchert, A.N. Christensen: Phys. Rev. B **17**, 111 (1978)
8.100 G. Benedek, M. Miura, W. Kress, H. Bilz: Phys. Rev. Lett. **52**, 1907 (1984)
8.101 G. Benedek, M. Miura, W. Kress, H. Bilz: Surf. Sci. **148**, 107 (1984)
8.102 K.H. Rieder, W. Drexel: Phys. Rev. Lett. **34**, 148 (1975)
8.103 M. Wuttig, C. Oshima, T. Aizawa, R. Souda, S. Otani, Y. Ishizawa: Surf. Sci. **192**, 573 (1987)
8.104 H.G. Smith, W. Gläser: Phys. Rev. Lett. **25**, 1611 (1970)
8.105 F. Gompf, L. Pintschovius, W. Reichardt, B. Scheerer: In *Proc. Conf. Neutron Scattering*, ed. by R. M. Moon, (US Department of Commerce, CONF-760601-Pl, Springfield 1976) Vol.1, p.129
8.106 C. Oshima, R. Souda, M. Aono, S. Otani, Y. Ishizawa: Phys. Rev. Lett. **56**, 240 (1986)
8.107 H. Ishida, K. Terakura: Phys. Rev. B **34**, 5719 (1986)
8.108 C. Oshima, R. Souda, M. Aono, S. Otani, Y. Ishizawa: Solid State Commun. **57**, 283 (1986)
8.109 C. Oshima: Phys. Rev. B **34**, 2949 (1986)
8.110 J. Müller, J.L. Olsen (eds.): *Proc. Int. Conf. on High Temperature Superconductors and Materials and Mechanisms of Superconductivity*, Physica C **153-155** (1988); R.N. Shelton, W.A. Harrison, N.E. Phillips (eds.): *Proc. Int. Conf. on Materials and Mechanisms of Superconductivity; High Temperature Superconductors* II, Physica C **162–164** (1989)
8.111 N. Lehner, H. Rau, K. Strobel, R. Geick, G. Heger, J. Bouillot, B. Renker, M. Rousseau, W.G. Stirling: J. Phys. C **15**, 6545 (1982)
8.112 R. Reiger, J. Prade, U. Schröder, F.W. de Wette, W. Kress.: J. Electron Spec. and Related Phenomena **44**, 403 (1987)
8.113 R. Reiger, J. Prade, U. Schröder, F.W. de Wette, A.D. Kulkarni, W. Kress: Phys. Rev. B **39**, 7938 (1989)
8.114 K. Gesi, J.D. Axe, G. Shirane, A. Linz: Phys. Rev. B **5**, 1933 (1972)
8.115 J. Prade, A.D. Kulkarni, F.W. de Wette, R. Reiger, U. Schröder, W. Kress: Surf. Sci. **211/212**, 329 (1989)
8.116 R.A. Cowley: Phys. Rev. Lett. **9**, 159 (1962)
8.117 R.A. Cowley: Phys. Rev. **134**, A981 (1964)
8.118 R.A. Cowley, W.J.L. Buyers, G. Dolling: Solid State Commun. **7**, 181 (1969)
8.119 G. Shirane, Y. Yamada: Phys. Rev. **177**, 859 (1969)
8.120 W.B. Stirling: J. Phys. C **5**, 2711 (1972)

8.121   M. Iizumi, K. Gesi, J. Harada: J. Phys. C **6**, 3021 (1973)
8.122   G. Shirane: Rev. Mod. Phys. **46**, 437 (1974)
8.123   W.G. Stirling, R. Currat: J. Phys. C **9**, L 519 (1976)
8.124   R. Migoni, H. Bilz, D. Bäuerle: Phys. Rev. Lett. **37**, 1155 (1976)
8.125   R. Migoni, K.H. Rieder, K. Fischer, H. Bilz: Ferroelectrics **13**, 377 (1976)
8.126   R.A. Cowley: In *Lattice Dynamics 1977*, ed. by M. Balkanski (Flammarion, Paris 1978) p.625.
8.127   J.E. Black, D.A. Campbell, R.F. Wallis: Surf. Sci. **115**, 161 (1982)
8.128   J. Prade, U. Schröder, W. Kress, A.D. Kulkarni, F.W. de Wette: In *Phonons 1989*, ed. by S.Hunklinger, W. Ludwig, G. Weiss (World Scientific, Singapore 1990) p. 841
8.129   N. Nickel, G. Schmidt, K. Heinz, K. Müller: Phys. Rev. Lett. **62**, 2009 (1989)
8.130   W. Kress: *Phonon Dispersion Curves, One-Phonon Densities of States and Impurity Vibrations of Metallic Systems*, Physics Data, Vol. 26-1, Fachinformationszentrum Karlsruhe, Karlsruhe (1987)
8.131   J.E. Black, V.T. Huynh, D.J. Cheng, R.F. Wallis: Surf. Sci. **192**, 541 (1987)
8.132   J.A. Stroscio, M. Persson, C.E. Bartsch, W. Ho: Phys. Rev. B **33**, 2879 (1986)
8.133   S.H. Chen, B.N. Brockhouse: Solid State Commun. **2**, 73 (1964)
8.134   A. Larose, B.N. Brockhouse: Can. J. Phys. **54**, 1819 (1976)
8.135   M.K. Debe, D.A. King: Phys. Rev. Lett. **39**, 708 (1977)
8.136   H.J. Ernst, E. Hulpke, J.P. Toennies: Phys. Rev. Lett. **58**, 1941 (1987)
8.137   J.P. Toennies: Physica Scr. T **19**, 39 (1987)
8.138   C.Z. Wang, E. Tosatti, A. Fasolino, M. Parinello: Surf. Sci. **189/190**, 679 (1987)
8.139   C.Z. Wang, A. Fasolino, E. Tosatti: Europhys. Lett. **7**, 263 (1988)
8.140   C.L. Fu, A.J. Freeman, E. Wimmer, M. Weinert: Phys. Rev. Lett. **54**, 2261 (1985)
8.141   X. W. Wang, W. Weber: Phys. Rev. Lett. **58**, 1452 (1987)
8.142   R. Stedman, G. Nilsson: Phys. Rev. **145**, 492 (1966)
8.143   K. Lam. M.L. Cohen: Phys. Rev. B **25**, 6139 (1982)
8.144   D.R. Hamann, M. Schlüter, C. Chiang: Phys. Rev. Lett. **43**, 1494 (1979)
8.145   A. Lock, J.P. Toennies, Ch. Wöll, V. Bortolani, A. Franchini, G. Santoro: Phys. Rev. B **37**, 7087 (1988)
8.146   K.M. Ho, K.P. Bohnen: Phys. Rev. Lett. **56**, 934 (1986)
8.147   K.M. Ho and K.P. Bohnen: Phys. Rev. B **32**, 3446 (1985); **38**, 12897 (1988)
8.148   J.P. Toennies, Ch. Wöll: Phys. Rev. B **36**, 4475 (1987)
8.149   A.G. Eguiluz, A.A. Maradudin, R.F. Wallis: Phys. Rev. Lett. **60**, 309 (1988)
8.150   K.P. Bohnen, K.M. Ho: Surf. Sci. **207**, 105 (1988)
8.151   R.J. Birgeneau, J. Cordes, G. Dolling, A.D.B. Woods: Phys. Rev. **136**, A1359 (1964)
8.152   G.A. de Witt, B.N. Brockhouse: J. Appl. Phys. **39**, 451 (1968)
8.153   S. Lehwald, J. Szenftel, H. Ibach, T.S. Rahman, D.L. Mills: Phys. Rev. Lett. **50**, 518 (1983)
8.154   M. Rocca, S. Lehwald, H. Ibach, T.S. Rahman: Surf. Sci. **138**, L 123 (1984)
8.155   Xu Mu-Liang, B.M. Hall, S.Y. Tong, M. Rocca, H. Ibach, S. Lehwald, J.E. Black: Phys. Rev. Lett. **54**, 1171 (1985)
8.156   S. Lehwald, M. Rocca, H. Ibach, T.S. Rahman: J. Electr. Spectr. Related Phenom. **38**, 29 (1986)
8.157   M. Rocca, S. Lehwald, H. Ibach, T.S. Rahman: Surf. Sci. **171**, 632 (1986)
8.158   V. Bortolani, A. Franchini, F. Nizzoli, G. Santoro: J. Electron Spectr. Relat. Phenom. **29**, 219 (1983)
8.159   A.P. Miller, B.N. Brockhouse: Phys. Rev. Lett. **20**, 798 (1968)
8.160   A.P. Miller, B.N. Brockhouse: Can. J. Phys. **49**, 704 (1971)
8.161   A.P. Miller: Can. J. Phys. **53**, 2491 (1975)
8.162   A.O.E. Animalu: Phys. Rev. B **8**, 3555 (1973)
8.163   K. Mohammed, M.M. Shukla, F. Milstein, J.L. Merz: Phys. Rev. B **29**, 3117 (1984)
8.164   R.D. Diehl, M. Lindroos, A. Kearsley, C.J. Barnes, D.A. King: J. Phys. C **18**, 4069 (1985)
8.165a A.M. Lahee, J.P. Toennies, Ch. Wöll: Surf. Sci. **191**, 529 (1987)
8.165b A.G. Eguiluz: private communication (1989)
8.165c J.A. Gaspar, A.G. Eguiluz: Phys. Rev. B **40**, 11976 (1989)
8.166   G. Binnig, H. Rohrer, Ch. Gerber, E. Weibel: Surf. Sci. **131**, L379 (1983)
8.167   H.D. Dutton, B.N. Brockhouse, A.P. Miller: Can. J. Phys. **50**, 2915 (1972)
8.168   U. Harten, J.P. Toennies, Ch. Wöll, G. Zhang: Phys. Rev. Lett. **55**, 2308 (1985)
8.169   K. Kern, R. David, P.L. Palmer, G. Comsa, T.S. Rahman: Phys. Rev. B **33**, 4334 (1986)

8.170   D. Neuhaus, F. Joo, B. Feuerbacher: Surf. Sci. **165**, L90 (1986)
8.171   V. Bortolani, A. Franchini, G. Santoro, J.P. Toennies, Ch. Wöll, G. Zang: Phys. Rev. B. **40**, 3524 (1989)
8.172   R. Nicklow, G. Gilat, H.G. Smith, L.J. Raubenheimer, M.K. Wilkinson: Phys. Rev. **164**, 922 (1967)
8.173   E.C. Swensson, B.N. Brockhouse, J.M. Rowe: Phys. Rev. **155**, 619 (1967)
8.174   G. Nilsson, S. Rolandson: Phys. Rev. B **7**, 2393 (1973)
8.175   A. Larose, B.N. Brockhouse: Can. J. Phys. **54**, 1990 (1976)
8.176   G. Nilsson: In Proc. Conf. Neutron Inelastic Scattering. (International Atomic Energy Agency, Vienna 1968) Vol. 1, p.187
8.177   G. Nilsson, S. Rolandson: Phys. Rev. B **9**, 3278 (1974)
8.178   A. Loidl: J. Phys. F **7**, L 57 (1977)
8.179   U. Harten, J.P. Toennies, Ch. Wöll: Faraday Discuss. Chem. Soc. **80**, 137 (1985)
8.180   V. Bortolani, A. Franchini, G. Santoro: In *Electronic, Dynamic, and Quantum Structural Properties of Condensed Matter*, ed. by J.T. Devreese, P.E. van Camp, p.401
8.181   C.S. Jayanthi, H. Bilz, W. Kress: In *Proceedings of the Second International Conference on Phonon Physics*, ed. by J. Kollar et al. (World Scientific, Singapore 1985) p.630
8.182   C.S. Jayanthi, H. Bilz, W. Kress, G. Benedek: Phys. Rev. Lett. **59**, 795 (1987)
8.183   B.F. Manson, K. McGreer, B.R. Williams: Surf. Sci. **130**, 282 (1983)
8.184   J.A. Stroscio, M. Persson, S.R. Bare, W. Ho: Phys. Rev. Lett. **54**, 1428 (1985)
8.185   M. Wuttig, R. Franchy, H. Ibach: Solid State Commun. **57**, 445 (1986)
8.186   M. Wuttig, R. Franchy, H. Ibach: Z. Phys. B **65**, 71 (1986)
8.187   B.F. Manson, B.R. Williams: Phys. Rev. Lett. **46**, 1138 (1981)
8.188   W.A. Kamitakahara, B.N. Brockhouse: Phys. Lett. A **29**, 639 (1969)
8.189   W. Drexel: Z. Phys. **255**, 281 (1981)
8.190   W. Bührer: Helv. Phys. Acta **51**, 15 (1978)
8.191   J.W. Lynn, H.G. Smith, R.M. Nicklow: Phys. Rev. B **8**, 3493 (1973)
8.192   R.B. Doak, U. Harten, J.P. Toennies: Phys. Rev. Lett. **51**, 578 (1983)
8.193   V. Bortolani, G. Santoro, U. Harten, J.P. Toennies: Surf. Sci. **148**, 82 (1984)
8.194a   G. Bracco, R. Tatarek, F. Tommasini, U. Linke, M. Persson: Phys. Rev. B **36**, 2928 (1987)
8.194b   W.R. Lambert, P.L. Trevor, R.B. Doak, M.J. Cardillo: J. Vac. Sci. Technol. A **2**, 1066 (1984)
8.195   R.E. Allen, G.P. Alldredge, F.W. de Wette: Phys. Rev. B **4**, 1661 (1971)
8.196   D. Castiel, L. Dobrzynski, D. Spanjaard: Surf. Sci. **59**, 252 (1976)
8.197   U. Harten, A.M. Lahee, J.P. Toennies, Ch. Wöll: Phys. Rev. Lett. **54**, 2619 (1985)
8.198   G. Santoro, A. Franchini, V. Bortolani, U. Harten, J.P. Toennies, Ch. Wöll: Surf. Sci. **183**, 180 (1987)
8.199   W. Moritz, D. Wolf: Surf. Sci. **88**, L29 (1979)
8.200   M. Manninen, J.K. Norskov, C. Umrigar: Surf. Sci. **119**, L393 (1982)
8.201   W. Moritz, D. Wolf: Surf. Sci. **163**, L655 (1985)
8.202   M. Copel, T. Gustafsson: Phys. Rev. Lett. **57**, 723 (1986)
8.203   J. Möller, K. J. Snowdon, W. Heiland, H Niehus: Surf. Sci. **178**, 475 (1986)
8.204   T. Hasegawa, N. Ikarashi, K. Kobayashi, K. Takayanagi, K. Yagi: In *Structure of Surfaces II*, ed. by J. F. van der Veen, M. A. Van Hove (Springer, Berlin 1988) p.43
8.205   K.M. Ho, K.P. Bohnen: Europhys. Lett. **4**, 345 (1987)
8.206   A.M. Lahee, J.P. Toennies, Ch. Wöll, K.P. Bohnen, M. Ho: Europhys. Lett. **10**, 261 (1989)
8.207   B. Voigtländer, S. Lehwald, H. Ibach, K.P. Bohnen, K.M. Ho: Phys. Rev. B **40**, 8068 (1989)
8.208   X.Q. Wang, G.L. Chiarotti, F. Ercolessi, E. Tosatti: Phys. Rev. B **38**, 8131 (1988)
8.209   G. Dolling: In *Inelastic Neutron Scattering*, (IAEA, Vienna 1963) Vol. 2, p. 37
8.210   G. Nilsson, G. Nelin: Phys. Rev. B **6**, 3777 (1972)
8.211   R.M. Martin: Phys. Rev. **186**, 871 (1969)
8.212   W. Cochran: Crit. Rev. Solid State Sci. **2**, 1 (1971)
8.213   R. Tubino, L. Piseri, G. Zerbi: J. Chem. Phys. **56**, 1022 (1972)
8.214   S.K. Sinha: Crit. Rev. Solid State Sci. **3**, 273 (1973)
8.215   F.A. Johnson, K. Moore: Proc. Roy. Soc. (London) A**339**, 85 (1974)
8.216   D.L. Price, S.K. Sinha, R.P. Gupta: Phys. Rev. B **9**, 2573 (1974)
8.217   W. Weber: Phys. Rev. B **15**, 4789 (1977)
8.218   C. Falter, M. Selmke, W. Ludwig, K. Kunc: Phys. Rev. B **32**, 6518 (1985)
8.219   W. Goldammer, W. Ludwig, W. Zierau, K.H. Wanser, R.F. Wallis: Phys. Rev. B **36**, 4624 (1987)

8.220 C. Falter, H. Rakel, W. Ludwig: Phys. Rev. B **38**, 3986 (1988)
8.221 C. Falter: Physics Rep. **164**, 1 (1988)
8.222 M. Klenner, C. Falter, W. Ludwig: Phys. Stat. Sol. B **151**, 503 (1989)
8.223 K.C. Pandey: Phys. Rev. Lett. **47**, 1913 (1981); **49**, 223 (1982)
8.224 U. Harten, J.P. Toennies, Ch. Wöll: Phys. Rev. Lett. **57**, 2947 (1986)
8.225 H. Ibach: Phys. Rev. Lett. **27**, 253 (1971)
8.226 O.L. Alerhand, D.C. Allan, E.J. Mele: Phys. Rev. Lett. **55**, 2700 (1985)
8.227 O.L. Alerhand, E.J. Mele: Phys. Rev. Lett. **59**, 657 (1987)
8.228 O.L. Alerhand, E.J. Mele: Phys. Rev. B **37**, 2536 (1988)
8.229 U. del Pennio, M.G. Betti, C. Mariani, S. Nannarone, I. Abbati, L. Braicovich, A. Rizzi: Surf. Sci. **189/190**, 689 (1987)
8.230 U. del Pennio, M.G. Betti, C. Mariani, S. Nannarone, C.M. Bertoni, I. Abbati, A. Rizzi: Phys. Rev. B **39**, 10380 (1989)
8.231 L. Miglio, L. Santini, P. Ruggerone, G. Benedek: Phys. Rev. Lett. **62**, 3070 (1989)
8.232 W. Goldammer, J. Backhaus, W. Ludwig: Physica Scripta **38**, 155 (1988)
8.233 J.L.T. Waugh, G. Dolling: Phys. Rev. **132**, 2410 (1963)
8.234 K. Kunc, H. Bilz: Solid State Commun. **19**, 1027 (1976)
8.235 K. Kunc, H. Bilz: In *Proc. Conf. Neutron Scattering*, ed. by R. M. Moon (US Department of Commerce, CONF-760601-Pl, Springfield 1976) Vol.1, p.195
8.236 U. Harten, J.P. Toennies: Europhys. Lett. **4**, 833 (1987)
8.237 R. Matz, H. Lüth: Phys. Rev. Lett. **46**, 500 (1981)
8.238 R. Nicklow, N. Wakabayashi, H.G. Smith: Phys. Rev. B **5**, 4951 (1972)
8.239 Y. Sato: J. Phys. Soc. Jpn. **24**, 489 (1968)
8.240 F. Tuinstra, J.L. Koenig: J. Chem. Phys. **53**, 1126 (1970)
8.241 R.J. Nemanich, G. Lucovsky, S.A. Solin: Solid State Commun. **23**, 117 (1977)
8.242 E. de Rouffignac, G.P. Alldredge, F.W. de Wette: Phys. Rev. B **23**, 4208 (1981); G.P. Alldredge, E. de Rouffignac, B. Firey, F.W. de Wette: Phys. Rev. B **29**, 3712 (1984)
8.243 C. Oshima, T. Aizawa, R. Souda, Y. Ishizawa, Y. Sumiyoshi: Solid State Commun. **65**, 1601 (1988)
8.244 G. Benedek, G. Brusdeylins, C. Heimlich, J.P. Toennies, U. Valbusa: Surf. Sci. **178**, 545 (1986)
8.245 D.E. Moncton, J.D. Axe, F.J. DiSalvo: Phys. Rev. B **16**, 801 (1977)
8.246 J.A. Holy, M.V. Klein, W.L. McMillan, S.F. Meyer: Phys. Rev. Lett. **37**, 1145 (1976)
8.247 J.C. Tsang, J.E. Smith, M.W. Shafer: Solid State Commun. **27**, 145 (1978)
8.248 N. Wakabayashi, H.G. Smith, K.C. Woo, F.C. Brown: Solid State Commun. **28**, 923 (1978)
8.249 D.E. Moncton, F.J. di Salvo, J.D. Axe: In *Lattice Dynamics*, ed. by M. Balkanski (Flammarion, Paris 1978) p.561
8.250 G. Benedek, L. Miglio, G. Brusdeylins, C. Heimlich, J.G. Skofronick, J.P. Toennies: Europhys. Lett. **5**, 253 (1988)
8.251 G. Benedek, G. Brusdeylins, C. Heimlich, L. Miglio, J.G. Skofronick, J.P. Toennies, R. Vollmer: Phys. Rev. Lett. **60**, 1037 (1988)
8.252 S. Jandl, J.L. Brebner, B.M. Powell: Phys. Rev. B **13**, 686 (1976)
8.253 G. Brusdeylins, R. Rechensteiner, J.G. Skofronick, J.P. Toennies, G. Benedek, L. Miglio: Phys. Rev. B **34**, 902 (1986)
8.254 G. Benedek: Surf. Sci. **61**, 603 (1976)
8.255 G. Brusdeylins, R.B. Doak, J.P. Toennies: Phys. Rev. Lett. **46**, 437 (1981)
8.256 W. Ludwig in: Festkörperprobleme (Advances in Solid State Physics) **29**, Vieweg, Braunschweig 1989, p.107
8.257 I.S. Braude: Sov. Phys.-Solid State **13**, 327 (1971)
8.258 J.E. Black, A. Franchini, V. Bortolani, G. Santoro, R.F. Wallis: Phys. Rev. B **36**, 2996 (1987)
8.259 F.J. Himpsel, P.M. Marcus, R. Tromp, I.P. Batra, M.R. Cook, F. Jona, H. Liu: Phys. Rev. B **30**, 2257 (1984)
8.260 U. Harten, J. P. Toennies, Ch. Wöll, L. Miglio, P. Ruggerone, L. Colombo, G. Benedek: Phys. Rev. B **38**, 3305 (1988)

Subject Index

α-quartz 32
Ab initio calculations 85, 93
Absorption, one-phonon 176
Acoustic bulk waves 5
Acoustic field 28
Acoustic surface waves 5 ff, 9, 11, 14 ff, 28
Acoustic wave device 5
Acousto-optic detection of surface waves 28
Acousto-optic interaction 28
Adiabatic approximation 181, 197
Adiabatic collisions 146
Adsorbate 196, 202
− covered surfaces 160, 279
Ag 92, 93, 257
− on Cu(100) 161
− on Ni(001) 161
Ag(001) 272, 273
Ag(110) 161, 272, 273
Ag(110) + O (2 × 1) 161
Ag(111) 50, 136, 145, 154, 155, 158, 161, 182, 270, 271
Ag(111) + Kr and Xe 161
Al 92, 111, 137, 257
Al(001) 260 ff
Al(110) 93 ff, 161, 259 ff
Al(111) 137 ff, 154, 155, 157, 161, 258, 259
Alkali halides 48, 55, 89 ff, 210 ff
Amplitude of vibration 71
Analyser, LiF crystal 167
Angle bending interaction 88
Angle resolved photoemission 198
Angular distribution of scattered electrons 194
Angular resolution of EELS 142
Anisotropic elastic media 5, 15
Anisotropic interaction 283
Annihilation, phonon 118, 134, 175
Anomalies
− interband 51
− Kohn 51, 263, 265, 266
− phonon 62, 63, 64, 157, 236, 240 ff
Anti-resonance 41
Anticrossing 216, 218, 226

Antidistortive phase transition 248
Applications 5, 6, 11, 24, 30
Area, effective surface 133, 143, 144
Aspect ratio 84, 101
Atom scattering 111
Atom–surface interaction 175 ff
Atom–surface matrix element 177
Atom–surface potential 175
Atom–surface scattering 167 ff
Attenuation
− curves 78, 79, 96
− of a pseudosurface wave 15, 16
− of a surface wave 29, 30, 32, 33
Attractive potential 183, 187
Au 92, 93, 154, 257
Au(110) 161, 275, 276
Au(111) 155, 161, 273 ff
Auger electron spectroscopy 131, 140
Axial-inversion symmetry 73
Axilrod–Teller three body interaction 183
Axis of easy magnetization 33

Ba 255
Background, thermal diffuse 195
Bandpass filters 11
bcc metals 92, 255, 256
BCS theory 32
Beam source, He nozzle 126 ff
Beeby correction 188
Bleustein-Gulyaev surface acoustic waves 21, 23, 24
Block diagonalization of the GF matrix 46
Boltzmann equation 128
Bond charge model 277, 279
Born approximation, distorted wave 115, 144
Born approximation, first order 41
Born–Mayer potentials 86, 87, 182, 186
Born–Oppenheimer (adiabatic) approximation 181
Born–von Kármán model 88, 94
Bose–Einstein function 176
Bose factor 146 ff
Bound states, surface- 186, 188

Boundary conditions 7, 9, 11, 12, 19, 23, 24, 26, 74
Breathing shell model 55, 211, 218
Brillouin scattering 25, 28, 29
Brillouin zone, surface 72
Broken bonds 201, 202
Bromides 215, 219 ff, 226, 227, 230 ff
Bulk acoustic waves 5
Bulk bands 74, 77, 112
Bulk mode 72

C on Ni(100) 161
Ca 257
Carbides 236 ff, 243 ff
Cauchy relation 86 ff
Causal Green's function 39, 41
Causality 39
CdS 20
Ce 257
Central force model 205
Charge density, electronic 50
Charge density wave (CDW) 51, 284 ff
Charge overlap 185
Charge transfer 210, 214
Chlorides 215, 218 ff, 223 ff, 230
Chopper 123, 124, 131
Classification of surface modes 57
Closed electronic shell configurations 85, 186, 211
Cluster deformation model 63, 241
Co 255
– on Ni(100) 161
CO on Pt(111) 158, 161
Collision, atom–surface 167 ff
Conservation laws 47, 116, 117, 167, 199
Constant Q scan 119, 121
Conversion of enthalpy 128
Cooper pairs 32
Correlation functions 83, 97 ff, 180
Corrugated surface 28, 147, 178, 179, 182, 186, 188
Coulomb interactions 37, 51, 52, 86 ff
Coupled channels 178
Cr 92, 255
Creation, phonon 118, 134, 175
Cross section 114, 172, 173, 194, 197, 204 ff
Crossing mode 57
Crystal analyser, LiF 167
Crystal grating spectrometer 123
Cu 92, 93, 154, 160, 257
Cu(100) 123, 158, 161, 204, 267 ff
Cu(100) + Ag(111) 161
Cu(100) + N c(2 × 2) 161

Cu(110) 161, 267, 268
Cu(111) 139, 145, 147, 148, 155, 161, 182, 266, 267
Cu–Pb sandwich samples 32
Curved surfaces 6
Cutoff effect 146
Cutoff factor 186, 187
Cylindrical isotropic solid surface 15

D on Pt(111) 161
Dangling bonds 277
Debye function 103
Debye model 169
Debye temperature 97, 103, 149
Debye–Waller factor 97, 146 ff, 169, 173, 174
Deceptons 129, 130, 134
Deflection angle 195
Delay lines 11
Density, electronic 181
Density functional theory 186, 259 ff, 275
Density of states 39
– surface projected 92
Density response function, surface 183
Detection of surface waves 6
– acousto-optic 28
– by Brillouin scattering 25
– by light scattering 25
– in piezoelectric media 25
Detector for scattered He atoms 132
Devices 5, 6, 11, 24, 30, 31
Diagonalization techniques 81
Diamond structure 277 ff
Dielectric function 53, 183, 187
Dielectric response theory 258, 260, 261
Dielectric tensor 17
Differential cross section 172, 173
Differential reflection coefficient 113, 142, 144 ff, 147, 148, 150, 173, 178, 179
Dipole (multipole expansion of the electronic charge density) 50
Dipole scattering 195, 196, 201
Discontinuities, geometrical, material 6
Dispersion curve 74, 61, 117, 119, 149, 193, 196, 203 ff, 209 ff
Dispersion in surface wave velocity 33
Dispersion relation 29, 44
Dispersive part 38
Dispersive-linear-chain model 287
Displacement correlation function 180
Displacement field (piezoelectric surface wave) 20
Displacement field (Rayleigh wave) 13
Displacive phase-transition 248

Distorted wave Born approximation (DWBA)
115, 144, 146, 173 ff
Distortions, ion 182
Double shell model 241, 245
DWBA (distorted wave Born approximation)
115, 144, 146, 173 ff
Dynamic repulsion 186
Dynamical equation 50, 51
Dynamical matrix 44, 71, 87, 200
Dyson equation 40

EELS (electron energy loss spectroscopy) 62,
112, 140 ff, 158, 159, 193 ff
Effective surface area 133, 143, 144
Eigensolutions 200
Eigenvalue 37, 38, 200
– equation 71
– spectrum 44
Eigenvector 38, 39, 40, 199, 201, 204
Einstein oscillator 180
Elastic media 6 ff
– anisotropic 15
– isotropic 13
Elastic modulus, piezoelectrically stiffened
22
Elastic modulus tensor 7, 21
Elastic relaxation 43, 52, 62
Elastic scattering 117, 142, 196, 197
Electric field 16, 20, 195
Electrical boundary conditions 26
Electron detector (in EELS) 140, 141
Electron diffraction 111
Electron energy loss spectra 142, 143
Electron energy loss spectroscopy (EELS) 62,
112, 140 ff, 158, 159, 193 ff
Electron optics (in EELS) 140, 141
Electron–phonon interaction 50 ff, 62 ff, 95,
116, 236
Electron scattering 111, 112, 114 ff, 121, 140 ff,
150, 158, 160
Electron source (in EELS) 140
Electron wave function 199
Electronic charge density 50, 181
Electronic contributions to surface dynamics
50
Electronic perturbation 52
Electronic polarizability 183
Electronic relaxation 52, 62
Electro-optic effects 25, 28
Electrostatic approximation 17
Elliptical polarization 58
Emission, one-phonon 176
Empty states 182

Energy 104, 167
– of atoms in a He beam 127 ff
– conservation 116, 117, 167
– resolution of a TOF spectrometer 135
– resolution of EELS 140
Enthalpy conservation 127
Entropy 104, 106
Equation of motion 7, 17, 42, 43, 50, 51, 71, 86
Equilibrium condition 87
Ewald diagram 117 ff
Exchange of vibrational character 216, 218,
226
Excitation of surface phonons (waves) 24, 198

Fabry-Perot interferometer 28, 29
fcc metals 92, 257 ff
fcc slab (monotomic) 74
Fe 92
Fe(111) 255
Fe–Pb sandwich samples 32
Ferroelectrics 51, 248
Field, macroscopic 53
Field penetration (in EELS) 140
Filters, bandpass 11
First order phase transition 32
First principles calculations 64, 85, 93, 210,
211, 254, 258 ff, 275
First sound – second sound 266, 270
Fluorides 211 ff, 222 ff, 229, 248 ff
Focussing, kinematic 121, 122, 137
Folding mechanism 66, 216
Force constant 52, 86, 87, 155, 210
– changes at the surface 62, 89, 93, 94, 201
– model 88, 94, 159
Forces, three body 61
Fraunhofer geometry 171, 172
Free electrical boundary conditions 24
Free energy 104, 106
Frequency distribution, surface 105
Fresnel geometry 171, 172, 173
Frozen phonon calculations 85, 210, 254, 259,
260, 275
Fuchs-Kliewer mode 55, 62

GaAs(110) 161, 280, 281
GaAs(111) 157
Gap energy 32
Gap modes 80
GaSe(001) 161, 286, 287
Gaussian-quadrature 83, 84
Ge(111) 62
Generalized Rayleigh wave 6, 14, 79
Generating coefficients 62

Generation of surface waves in piezoelectric
  media 25
Gilat-Dolling-Raubenheimer interpolation 84
Glauber–van Hove trick 180
Glue model 276
Golden rule 113
Grafoil 102
Graphite 95, 102, 182
Graphite(0001) 161, 281 ff
Grating spectrometer 123
Green's function 37, 38, 39, 56, 116, 197, 201, 203
Group velocity vector of a surface acoustic
  wave 15
Guided waves 5

H on Pt(111) 161
H on Si(111) 161
H on W(001) 157, 161
Hard wall model 182, 186, 188
Harmonic approximation 175, 202, 203
Hartree–Fock approximation 181
He nozzle beam source 126 ff
He scattering 61, 62, 111 ff, 123 ff, 144 ff, 160, 167 ff
He–solid interaction 114, 145, 149, 155
Heine-Abarenkov pseudopotential 258, 260 ff
Helmholtz free energy 104, 106
Hermiticity condition 82
Heterodyne techniques 28
HfC(001) 243, 244
High frequency signal processing 5
Homopolar crystals 210, 277 ff
Hybridization 58, 66
Hydrides 211, 215

Impact energies 204
Impact regime 196, 197
In films 32
Incoherent diffuse elastic scattering 134
Inelastic cross section 169, 188
Inelastic matrix elements 186
Inelastic neutron scattering 211
Inelastic scattering 116, 152
Influence factor 187
Insulators 210 ff, 248 ff
Intensities 121, 142, 152
Intensity enhancement by kinematic focussing 121
Interaction
– atom–surface 175 ff
– Axilrod–Teller three body 183
– long range Coulomb 37

Interaction models 85
Interaction potential 70, 209, 210
Interatomic force constants, effective 52
Interband anomalies 51
Interdigital transducers 5, 24, 25, 27, 30, 31
Interdigitated structures 28
Interface
– solid–liquid 6, 15, 28
– solid–vacuum 15
Interferometer, Fabry-Perot 28, 29
Interlayer spacing 205
Interplanar force constants 275
Intra-molecular vibrations 193
Intrinsic perturbation 46
Invariance conditions 37, 43, 47 ff
Iodides 215, 221, 222, 227 ff, 232, 233
Ionic crystal 37, 44, 86
Ionizer 132
Ir 92
Irreducible element 73
Isotropic media 13, 14

Jellium model 181, 184, 186

K 92, 255
KBr(001) 55, 226, 227
KCl(001) 55, 57, 89, 90, 161, 223 ff
Kellermann 105
KF(001) 55, 56, 90, 222 ff
KI(001) 55, 227 ff
Kinematic focussing 121, 122, 137, 170, 171
Kinematic smearing 135
Kinematics 116 ff, 167, 168
KMnF$_3$ 33, 250 ff
Kohn anomalies 51, 263, 265, 266
Kr on Ag(111) 161
Kr on Pt(111) 158, 161
Kramers-Kronig relations 39
KZnF$_3$ 248 ff

La 257
Lamb waves 15
Landau-Peierls divergence 83
Lateral variation of the potential 185
Lattice vectors 70
Layer projected spectral density 93
Layered structures 95, 281 ff
Layered systems 5
Leaky surface acoustic waves 6, 7, 15, 16
LEED (low energy electron diffraction) 116, 131, 140, 151, 197
Lennard-Jones potential 80, 81, 85, 92
Li 255

LiBr(001) 215
LiCl(001) 215
LiF 123, 139
LiF(001) 55, 56, 58, 59, 61, 62, 74 ff, 92, 97 ff,
   106, 111, 117, 122, 129, 130, 134, 145, 147,
   161, 182, 188, 211 ff
LiF crystal analyser 167
Lifetime of a resonance 203
Light scattering 6, 25
LiH(001) 211
LiI(001) 215
LiNbO₃ 32
Lindhard screening 237
Linear dielectric response theory 95
Linear polarization 58
Lippmann-Schwinger equation 39, 40
Liquid He films on LiNbO₃ 32
Local density approximation 186
Local-density-functional formalism 259 ff, 275
Long wavelength phonons 195
Long-range Coulomb terms 52
Long-range forces 44
Longitudinal modes 58, 112, 113, 160
Longitudinal resonance 62, 92, 93, 155
Low energy electron diffraction (LEED) 197
Lucas mode 55, 57

Macroscopic field 53
Macroscopic modes 53, 54, 78
Madelung constant 88
Magnetic field dependent attenuation of surface
   waves 33
Magnetic media 6
Magnetization, axis of easy 33
Many-body contributions 180, 183, 210, 214
Mass spectrometer 131, 132
Matrix element
- atom–surface 177
- phonon 175
Maxwell equations 54
Maxwellian velocity distribution (He beam)
   129, 130
Mean-square amplitude of vibration 83, 97 ff
Mean-square displacements 169
Metal oxides 233 ff
Metallized surfaces, Bleustein-Gulyaev waves
   on 21, 23, 24
Metals 51, 85, 88, 181, 210, 236 ff, 254 ff
Metastable detection scheme 158
MgO(001) 55, 57, 161, 233 ff
Microscopic models 64
Microscopic modes 53, 54, 78
Microscopic perturbation 53

Missing row reconstruction 275
Mixed mode 72, 78, 79
Mixed valence compounds 51
Mo 92, 255
Mode softening 157
Mode, Fuchs-Kliever 62
Models, interaction 85, 205
Modes
- longitudinal 58
- microscopic 53, 58, 62
- parallel 56, 57
- pseudosurface 45
- resonant 45
- sagittal 56, 57, 58
- shear horizontal 56
- surface 45
- transverse 58
Molecular crystals 85
Molecular dynamics calculations 256, 257, 275
Momentum conservation 116, 117, 167
Momentum transfer 116, 117, 134
Monoclinic crystals 24
Monopole 50
Muffin tin 151, 197
Multiphonon processes 168, 169
Multiphonon scattering 123, 148, 149
Multiple scattering 114, 116, 151, 152, 197,
   204
Multipoles 50, 184

N on Cu(100) 161
N on Ni(100) 161
Na 92, 255
NaBr(001) 90, 219 ff
NaCl(001) 49, 55, 57, 89, 90, 105, 161, 182,
   218 ff
NaF(001) 55 ff, 60 ff, 89, 90, 122, 161, 215 ff
NaH(001) 215
NaI(001) 55, 221 ff
Nb 255
NbC(001) 63, 161, 245, 246
Nb₃Ge films 32
NbN films 32
Nb₃Sn films 32
Ne-atom scattering 167
Neutron scattering 111, 211
Ni 92, 93, 257
Ni films 33
Ni(100) 111, 142, 143, 151 ff, 158 ff, 204 ff,
   262
Ni(100) + Ag(100) 161
Ni(100) + C p4g(2 × 2) 161
Ni(100) + CO 161

Ni(100) + N c(2 × 2) 161
Ni(100) + O c(2 × 2) 161
Ni(100) + O p(2 × 2) 161, 201 ff
Ni(100) + S c(2 × 2) 161, 204 ff
Ni(110) (1 × 2) 161
Ni(111) 182
Nitrides 236, 240 ff
Noble gas crystals 89
Noble gas overlayers 182
Noble metals 50, 183
Nonpiezoelectric substrates 24
Noncentral forces 205
Noncentral potential 214
Nondispersive waves 20
Nonlinear interactions 6
Normal modes 74, 175, 200
Normalization 173
Nozzle, He beam source 126 ff

O^{2-} ion 233
O on Ag(110) 161
O on Ni(100) 161
O on Pt(111) 161
Off-specular scattering 196
One-phonon absorption 176
One-phonon emission 176
One-phonon scattering 149, 169 ff
Order parameter, changes in 33
Oscillators 11

π-bonded chain model 277
p–wave scattering 182
Pair potentials 85
Parallel modes 56, 57
Parallel polarization 45
Particle displacement vector 13, 20
Pb 257
Pb(110) 161
Pb–Ag sandwich samples 32
Pb films 32
Pd 92, 154, 257
Pd(110) 155, 160, 161, 263, 264
Peeling-off modes 80
Penetration depth of electrons 150
Perovskites 248 ff
Perturbation
– effective surface 51
– electronic 52
– localized 39
– matrix 47, 48, 51, 52
– short range 37
– surface as a 43
Perturbed Green's function 40

Perturbing field 38
Phase shift 197
Phase transition 6, 32, 33, 34, 157, 248, 285, 286
Phase velocity 9, 15
Phonions 129, 130
Phonon annihilation (absorption, phonon loss) 118, 151, 168
Phonon anomalies 62, 236, 240 ff
Phonon creation (emission) 118, 168
Phonon density, surface-projected 56
Phonon dispersion curves, surface 149
Phonon eigenvector 199, 201
Phonon frequency 168
Phonon matrix element 176
Phonon softening 236, 240 ff
Piezoelectric field 20
Piezoelectric media 5, 6, 16, 17, 25, 26
Piezoelectric surface waves 16, 18 ff
Piezoelectric tensor 17, 21
Piezoelectrically stiffened elastic modulus 22
Planar isotropic solid surface 15
Plasma frequency 55
Polariton, surface phonon 53, 54, 57
Polarizability 61, 88, 183, 214
Polarization 11, 112, 113, 181
– elliptical 58
– linear 58
– parallel 45
– sagittal 45
– vector 71, 72
Potential 70, 80, 81, 85, 86, 209, 210
– atom–surface 175
– attractive 183
– Born–Mayer 182, 186
– of He-solid interaction 114
– repulsive 181, 182, 186
– total static 185
– two-body 177, 180
– van der Waals 181, 183
– Yukawa 182, 186
Power, surface 28
Projected Green's function 42, 45
Projected phonon density, surface 45
Propagator 116
Proximity effects 32
Pseudo surface-mode 6, 15, 45, 78, 79, 155
Pseudoatoms 50
Pseudoparticle model 94, 95, 266, 267, 270, 271, 275
Pseudopotential 155, 181, 258, 260 ff
Pseudorandom chopper 124
Pt 92, 93, 257

Pt(111)   154, 155 ff, 161, 264 ff
Pt(111) + CO c(4 × 2)   161
Pt(111) + CO p($\sqrt{3}$ × $\sqrt{3}$)   161
Pt(111) + D (1 × 1)   161
Pt(111) + H (1 × 1)   161
Pt(111) + Kr and Xe   161
Pt(111) + O (1 × 1)   161
Pt(111) + O p(2 × 2)   161
Pulsed nozzle   131
Pumping stages   133

Quadrupole   50
Quantization box   171
Quantum effects in a He beam   128
Quantum number   74
Quasi–harmonic approximation   71
Quasi-particle model   50, 62
Quasi-static approximation   17

Radar system, signal processing   31
Radial force constants   86, 87
Rayleigh mode (wave)   5, 6, 13, 14, 20, 33, 34,
    57, 61 ff, 78, 79, 92, 119, 122, 136 ff, 142,
    143, 145, 151, 155 ff, 159, 170, 204
Rb   255
RbBr(001)   91, 230 ff
RbCl(001)   55, 230
RbF(001)   55, 90, 229
RbI(001)   91, 232, 233
Reciprocal lattice vectors   70
Reconstruction   62, 64, 157, 248, 254, 255, 273,
    275, 277 ff
Reduced subspace   46, 47
Reflection coefficient   113, 121, 142, 144, 145,
    147, 148, 150, 173, 178, 180
Refractory compounds   63, 236 ff
Relaxation   52, 61, 62, 80, 81, 89 ff, 94, 155,
    159, 160, 210, 211, 213, 216, 218, 219, 221 ff,
    227, 229, 230, 232, 234, 238, 241, 243, 248 ff,
    258, 263, 264, 275
Repulsive potential   181, 182, 186
Resolution of a TOF spectrometer   135 ff
Resonance (resonant modes)   41, 45, 59, 62, 72,
    77, 112, 113, 142, 143, 151, 159, 160, 202
Resonance enhancement   42, 204
Resonant scattering   188
Resonators   11
Response   38
Response function, surface density   183
Rh   92
Rhombic crystals   24
Rigid ion model   51, 105
Rippling   25, 187

Rock salt structure   210 ff
Root sampling method   83
Rotation invariance condition   43, 47 ff
Rough surfaces   6
Rumpling, surface   253

S on Ni(100)   161
Sagittal modes   56, 57, 58, 73
Sagittal plane   5, 11, 20, 22, 112, 113
Sagittal polarization   45
Sagittal resonance   57
Sampling   82
Sandwich samples   32
SBZ (surface Brillouin zone)   72
Scan curve   119 ff, 134, 142, 168
Scattering
– amplitude   116, 197, 198
– atom–surface   167 ff
– cross section   114, 204
– dipole   195
– function S   199
– geometry   204
– He-atom   167 ff
– intensities   142, 143, 153, 154
– mechanism, corrugation   28
– Ne-atom   167
– p–wave   182
– potential   114
Second order phase transition   32, 34
Second sound – first sound   266, 270
Secular equation   40
Seismic waves   5, 30
Selective adsorption   149
Self–interaction terms   71
Semi-infinite lattice   43, 201
Semiconductors   51, 85, 88
Shear horizontal (SH) modes   24, 56, 57, 74,
    80, 112, 113, 118
Shear vertical (SV) modes   80
Shell model   50, 86, 211
Short circuit boundary conditions   23
Si(111)   62, 157, 277 ff
Si(111) (2 × 1)   161
Si(111) + H (1 x 1)   278 ff
Signal processing   5, 30, 31
Skimmer   127
Slab adapted bulk spectra   75
Slab method   56, 67 ff, 203
Slab shape   84
Slab thickness   69
Small angle scattering   195, 197, 201
Soft mode   157, 248
Soft wall model   182, 186

Softening of force constants 62, 155
Softening of phonon frequencies 51
Solid–liquid interface 15
Solid–solid interface 15
Solid–vacuum interface 15
SP modes 73, 78
Specific heat, surface-excess 104 ff
Spectral density 93, 199 ff
Spectral part 38, 39
Specular lobe 195
Specular peak 196
Speed of a surface acoustic wave 12
Speed of surface piezoelectric waves 18 ff
Speed ratio 127, 128
Spin reorientation phase transition 34
Spurions 129, 130
Sr 257
SrTiO$_3$ 33, 251 ff
Stability, microscopic 82
Standing waves 74
State-to-state cross section 173
State-to-state reflection coefficients 173, 178
State-to-state transition rate 174
Static potential, total 185
Stepped surfaces 62
Stress 7, 17, 267
Structural phase transitions 33
Subspace, reduced 46, 47
Superconductivity 6, 32, 51, 63, 64, 236, 240, 245 ff
Surface acoustic waves 5, 6, 7, 15, 16, 17, 18, 21, 22, 24, 30, 32
Surface area, effective 133, 143 ff
Surface bound states 186, 188
Surface Brillouin zone 72
Surface Debye temperature 149
Surface density response function 183
Surface-excess distribution of frequencies 83
Surface-excess thermodynamic quantities 83
Surface force constants 62, 155
Surface geometry 205
Surface localized modes (surface modes, surface vibrations) 45, 57, 59, 72, 74, 77, 112
Surface perturbation matrix 51, 52
Surface phonon anomalies 62
Surface phonon density 241, 242
Surface phonon density of states 83
Surface phonon excitation 198
Surface phonon polaritons 53, 54, 57
Surface phonons 6, 29, 193, 194, 202, 203
Surface piezoelectric waves 18 ff
Surface power 28

Surface projected bulk phonons 168, 170, 179
Surface projected density of states 45, 56, 92
Surface resonance 45, 72, 77, 79, 194, 196, 202, 203
Surface scattering 173
Surface stress 267
Surface structure 194
Surface superconductivity 32
Surface to volume ratio 101
Surface wave device technology 6
Surface waves 6, 24, 25, 27, 33
Surface waves, excitation of 24
Surface waves on piezoelectric media 25
Susceptibility, dielectric 53
SV modes 80
Symmetry 73
– breaking 48
– transformation 46

T-matrix 116
Ta 255
TaC(001) 161, 246, 247
Tangential force constants 86, 87
Target chamber 131
TaS$_2$(0001) 161
TaSe$_2$(0001) 161, 283 ff
Technology, surface wave device 6
Thermal diffuse background 195
Thermal motion 197
Thermodynamic limit 84, 101
Thermodynamic quantities, surface-excess 83, 104
Thin-film plate transducers 27
Three-body interaction, Axilrod–Teller 183
Three-body forces 61
TiC(001) 63, 161, 237 ff
TiC(110) 238 ff
TiC(111) 240, 241
Time dependent Green's function 39
Time-of-flight distribution 135, 136
Time-of-flight (TOF) measurements 167, 168
Time-of-flight spectrometer 123 ff, 126
Time-of-flight spectrum 61, 120, 121
TiN(001) 63, 240 ff
Transducers, interdigital 24, 25
Transformation of the elastic moduli 7, 8, 10
Transition matrix element 174
Transition metal compounds 236 ff
Transition rate 174
Translational invariance 43, 47 ff, 82
Translational symmetry 74
Transverse force constants 86, 87
Transverse modes 58, 90, 112, 113

Travelling waves 74
Turning point 182, 187
Two-body potential 177, 180

Umklapp processes 178

V 92, 255
Valence force field 280
Van der Waals interactions 281
Van der Waals potential 149, 181
Velocity of atoms in a He beam 127 ff
Velocity, surface wave 26, 30, 32, 33
Vibrational correlation functions 97
Voigt notation 7
Vortex pairs 32

W 92
W(001) 157, 161, 255 ff
W(100) + H(1 × 1) 161, 196

Wave functions 171, 177, 178, 199
Wave vector 167
– conservation 198, 199
Waves, bulk acoustic 5
Waves, surface acoustic 5, 6, 7
Weare criterion 149
Work function, surface 182

X-ray scattering 221
Xe on Ag(111) 161
Xe on Pt(111) 158, 161

Yb 257
Yukawa potential 182, 186

Zero energy resonance 128
Zinc blende structure 280, 281
Zn films 32
ZrC(100) 161